Organic Chemistry from Retrosynthesis to Asymmetric Synthesis

Vitomir Šunjić · Vesna Petrović Peroković

Organic Chemistry from Retrosynthesis to Asymmetric Synthesis

Springer

Vitomir Šunjić
Croatian Academy of Sciences and Arts
Zagreb
Croatia

Vesna Petrović Peroković
Faculty of Science
University of Zagreb
Zagreb
Croatia

ISBN 978-3-319-29924-2 ISBN 978-3-319-29926-6 (eBook)
DOI 10.1007/978-3-319-29926-6

Library of Congress Control Number: 2016935567

Printed on acid-free paper

This Springer imprint is published by Springer Nature
The registered company is Springer International Publishing AG Switzerland

Preface

There is a substantial didactic difference between retrosynthetic analysis and asymmetric synthesis. This difference refers to the chiral target molecules. They are regarded as racemic in "two-dimensional" retrosynthesis, but one enantiomer is the target in asymmetric synthesis. Retrosynthesis without considering the absolute configuration anticipates the synthesis of racemic target molecules, while asymmetric synthesis leads to the preferred enantiomers. Concerning the conceptual and practical difference between retrosynthesis without consideration of the stereochemistry and asymmetric synthesis of optically pure target molecules, we underline considering asymmetric synthesis as a "departure to the third dimension."

This book is an attempt to bridge these two aspects of teaching and practicing synthetic organic chemistry. Retrosynthetic analysis is based on the method developed by S. Warren in his monographs as a creative mnemonic tool and a specific departure from the computer-designed multistep syntheses. The attractiveness and pragmatic value of Warren's approach rest in the adoption of the basic principles of retrosynthetic analysis through application to the problems of the increasing complexity, attaching "computer-like" properties to the brain of synthetic chemists, in particular the capacity to see the target structures in a new, creative way.

The selected examples of asymmetric syntheses in this book are not regularly related to the target molecule of retrosynthetic analysis. Chiral target molecules are selected either to demonstrate the practicability of certain asymmetric syntheses in non-catalytic and catalytic mode, on the laboratory and industrial scale, or because of their scientific originality.

The book characterizes the framework consisting of *chapters* divided into *sections* and preceded by *abstracts* and *introductions*.

Chapter 1 sets the scene by presenting retrosynthetic analysis and a proposal for the synthesis of one simple racemic target molecule, immediately followed by presentation of asymmetric syntheses of one enantiomer of the same target. The aim of this endeavor is to present the substantial difference in the complexity of retrosynthetic analysis of racemates and completing the asymmetric synthesis of the selected enantiomer to the reader. The following chapters are basically organized around a discussion of the preferred C–C bond disconnections controlled by participation of one or more functional groups. Many aspects of organic reactions are discussed in relation to proposed retrosynthetic steps such as the Diels-Alder cyclization, Birch reduction, Heck reaction, Jones oxidation, Nef reaction, Pfitzner-Moffat oxidation, Pictet-Spengler cyclization, Strecker addition, Suzuki coupling, Wittig condensation and some others. For many important raw materials or building blocks, such as adipic acid, acrolein, *n*- and *iso*-butyric acid, methyl vinyl ketone, cyclohexanone, pyridine, caprolactame, *n*-hexanol, ethyl acetoacetate and phenylacetic acid, a short description of the industrial production method is given. The final chapters are devoted to specific topics, the retrosynthetic approach to heterocyclic structures, rearrangement reactions, retrosyntheses and asymmetric synthesis of complex biologically active compounds. Specific *sections* are devoted to selected topics such as the environmental aspects of organic synthesis, feasibility of the Wittig reaction on the industrial scale, disconnection of the C–C bond correlated to the C–H acidity scale of organic compounds, the Baldwin rules in cyclization reactions, etc. *Examples* are the soul of the book. Most require completion of the retrosynthetic analysis and a proposal for the synthesis. Some of them discuss only synthetic aspects of complex molecules and specific methods of their preparation. *Notes* are inserted into the retrosynthetic discussion and are devoted to a concise description of specific topics, production methods of commodities, explanation of the mechanisms of important reactions, etc.

Nowadays retrosynthetic packages, workshops and retrosynthesis competitions are being developed by collaborations of industry and academy. The authors expect this book to contribute to this trend and bridge retrosynthetic analysis and asymmetric synthesis of chiral target molecules in the optically pure form.

We are very grateful for the support and assistance provided by the publisher, Springer, particularly that from Dr. Hans-Detlef Klueber, Dr. Jutta Lindenborn and Ms. Abirami Purushothaman.

Zagreb, Croatia
December 2015

Vitomir Šunjić
Vesna Petrović Peroković

Contents

Abbreviations and Acronyms

$(AcO)_2O$	Peracetic acid anhydride
$(Ipc)_2BCl$	Chlorodiisopinocampheyl borane
1,*n*-CO	1,*n*-Dioxygenated pattern in target molecule
3-KB	3-Chlorobutan-2-one
Ac_2O	Acetic anhydride
API	Active pharmaceutical ingredient
BINAP	2,2′-*bis*-Diphenylphosphino-1,1′-binaphtyl
BINAPO	2,2′-*bis*-Diphenylphosphinoxy-1,1′-binaphtyl
Bmim	1-Butyl-3methylimidazolium hexachlorophosphate (ionic liquid)
Bmpy	1-Butyl-3-methyl pyridinium bromide (ionic liquid)
Boc	*tert*-Butoxycarbonyl
BPPM	Diphosphine ligand derived from 3-hydroxy-D-proline
CAMP	Cyclohexyl anisyl methyl phosphinoxide
CIP	Cahn-Ingold-Prelog
CNDO/2	Complete neglect of differential overlap, one of the first semi empirical methods
COD	Cyclooctadiene
CO-pattern	Oxygenated pattern (oxygen functionality)
CSP	Chiral stationary phase
Cy	Cylopentadiene
D.-A.	Diels-Alder
d.e.	Diastereomeric excess
DAIB	3-*exo*-Dimethylaminoisoborneol
DBU	1,8-Diazabicycloundec-7-ene
DCC	Dicyclohexylcarbodiimide
DCE	Dichloroethane
DCM	Dichloromethane
DIBAL-H	Diisobutylaluminium hydride
Dimsyl	Dimethylsulfoxide carbanion
DIOP	Diphosphine acetonide, ligand derived from tartaric acid
DIS	Disconnection

DMAP	4-Dimethylamino pyridine
DMPU	1,3-Dimethyl-3,4,5,6-tetrahydro-2(1H)-pyrimidinone
DMSO	Dimethylsulfoxide
DPPA	Diphenyl phosphoryl azide
DuPHOS	(1,2-bis(2,5-Diisopropylphospholano)benzene), chiral ligand
e.e.	Enantiomeric excess
EBTH	Ethylienebistetrahydroindenyl ligand
EDG	Electron-donor group
E-factor	Environmental factor
Emim	1-Ethyl-3-methlyimidazolium chloride
EQ	Environmental quotient
EWG	Electron-withdrawing group
F.-C.	Friedel-Crafts
FGA	Functional group addition
FGE	Functional group elimination
FGI	Functional group interconversion
HIV	Human immunodeficiency virus
HMPTA	Hexamethylphosphortriamide
HOMO	Highest occupied molecular orbital
HSAB	Hard and soft acids and bases
HTS	High throughput screening
Hydemin	1-Hydroxyethyl-3-methylimidazolium hexafluorophosphate
Josiphos	Chiral diphosphine ligand derived from ferrocene
LC	Lead compound
LDA	Lithium diisopropylamide
LUMO	Lowest unoccupied molecular orbital
MCPBA	*meta*-Chloroperbenzoic acid
MCR	Multicomponent reaction
MEK	Methyl ethyl ketone
Nafion	Sulphonated polytetrafluoroethylene
NDE	New drug entity
NMP	*N*-Methylpyrrolidone
NNBP	Non-nucleoside binding pocket
PAS	*para*-Acetylaminobenzensulfonyl chloride
PCC	Pyridinium chlorochromate
Phen	1,10-Phenantroline
PMP	*para*-Methoxyphenyl
PTC	Phase-transfer catalysis
p-TsCl	*para*-Toluenesulfonyl chloride
p-TsOH	*para*-Toluenesulfonic acid
r.t.	Room temperature
RaNi	Raney-nickel catalyst
RCN	Reconnection
SAR	Structure activity relationship

SCRAM	Dimer of pentamethylcyclopentadienyl Ir(III) diiodide, catalytic complex
SET	Single electron transfer
SMB	Simulated moving bed
SPR	Structure-property relationship
TBAF	Tetrabutylammonium fluoride
TBDPS	*tert*-Butyldiphenylsilyl
TEA	Triethylamine
TFA	Trifluoroacetic acid
TfO	Triphlate (triphloroacetyl) group
THF	Tetrahydrofurane
TIPS	Triisopropylsilyl
TM	Target molecule
TPPTS	Triphenylphosphine-*para*-toluenesulphonate
TS	Transition state
VAPOL	Phosphate ester of 2,2′-biphenyl-[3,3′-biphenylanthren]-4,4′-diol, chiral ligand
ZSM-5	Zeolite-based Si/Al catalyst

Chapter 1
Disconnection, Synthons, Introductory Example

Abstract Retrosynthetic analysis as an imaginative process is introduced. Disconnection and functional group interconversion are discussed. 1-(Pyridine-3-yl) propan-1-ol is selected as an exemplary target molecule for retrosynthetic analysis and its (*S*)-enantiomer for asymmetric synthesis. Interconversions of oxygen functionalities are overviewed. The acidity of the C–H bond as a key property for C–C disconnections is indicated. Some historical and environmental aspects of organic synthesis are concisely presented.

1.1 Introduction

A proposal for the synthesis of the target molecule, irrespective of its complexity, can be elaborated by *retrosynthetic analysis* based on the *disconnection approach*. For chiral molecules this approach results in a proposal for the synthesis of racemic target molecules. Preparation of one enantiomer, or optically pure target molecule, enables *asymmetric synthesis*. 1-(Pyridine-3-yl)propan-1-ol is selected to demonstrate various retrosynthetic approaches to this relatively simple target molecule and to show the complexity of asymmetric syntheses of the preferred enantiomer. An introductory example is elaborated in some detail to familiarize the reader with the philosophy behind retrosynthetic analysis and to underline the need for *chiral information* in the reacting system to complete *asymmetric synthesis*. Today *chiral variants* of synthetic reactions are the subject of intensive research, and it is said that their number is limited only by the creativity of the organic chemist.

To demonstrate the progress of organic synthesis, some historical signposts are presented and the environmental aspect of industrial syntheses briefly discussed.

V. Šunjić and V. Petrović Peroković, *Organic Chemistry from Retrosynthesis to Asymmetric Synthesis*, DOI 10.1007/978-3-319-29926-6_1

1.2 General Aspects of Retrosynthetic Analysis

As mentioned in the introduction, *retrosynthetic analysis is an imaginative process* in which the *target molecule* (**TM**) is *disconnected* into less complex structures, the next generation of target molecules. This procedure is repeated down to simple, easily available starting compounds. Methodical breaking apart of the target molecule leading to more simple structures that can be prepared by known or conceivable reactions is the basis of *retrosynthetic analysis*. Application of this procedure requires a basic knowledge of organic chemistry and rather strict adherence to certain rules.

First we accentuate that the most important retrosynthetic rule is related to the basic property of the C–C bond, electronic structure and electronic charges of the fragments that emerge on *disconnection* of this bond. The rule states that *disconnection should follow the correct mechanism*. Products of disconnection are *synthons*—anionic or cationic fragments or radicals. Behind synthons, however, real molecules should exist, denoted as *reagents* or *synthetic equivalents*.

Before consideration of the electronic structure of synthons and properties of their acceptable synthetic equivalents, let us see the general scheme that illustrates retrosynthetic analysis (Scheme 1.1).

In Scheme 1.1 the waved line and bent arrow over the line representing the critical C–C bond indicate the *site of disconnection*. The broad arrow indicates the *disconnection process* from target molecule **TM I** to the charged species, *anionic synthon* A and *cationic synthon* B. The dashed arrows then indicate the conceptual connection of *synthons* with real compounds, *reagents or synthetic equivalents*. Real compounds are denoted as **TM Ia** and **TM Ib** and called *target molecules of the second generation*. Their disconnection represents the second retrosynthetic steps 2A and 2B, and this process continues until simple, commercially available compounds are reached. Consequently, any new target molecule along the retrosynthetic scheme should have more easily available synthetic equivalents than the previous one.

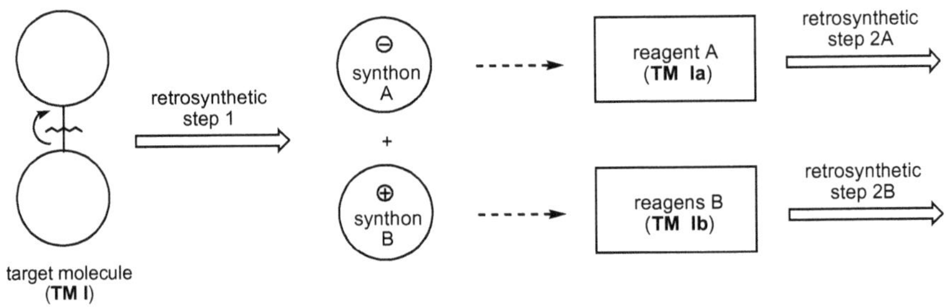

Scheme 1.1 General scheme of retrosynthetic analysis

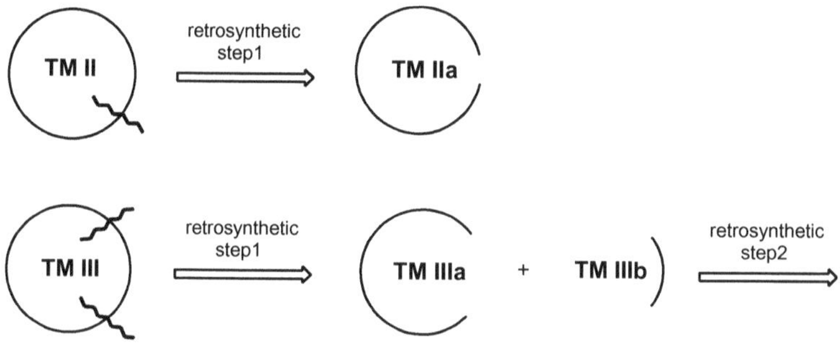

Scheme 1.2 Retrosynthetic analysis of cyclic target molecules

Let us now consider disconnections of the C–C bond in cyclic target molecules (Scheme 1.2). By disconnection of one bond in **TM II**, we open the ring, and only one synthon is formed, respectively one new target molecule **TM IIa** is envisaged. Of course, the complex open-chain structure requires further retrosynthetic consideration.

If in the cyclic **TM III** we contemporaneously disconnect two C–C bonds, two synthons, **TM IIIa** and **TM IIIb**, are obtained. Again, if one or both **TM**s of this second generation are complex structures, retrosynthetic consideration continues.

Disconnection resulting in more than two synthons represents a multicomponent reaction in the *synthetic direction*. In such a reaction three or more reagents form three or more new bonds in a one-pot reaction. Some long-known reactions belong to this group and are discussed in the Sects. 6.2.1 and 6.2.2.

1.2.1 Disconnection Versus Interconversion of the Functional Group

When disconnection of a C–C bond is completed in a way to obtain synthons with a charge stabilized by the neighboring group, we talk about *logical disconnection* or disconnection that follows the *correct mechanism*. As to the electronic properties, synthons resulting from correct disconnection accommodate a stabilized charge, negative or positive, or an electronic sextet in carbene.

Interconversion of a functional group (FGI) is one of the possible *transformations of the functional group* in the target molecule and includes change of the oxidation state or exchange of a heteroatom in this group. To the first group belongs, for example, interconversion of an ester to aldehyde or alcohol or oxidation of *sec* alcohol to ketone. Remember that the formal oxidation state of the C atom in organic compounds varies from −4, e.g., in CH_4, to +4 in CO_2. For the heteroatoms most frequently present in organic compounds, the oxidation state varies from −2 to +2 for oxygen, from −3 to +5 for nitrogen and from −2 to +6 for sulfur. Such diversity of the oxidation states of heteroatoms, primarily of O, N, S and P, enables various redox reactions of functional groups supporting FGI as one of the basic retrosynthetic steps.

In the second group we consider interconversions of functional groups with the exchange of hetereoatoms, breaking of old and formation of new C–heteroatom bonds. Examples of these transformations are interconversion of an amide to ester, thioketone to ketone or alkylhalide to alcohol. They are related to synthetic reactions: formation of amide from ester, thioketone from ketone or haloalkane from alcohol. Characteristic of all the above interconversions is the disconnection (imaginative process!) of the C–heteroatom bond, C–N or C–O. In the synthetic direction C–N, C–S and C–Hal bonds are formed. Therefore, such FGIs are also denoted as DIS–C–X, where X stands for heteroatom.

There are three main transformations of the functional group: *interconversion* (FGI), *elimination* (FGE) and *addition* (FGA), assigned by double arrows (Table 1.1). Another important retrosynthetic tool represents *reconnection* (RCN) of acyclic to cyclic structures. Reversal to a synthetic direction is the formation of an acyclic structure by ring opening. A typical example is reconnection of an α, ω-dicarbonyl compound to cycloalkene, e.g., 1,6-hexanedialdehyde to cyclohexene. In the synthetic direction ozonolysis of the double bond in cyclohexene affords 1,6-dialdehyde.

Most rearrangements well known to synthetic organic chemists as the Beckmann or Claisen rearrangement can be analyzed retrosynthetically as *retro-rearrangements*. They are characterized by *multiple disconnections-reconnections*. Detailed retrosynthetic consideration of some rearrangements is given in Chap. 8.

Concluding this section, it should be pointed out that the disconnection procedure is primarily governed by *the number, type and relative position of the functional groups in the target molecule.*

Table 1.1 Abbreviations and symbols used in retrosynthetic analysis

Retrosynthetic step	Abbreviation	Symbol
Disconnection	DIS	DIS ⟹
Reconnection	RCN	RCN ⟹
Functional group addition	FGA	FGA ⟹
Functional group elimination	FGE	FGE ⟹
Functional group interconversion	FGI	FGI ⟹

1.3 Retrosynthesis and Asymmetric Synthesis, Introductory Example

1.3.1 Retrosynthetic Analysis of 1-(Pyridine-3-yl) propan-1-ol

The target molecule of our first retrosynthetic analysis is racemic 1-(pyridine-3-yl) propan-1-ol, **TM 1**. In Sect. 1.3.2, we discuss some possibilities for obtaining optically pure enantiomer (*S*)-**TM 1** by application of *asymmetric synthesis*.

The proper starting material for alcohol **TM 1**, a target molecule of the *second generation*, is ketone **TM 1a**. Here we apply the formalism explained in Sect. 1.2.1; the double arrow indicates the imaginative process assigned as FGI revealing alcohol as a convenient precursor for the ketone.

Let us now consider the methodological aspect of the synthetic reaction proposed in the scheme. Besides complex hydride indicated as a reducing agent, there are many other non-catalytic and catalytic methods available for such reduction. All of them should be *chemoselective* to afford secondary alcohol without hydrogenation of heterocyclic ring.

We then focus on retrosynthetic consideration of the target molecule of the next generation, ketone **TM 1a**. Aryl-alkyl ketones are generally available by the well-known Friedel-Crafts (F.-C.) acylation of aromatic substrates. This reaction is discussed in some detail in Sect. 2.3.2. Here we meet an exception; the pyridine ring is an unreactive substrate for F.-C. acylation. Its inertness originates from the presence of the *N*-heteroatom in the ring, which under reaction conditions is either a protonated or coordinated Lewis acid as catalyst, usually to $ZnCl_2$ or $AlCl_3$. As the consequence of protonation or coordination of the *N*-atom, π-electrons are depleted, rendering the aromatic ring unreactive toward electrophilic acylating agents.

Note Unreactive means failing to react with specified chemical species under specified conditions. The term should not be used in place of stable, since a relatively more stable species may nevertheless be more reactive than some reference species towards a given reaction partner.

Mechanistic consideration of the synthetic reaction reveals *retro-F.-C. disconnection* as misleading (Scheme 1.4).

Here it is important to note the principle difference between the retrosynthetic steps in Schemes 1.3 and 1.4. In the first one we anticipate interconversion of the functional group (FGI), in the second one the disconnection of the C–C bond (DIS). They reflect the difference between two basic types of reactions in synthetic organic chemistry: the transformation of one functional group and formation of a new C–C bond. By far more synthetically important are C–C bond-forming reactions, which enhance the complexity of the carbon skeleton. A rather sharp difference between these two types of reactions in synthetic organic chemistry is reflected in retrosynthetic analysis. Disconnection of the C–C bond in **TM 1a** is presented in some detail (Scheme 1.5).

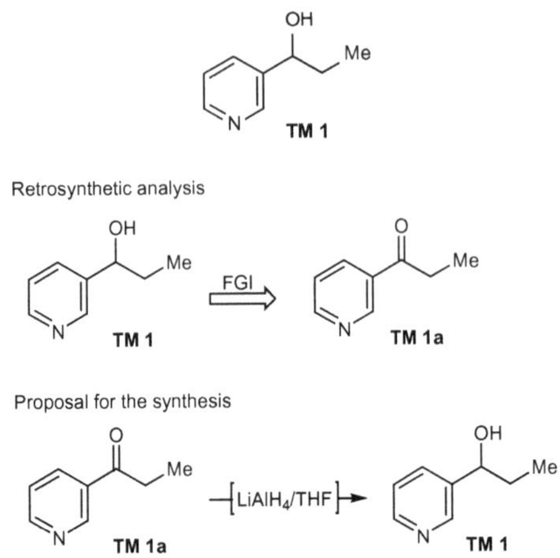

Scheme 1.3 The first step in retrosynthetic analysis of **TM 1** and proposal for the synthesis

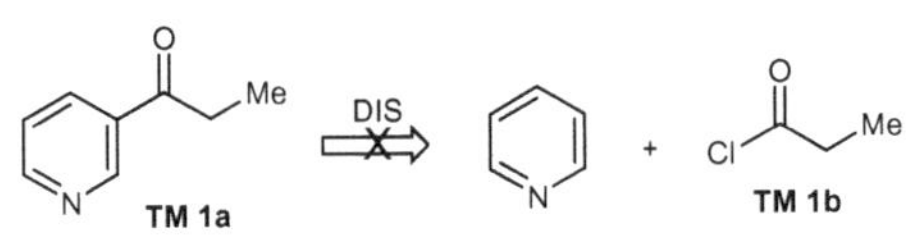

Scheme 1.4 Unacceptable disconnection of ethyl pyridine-3-yl ketone **TM 1a**

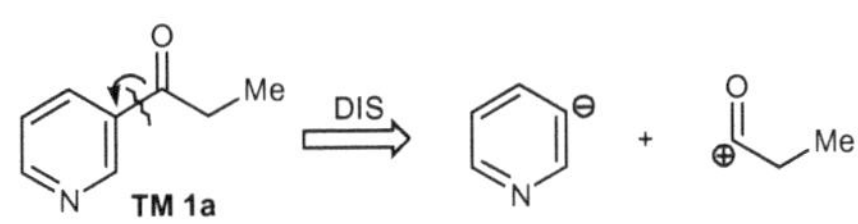

Scheme 1.5 Indication of synthons in the *retro*-F.-C. disconnection of **TM 1a**

Formalism in the presentation of C–C bond disconnection comprises, besides writing DIS over the broad arrow, an indication of the splitting by a waved line across the C–C bond, and a bent arrow indicates the electron flow to one of the C atoms. The appearance of the charged synthons in Scheme 1.5 deserves further comment. Anticipation of the chemical stability of synthons is based on their electronic properties, which are either free or coordinated to a specific stabilizing agent. The retrosynthetic step outlined in Scheme 1.5 is not acceptable since synthons correspond to reagents used in the Friedel-Crafts reaction, and we already argued why this reaction is not feasible with pyridine. It is premature to abandon this type of reaction, however, since there is an available synthetic route according to the retrosynthesis in Scheme 1.5! Before coming to this synthetic opportunity, let us consider the disconnection of the second C–C bond to carbonyl in **TM 1a** (Scheme 1.6).

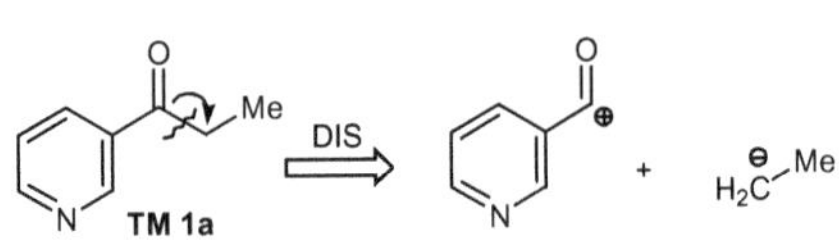

Scheme 1.6 Alternative C-C bond disconnection in **TM 1a**

At first sight this disconnection is no better than the one in Scheme 1.5. Two new synthons appear, stable carbocation and highly unstable carbanion. Now we consider what might be hidden behind these two synthons attempting to complete a synthetic step. An obvious reagent for cationic synthons is chloride of pyridine-3-carboxylic acid, known as nicotinic acid. A by far less obvious reagent is anionic synthon. This notoriously *unstable carbanion can be stabilized* within an organometallic complex, such as, e.g., EtMgBr. This quite acceptable, relatively stable organometallic species is the well-known Grignard reagent. Knowing the relative reactivity of various carbonyl compounds toward Grignard reagents, we assume difficulties. Namely, in the reaction of carboxylic ester with Grignard reagent, the first product ketone **TM 1a** is a more reactive species than the starting ester. Ketone is alkylated in the faster second step affording *tert* alcohol or carbinol and can be regarded as steady-state intermediate. This aspect of the Grignard reaction is considered in detail in handbooks of organic chemistry.

The question emerges whether retrosynthetic Scheme 1.6 is feasible in the synthetic direction. The answer is positive, but requires familiarity with properties of specific reagents in organic synthesis. In other words, besides understanding of the basic rules for disconnection, a proposal for a synthetic route requires knowledge of synthetic reactions.

The proposed synthesis that emerges from the retrosynthetic analysis in Scheme 1.6 helps obtain better insight into the reactivity of specific carboxylic acid derivatives in the Grignard reaction. Nitriles, considered as carboxylic acid derivatives, form an Mg complex of imine as a stable intermediate in the Grignard reaction. Starting from nitrile **1**, stable complex **2** is formed, which can be hydrolyzed to ketone **TM 1a** (Scheme 1.7).

Intermediary complex **2** owes its stability to strong N-Mg coordination preventing reaction with the second mole of Grignard reagent. Only on pouring of the reaction mixture into water does it hydrolyze to the targeted ketone **TM 1a**.

To completely master the retrosynthetic approach it is strongly suggested to write down a proposal of the synthetic scheme on completion of retrosynthetic analysis. The proposed scheme should contain reagents, solvents and specific reaction conditions. This concept is exemplified for **TM 1** in Scheme 1.8.

In writing the details, we often avoid the pitfalls of certain synthetic steps suggested by retrosynthetic analysis as a consequence of overlooking the competition of functional groups. Either activation of targeted functionality or protection of the competing groups is required to obtain *chemoselectivity*.

Scheme 1.7 Synthesis of ketone **TM 1a** from 3-cyanopyridine

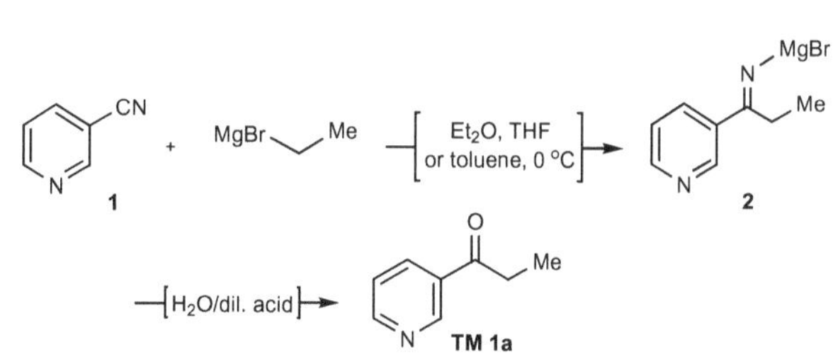

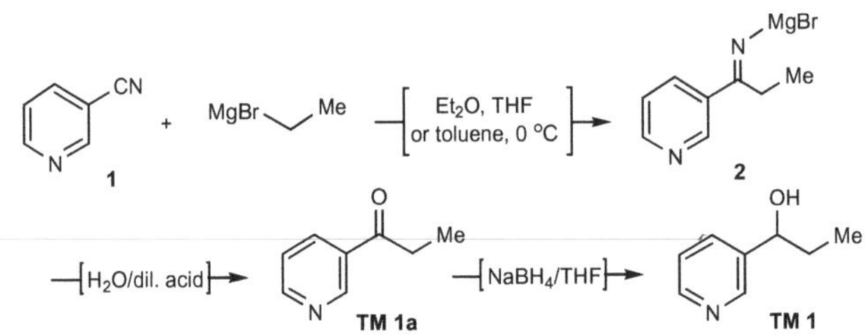

Scheme 1.8 Proposed synthesis of **TM 1**

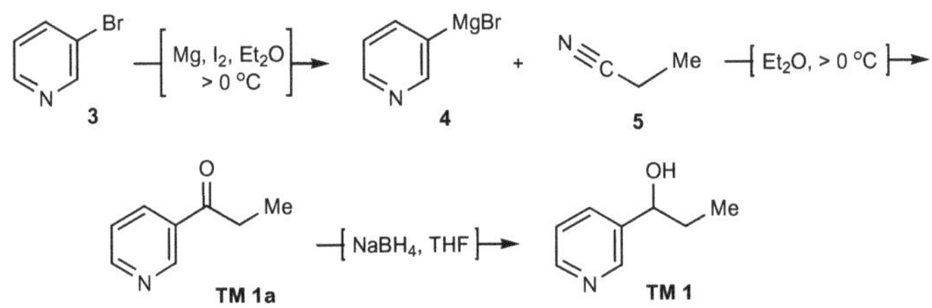

Scheme 1.9 Alternative synthesis of **TM 1**

Based on the completed retrosynthetic analyses in Schemes 1.3 and 1.6, the synthetic route in Scheme 1.8 is proposed. There is a legitimate question, however: are there some other workable routes to **TM 1**? In many real situations the answer is positive. Let us reconsider the disconnection in Scheme 1.5. We rejected this approach considering only *retro*-F.-C. disconnection. If we conceive of *retro*-Grignard disconnection as placing a cyano group and organomagnesium complex on the reagents opposite to those in Scheme 1.8, a workable synthetic route emerges (Scheme 1.9) [1].

The reagent for cationic synthon is propionitrile **5** and the reagent for anionic synthon Grignard reagent **4**, available from 3-bromopyridine.

Still, we did not exhaust the retrosynthetic routes to **TM 1**! Let us consider the disconnection and synthesis outlined in Scheme 1.10.

This scheme introduces one of the most important concepts in retrosynthetic analysis: *disconnection of the C–C bond with participation of the neighboring group*. Many examples of this concept are presented in the Chaps. 4 and 5. In the above scheme, disconnection of the C–C bond is accompanied by a contemporaneous shift of π-electrons from the O–H to C–O bond. Both processes are indicated by the bent arrows. *Participation of the neighboring O–H group* results in the formation of the carbonyl group and avoids the formation of highly unstable carbocation on a single C–O bond. A shift of π-electrons from the O–H bond is an energetically favorable concept since the hydrogen atom dissociates as a proton. The proton removed from the X–H group (X = heteroatom) is written in parentheses to satisfy the formal stoichiometry of the retrosynthetic scheme. This

Scheme 1.10 Alternative retrosynthetic analysis of **TM 1**

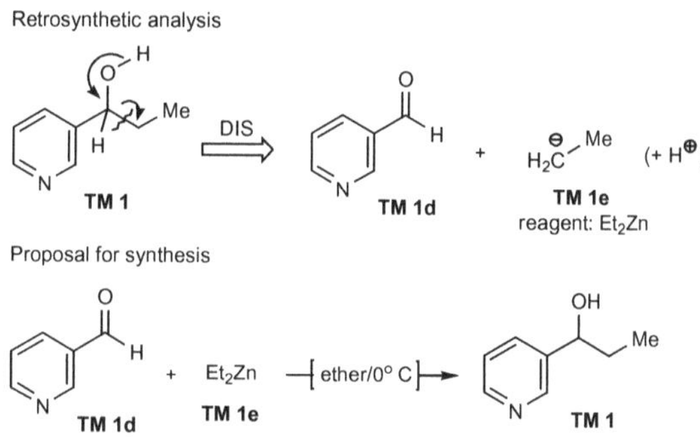

formalism is adopted in all future schemes for disconnections with the participation of the X–H group.

Note The disconnection in Scheme 1.10 results in the molecule of pyridine-3-aldehide **TM 1d** and unstable anionic synthon, which has its synthetic equivalent in organometallic compound Et_2Zn, **TM 1e**. Modern synthetic organic chemistry is increasingly based on reactions with organometallic compounds. We shall get acquainted with many of them in the following chapters, particularly with examples of asymmetric synthesis. In these complexes, the C-metal bond often exhibits a highly covalent character stabilizing the negative charge on the carbon atom. The synthetic proposal in Scheme 1.10 suggests Et_2Z in the aprotic solvent at low temperature, a usual working condition for organometallic complexes.

This introductory example served to illustrate the pallet of retrosynthetic proposals available for the synthesis for relatively simple target molecules. In the next section, we shall discover a huge difference in completing the synthesis of **TM 1** in the optically pure form; (*S*)-enantiomer is deliberately selected.

1.3.2 *Asymmetric Synthesis of (S)-1-(Pyridine-3-yl) propan-1-ol*

In the course of retrosynthetic consideration of **TM 1**, we completely neglected the stereochemistry. This is chiral molecule, and in praxis usually preparation of one enantiomer, denoted as *asymmetric synthesis* or *synthesis of an enantiomerically pure compound*, is targeted. When more stereogenic centers are present, expression of the *asymmetric synthesis of an optically pure compound* is preferred. Let as now assume that our target is (*S*)-**TM 1** (for *R*, *S* nomenclature and CIP convention, see Sect. 3.1).

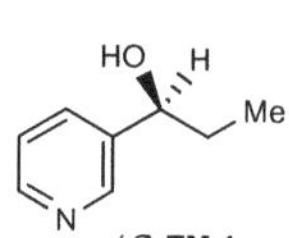

(*S*)-**TM 1**

Asymmetric synthesis of (*S*)-**TM 1** serves as an example of "*departure into the third dimension*" as we underline the substantial difference in complexity between two-dimensional retrosynthetic analysis and the completion of the asymmetric synthesis of the chiral target molecule in the optically pure form.

In order "*to depart to the third dimension*" to perform asymmetric synthesis in a way to obtain only or prevalently one enantiomer of the target molecule, there is a substantial prerequisite; *chiral information* has to be present in the reaction system. This is materialized as a chiral catalyst, chiral auxiliary agent or even chiral solvent. Detailed discussion of stereoisomerism as an introduction to stereoselective reactions is presented in Chap. 3. We suggest detailed study of the following examples after reading this chapter.

All reactions for obtaining **(*S*)-TM 1** are *asymmetric* and *enantioselective*. According to basic stoichiometry, they are analogous to those leading to racemic **TM 1** and therefore also known as their *chiral variants*. However, the stereochemical course of certain synthetic reactions is not amenable to retrosynthetic analysis since this methodology does not consider chirality or the absolute configuration of the target molecule. Many synthetic reactions proposed on the route to racemic target molecules can in principle be performed in its chiral variant as *asymmetric synthesis*. To complete multistep asymmetric syntheses, it is sufficient to envisage one *asymmetric reaction* or *enantioselective step* on the synthetic route.

This concept is exemplified by two approaches to (*S*)-**TM 1**. We have seen that the chiral center bearing the hydroxy group might be generated by

- reduction of the carbonyl group in the ketone **TM 1a**
- alkylation of the carbonyl group in the aldehyde **TM 1d**.

When performed in the presence of a *chiral catalyst with a defined absolute configuration*, the catalyst induces an *S*-configuration in the enantiomer of **TM 1**. Prediction of *asymmetric bias* or *direction asymmetry*, i.e., prevailing formation of an *R*- or *S*-enantiomer in an enantioselective process, is a difficult task.

Some details on selected asymmetric reduction and alkylation of the carbonyl group on the route to (*S*)-**TM 1** are given in Scheme 1.11. The catalytic complexes presented in this scheme are reported to produce an (*S*) configuration of the *sec* alcohols.

In order to obtain deeper knowledge of these reactions, it is suggested to consult handbooks on reactions of organometallic complexes and basic mechanisms of their catalytic activity [2–5].

Two chiral borate complexes are presented for reduction of **TM 1a**: (a) chloro-pinocampheylborane (−)-(1*R*,1′*R*)-$(Ipc)_2BCl$ and (c) borate complex derived from D-proline and dimethylsulfide-borohydride as sources of hydride. Ru-complex (b) transfers to ketone one mol of hydrogen created in situ by decarbonylation of formic acid into H_2 and CO_2 (see Sect. 3.7.1, Example 3.2).

As illustrative examples here we consider mechanisms of asymmetric reduction of **TM 1a** by (−)-(1*R*,1′*R*)-$(Ipc)_2BCl$ (Scheme 1.12) [6, 7] and asymmetric alkylation (Scheme 1.13) [8, 9].

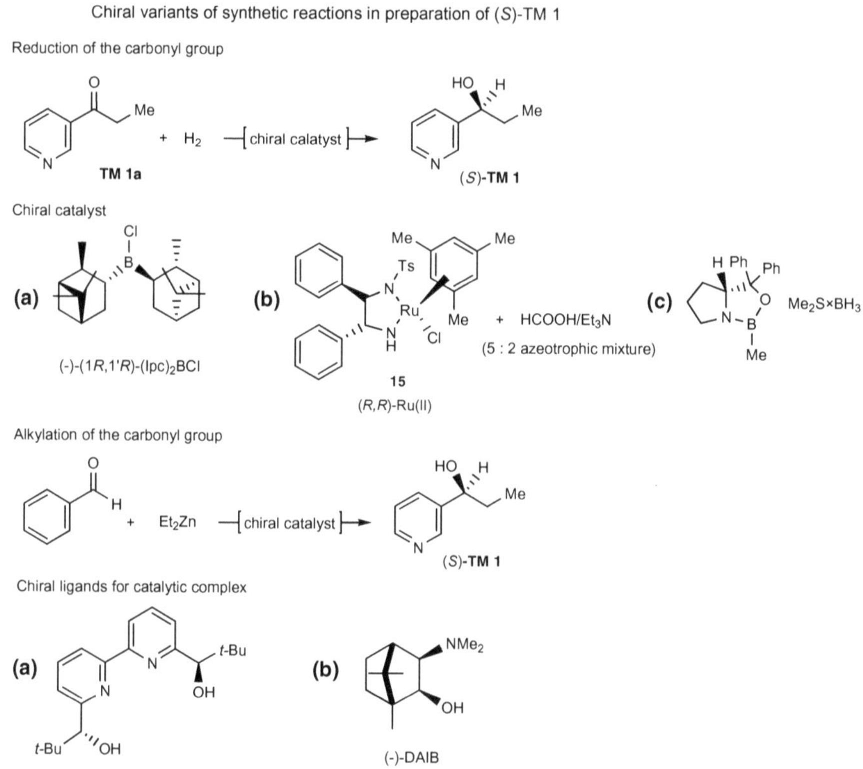

Scheme 1.11 Examples of asymmetric hydrogenation and alkylation on the route to (*S*)-**TM 1**

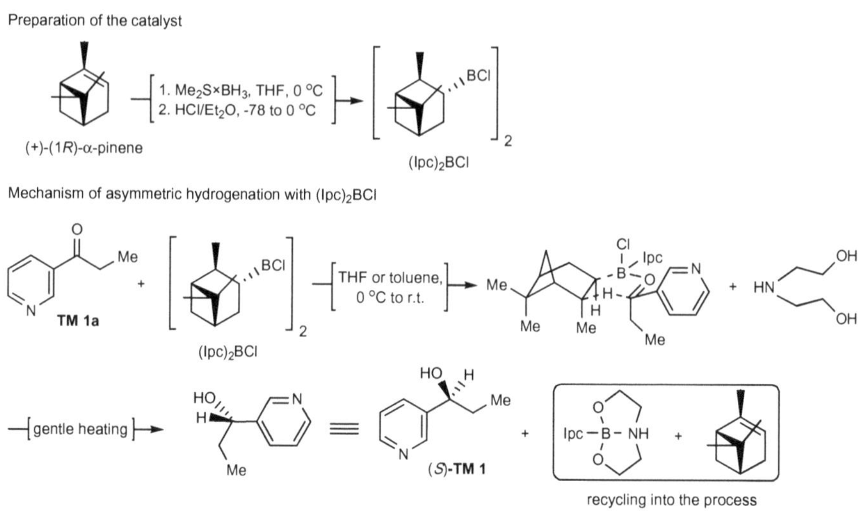

Scheme 1.12 Preparation of (−)-(Ipc)$_2$BCl and mechanism of asymmetric hydrogenation

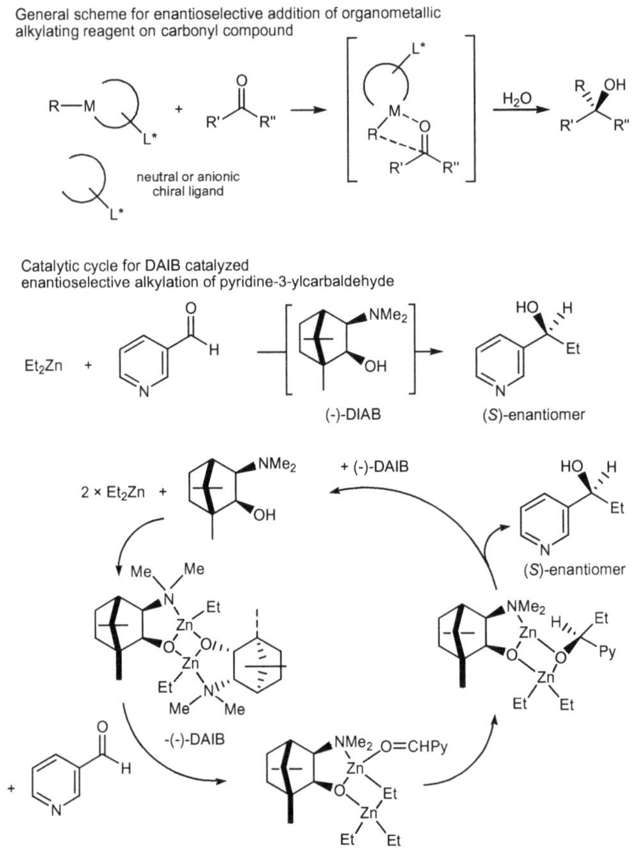

Scheme 1.13 Enantioselective alkylation and catalytic cycle for (*S*)-**TM 1**

The chiral reducing agent is available from α-pinene in two steps. It forms a six-membered complex wherein ketone **TM 1a** is coordinated to borone, extending a larger aryl group in the *pseudo*-equatorial position and a smaller alkyl group in the *pseudo*-axial position of the chelate ring. This arrangement determines the hydride ion transfer to the carbonyl C atom, forming a new stereogenic center with an *S*-configuration. Boronate is precipitated with diethanolamine and recycled in the process.

The catalytic cycle for enantioselective ethylation of aldehyde **TM 1e** is presented in Scheme 1.13.

Chiral bidentate O,N-ligand (−)-DAIB, (−)-3-exo-(dimethylamino)isoborneol, efficiently catalyzes the alkylation of aldehydes when present in a 1–2 % molar ratio. The ligand coordinates diethyl-zinc in the first step and forms a dimeric complex. An intermediary complex coordinates one mol of aldehyde and decomposes into monomeric species enabling the transfer of ethyl to the carbonyl group. In the last step, chiral alcohol is eliminated, and the ligand returns to the catalytic cycle [9].

1.4 Interconversion of Functional Groups and C–H Acidity

1.4.1 Interconversions of Oxygenated Functional Groups

Of utmost importance for the synthesis and reactivity of organic compounds are functional groups with oxygen as a heteroatom, the *oxygenated functionalities*. They are easily introduced in organic molecules and transformed into other functionalities, rendering oxygen the most widespread heteroatom in organic compounds. The hydroxy group is the oxygen functionality with the lowest oxidation state, and the carboxylic group has the highest oxidation state. The carbonyl group stays between them. Consequently, interconversion of these groups is a redox-process requiring a proper oxidizing or reducing agent.

Alcohols are frequent sources of organic compounds with other heteroatoms, and their interconversion is the subject of many retrosynthetic FGIs. Interconversion of various functional groups to primary or secondary alcohols is another frequent retrosynthetic step, since alcohols can be disconnected to available building blocks (Sect. 2.1).

A pallet of important transformations of alcohols is presented in Scheme 1.14.

Carboxylic acid derivatives deserve specific attention since they are easily interconverted, enhancing or lowering their reactivity (Scheme 1.15).

Transformation of carboxylic acids into more reactive derivatives, particularly halides and anhydrides, is an activation of the carboxylic group. Formation of more reactive carboxylic acid derivatives may be described as "moving the carboxylic group energetically uphill," from where it can be favorably transformed "downhill" into a less reactive congener. Activated carboxylic acid derivatives characterize the localized partial positive charge on the carbonyl C atom and reactivity against weak nucleophiles such as water, alcohols, ammonia or amines.

Scheme 1.14 Most important transformations of primary alcohols

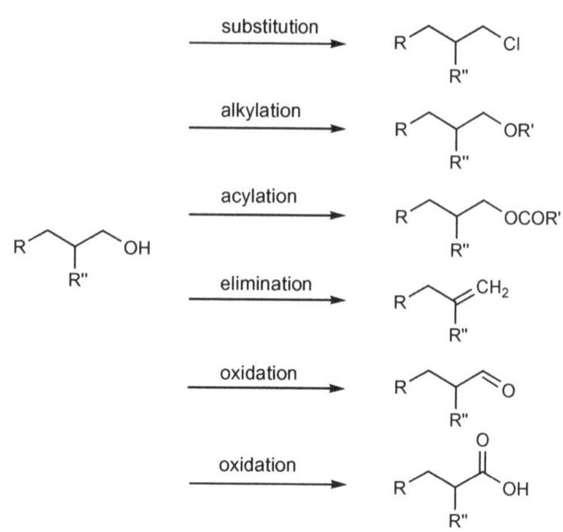

Scheme 1.15 Relative reactivity of carboxylic acid derivatives

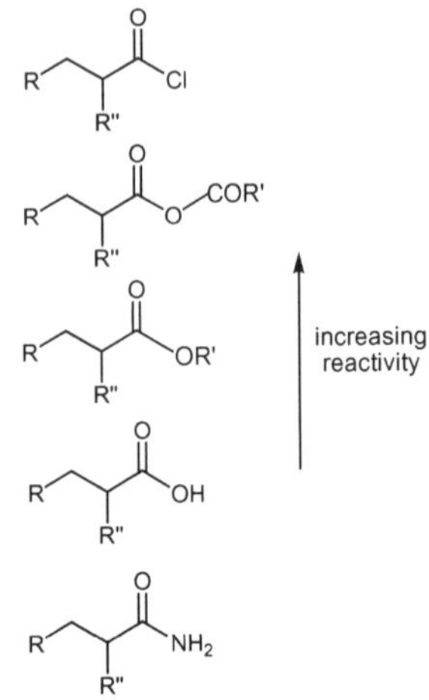

1.4.2 *Acidity of C–H Bond, Stabilization of Carbanions*

Stabilized carbanionic synthons have their synthetic equivalents in neutral reagents with an acid C–H bond. Unstable carbanions within organometallic complexes represent a specific group of reagents. Division of carbanions into *unstable and stabilized ones* is rather arbitrary and based on the pK_a value of the C–H bond, which is deprotonated. The weaker the C–H bond is, the higher the acidity; the pK_a value is lower, and a weaker base is needed for deprotonation. In the Table 1.2, the pK_a values of some C–H acid compounds and the bases needed for their deprotonation are given.

Note that *stabilized carbanions* derive from C–H acid organic compounds having $pK_a < 15.7$, lower than water. The rationale behind this classification lies in the C–H acids of these carbanions being stronger acids than water, and as a result the carbanions are stable in organic solvents in the presence of water, especially at lower temperatures. Therefore, in Table 1.2 we can define the carbanions of the last five C–H acids as stable.

Since alcohols are less strong acids than water, their conjugated bases, alkoxide ions, are stronger bases than hydroxyl ions. Generally, the stronger the C–H acid is, the weaker the base required for its deprotonation, and vice versa. Let us note that carbanions of very weak acids ($pK_a > 25$) are stabilized in the form of organometallic compounds or complex salts. Their use in C–C bond forming reactions is one of the key methods in organic synthesis. Disconnection of C–C bonds with formation of unstable carbanions as synthon, though unfavorable at the first glance, represents a useful retrosynthetic step when a stabilizing organometallic reagent is available.

Table 1.2 Some C–H acid compounds and the bases needed for their deprotonation

C–*H* acid compounds (*H* is an acidic proton)	pK_a	Base for deprotection	pK_a of conjugated acid	Base availability
Alk–*H*	42	*n*-BuLi RMgBr	42 42	Commercial RBr + Mg
Ph–*H*	40	PhLi	40	Commercial
$CH_2 = CHCH_3$	38			
$PhCH_3$	37	NaH R_2NLi	37 36	Commercial R_2NLi + BuLi
$MeSOCH_3$	35	$NaNH_2$	35	Na + NH_3
Ph_3CH	30	Ph_3Na	35	Na + Ph_3CH
HC≡C*H*	25	$NaNH_2$	35	Na + NH_3
CH_3CN	25	ibid.		
CH_3COOEt	25	ibid.		
CH_3COMe	20	ibid.		
CH_3COPh	19	*t*-BuOK	19	Commercial
$CH_3P^+Ph_3$	18	NaOMe, NaOEt	18	ROH + Na
$ClCH_2COMe$	17	ibid.		
$PhCH_2COPh$	16	NaOH	16	Commercial
$CH_2(COOEt)_2$	13	ibid.		
$MeCOCH_2CO_2Et$	11	PhONa	10	PhOH + NaOH
CH_3NO_2	10	Na_2CO_3, Et_3N, etc.	10	Commercial
$NCCH_2CO_2Et$	9	ibid.		
$Ph_3P^+CH_2CO_2Et$	6	$NaHCO_3$, NaAc	5	Commercial

1.5 Organic Synthesis and the Environment

There is a large difference in the approaches to laboratory- and industrial-scale organic synthesis, primarily concerning environmental protection. Deposition of the wastes exerts an important effect on the environment, and in this context everything besides the product is waste. In laboratory synthesis, the quantity of such wastes is not considered, or nearly neglected, until a high yield of the product and easy recovery are achieved. However, the quantity and nature of the waste, along with the energy consumption, are key factors in developing industrial synthesis. Therefore, in the last decades concepts such *green chemistry* and *environmentally friendly synthesis* have emerged.

Many "elegant syntheses" published in scientific journals produce large amounts of waste when applied on the industrial scale. This problem has been ignored for a long time, resulting in a large gap between the academic achievements in organic

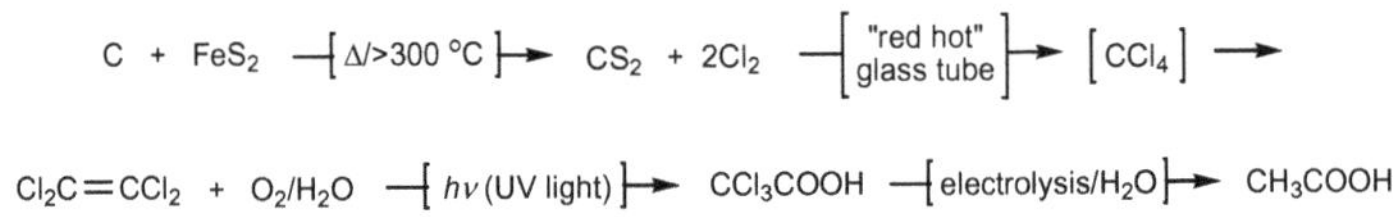

Scheme 1.16 Kolbe synthesis of acetic acid

synthesis and requirements of chemical industries. Many chemical industry problems fully correlate with or directly originate from social or community problems concerning protection of the environment and economic consumption of energy.

Wohler's synthesis of urea (1828) and Kolbe's synthesis of acetic acid (1845) moved from the scientific scene's *vis vitalis* theory. Kolbe synthesis is presented here because of its historical importance and also as an extraordinary introduction to the ecological issues of industrial organic synthesis (Scheme 1.16).

The scheme reveals the multistep character of this synthesis and its intensive energy consumption. Thermal, photochemical and electrical energy are consumed to produce, besides acetic acid, large quantities of sulfur, hydrogen chloride and salts! However, completing the synthesis of acetic acid from inorganic raw materials, a historical goal was achieved. Kolbe demonstrated that a vitalistic theory is untenable. To the contrary, the modern production method for acetic acid aims to demonstrate a drastic economy of material and energy (Scheme 1.18).

Economic use of material and energy in industrial synthesis has a direct impact on protection of the environment. Industrial synthesis is forced to minimalize or completely eliminate the production of waste. Today, "zero-emission plants" are being installed, and concepts such as zero emissions, environmentally friendly production and integrated monitoring of chemical processes have become key phrases in the chemical industry. The burden of this problem in various production plants is best illustrated by considering the quantity of waste products per kg of product. This ratio is defined as *environmental factor E* and varies greatly in different segments of the chemical industry (Table 1.3).

Note that the largest polluters are the pharmaceutical industry and producers of fine chemicals, those sectors of the chemical industry where organic synthesis has a dominant role.

The *economy of material* from the perspective of organic synthesis is properly represented by *atom utilization* or *atom selectivity*. This parameter is obtained by dividing the molecular weight of the product by the sum of molecular weights of all

Table 1.3 The *E*-factor for some industrial sectors

Industrial sector	Production per year (in tons)	*E*-factor (kg waste products per kg targeted product)
Oil refinery	10^7–10^{10}	cca 0.1
Chemical raw materials	10^5–10^7	<1–5
Fine chemicals	10^3–10^5	5–50
Pharmaceuticals	1–10^3	25–100+

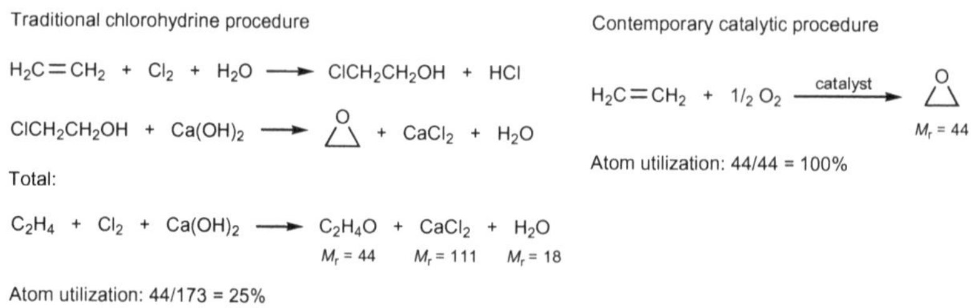

Scheme 1.17 Atom utilization in the production of ethylene oxide

compounds that are formed according to stoichiometry of the reaction equation. An example that properly illustrates this concept is the production of ethylene oxide (Scheme 1.17).

While for traditional processes the atom utilization is 25 %, in catalytic procedures it amounts to 100 %. According to this parameter, the former process resembles more to production of calcium chloride with ethylene oxide as the side product! In other words, this process affords 3 kg of calcium chloride per 1 kg of ethylene oxide by 100 % yield.

The second example, presented in Scheme 1.18, demonstrates how efficient modern technology changes facets of the Kolbe synthesis of acetic acid.

Rh(I) complex, which is soluble in methanol, serves as the catalyst in the presence of HI as promoter [10, 11]. Of approximately 5 million tons/year of acetic acid, about 50 % is produced by this process.

The next example demonstrates the difference in atom utilization and some other parameters between the traditional and catalytic process for the production of methyl methacrylate (Scheme 1.19).

Production of this monomer amounts to 20 million tons/year. Traditionally, production was based on cheap and toxic raw materials: acetone, a side product of the production of phenol, and hydrogen cyanide, which is highly toxic, a side product of the production of acrylonitrile. The process has now been abandoned because it produces 2.5 kg of ammonium hydrogen sulfate/kg of methyl methacrylate and is characterized by *E*-factor 2.5. The catalytic process, introduced by chemists of the Shell Co., is based on the methoxycarbonylation of methylacetylene (propyne). Besides 100 % atom utilization, this process is characterized by high chemical yield and selectivity, both over 99 % [12]. For the economy of any industrial process, the turnover is a particularly important technological parameter. Approximately 100,000 mols (ca. 4000 kg) of methylacetylene is

Scheme 1.18 Monsanto process for production of acetic acid

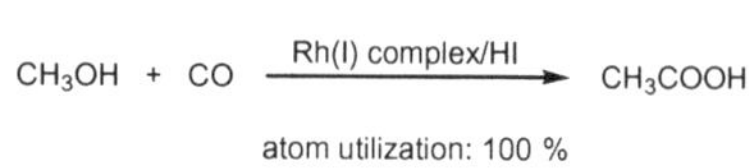

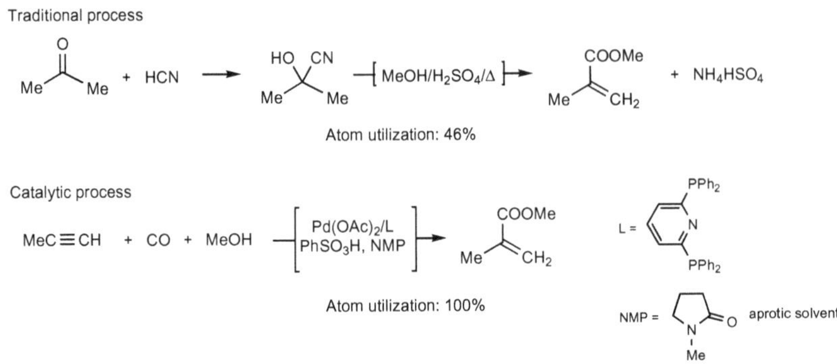

Scheme 1.19 Traditional and catalytic process for the production of methyl methacrylate

converted to methacrylate per gram of catalyst in 1 h, confirming the importance of catalysis in the development of clean technologies.

Comparison of technologies based only on the quantity of waste represents an oversimplification. The inherent nature of the waste is of prime importance, in particular its toxicity or generally negative impact on the environment. R. A. Sheldon has suggested the *environmental quotient (EQ)*, the product of the *E*-factor and the factor of *unacceptability for the environment Q* [13],

$$EQ = E \times Q$$

The scale ranging *Q* factors is not generally accepted since it depends on the position and technology of the production plant. Generally, the *Q* factor ranges from 1 for non-toxic wastes to 100 or 1000 for heavy metals or organic poisons.

The great German chemist, J. Liebig, said "*Progress of the method is progress of science.*" A similar expression applies for the chemical industry; progress of catalytic methods is progress of chemical technology. The development of the production technology for caprolactame, the starting material for the production of polymer nylon-6, serves as an excellent illustration.

Oxidation of cyclohexanone and rearrangement of oxime into caprolactame, according to traditional technology (a), result in large quantities of inorganic wastes (Scheme 1.20).

Hydroxylamine hydrochloride is prepared by catalytic oxidation of ammonia with oxygen to nitric acid and its partial hydrogenation in the presence of hydrochloric acid. In the last steps, the formation and rearrangement of oxime in the presence of a large excess of sulfuric acid require neutralization producing huge quantities of ammonium sulfate. This amounts to 4.5 kg/kg of caprolactame!

Alternative procedure (b), developed by Euchem Co., consists of catalytic oximation of cyclohexanone with a mixture of ammonia and hydrogen peroxide [14, 15]. The process is catalyzed by titanium silicate, which is a prototype of a new generation of solid, recyclable industrial catalysts, known as redox molecular sieves [16]. This step eliminates the large quantities of salts formed in the former method,

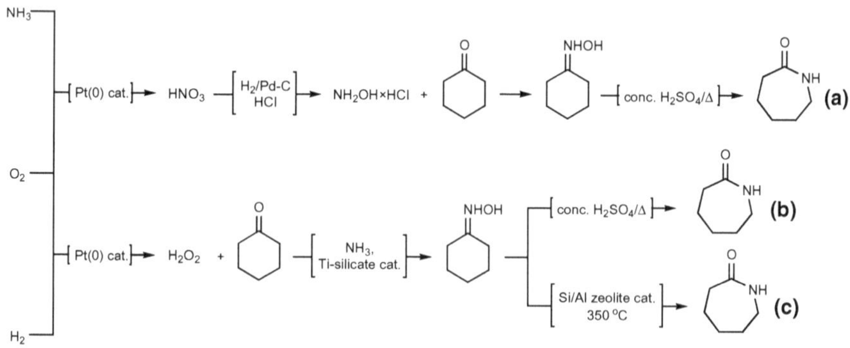

Scheme 1.20 Development of industrial methods for production of caprolactame

but uses relatively expensive hydrogen peroxide. In the last step, sulfuric acid is used as a promoter of the Beckmann rearrangement, as in method (a), and thus the formation of large qualities of inorganic waste is not completely avoided.

The rearrangement step is most efficiently performed by the original solid catalyst based on Si/Al zeolite and coded as ZSM-5, discovered by the research team of Sumitomo Co. (Scheme 1.20c) [17, 18]. It catalyzes the rearrangement to caprolactame with 95 % selectivity and 100 % yield, without side products.

References

1. Wibaut JP, van der Voort H, Markus R (2010) Rec Trav Chim Pays-Bas 69:1048–1052
2. Crabtree RH (2005) The organometallic chemistry of the transition metals. Wiley, New York
3. Astruc D (2007) Organometallic chemistry and catalysis. Springer, Basel
4. Steinborn D (2011) Fundamentals of organometallic catalysis. Wiley, New York
5. Hartwig JF (2010) Organotransition metal chemistry. From bonding to catalysis. University Science Books, Sausalito
6. Chandrasekharan L, Ramachandran PV, Brown HC (1985) J Org Chem 50:5446–5448
7. Labelle M, Belley M, Gareau Y, Frenette R, Guay D, Gordon R, Grossman SJ, Jones TR, Leblanc Y, McAuliffe M, McFarlane C, Masson P, Metters KM, Oimet N, Patrick DH, Piechuta H, Sawyer N, Yiang YB, Pickett CB, Ford-Hutchinson AW, Zamboni RJ, Young RN (1995) Bioorg Med Chem Lett 5:283–288
8. Jansen JFG, Feringa BL (1992) Tetrahedron: Asymm 3:581–582
9. Kitamura M, Okada S, Suga SS, Noyori RJ (1989) Am Chem Soc 111:4028–4032
10. Forster D (1980) US Patent 4,190,729 (for Monsanto Co.)
11. Jones JH (2000) Platinum Metals Rev 44:94–105
12. Ward SG, Taylor RC (1990) Eur Pat Appl 386833 (for Shell Int Res)
13. Sheldon RA (1994) ChemTech 3:38–47
14. Zechina A, Spoto G, Bordiga S, Geobaldo F, Leofanti G, Padovan M, Mantegazza M, Roffia P (1992) New frontiers in catalysis. Elsevier, Amsterdam
15. Romano U, Esposito A, Maspero F, Neri C, Clerici MG (1990) Chem Ind 72:610–616
16. Sheldon RA, Dakka J (1994) Catalyst Today 19:215–219

17. Sato H, Hirose K, Kitamura M, Nakamura Y (1989) A vapour phase Beckmann rearrangement over high-silicious ZSM-5, zeolites: facts, figures, future. Elsevier, Amsterdam, pp 1213–1266
18. Ichihasi H, Kitamura M, Nakamura Y, Suzukamo G (1990) US Patent 4,968,793 (for Sumitomo Co.)

Chapter 2
Retrosynthetic Analysis of the Compounds with One Functional Group

Abstract A problem-solving approach to retrosynthesis is introduced. Basic principles for good disconnections are postulated. Examples of interconversion and disconnection of carbinols, alkenes, ketones and nitro compounds are discussed. Concepts of *retro*-Diels–Alder and *retro*-Wittig disconnections are presented and the mechanisms of reactions explained. Application of the Wittig reaction on the industrial scale is exemplified by the synthesis of the analog of *bombykol*, the principal of pear odor and anti-appetizer *chlorphentermine*.

2.1 Introduction

A problem-solving approach to retrosynthesis is introduced with examples selected according to the functional group that participates in C–C bond disconnection or is interconverted. The key retrosynthetic steps suggest important synthetic reactions, such as Diels–Alder, cyanhydrin, Wittig and Nef reactions. An argumentation is presented for the feasibility of the Wittig reaction on the industrial scale and its acceptability from the environmental aspect.

The importance of nitroalkanes as building blocks and precursors of *prim*-amines and ketones is exemplified.

Retrosyntheses and syntheses of natural and commercial compounds of medium complexity are presented, such as the analog of the pheromone *Bambykol*, an unsaturated ester principle of pear odor, a key intermediate in the industrial synthesis of β-carotene, and the anti-appetizer *chlorphentermine*.

2.2 Disconnection of Carbinols

In the course of the retrosynthetic analysis of **TM 1**, we met the first example of disconnection with the *participation of one functional group* (Sect. 1.3.1, Scheme 1.10). Participation of the hydroxyl group enabled disconnection of the C–

V. Šunjić and V. Petrović Peroković, *Organic Chemistry from Retrosynthesis to Asymmetric Synthesis*, DOI 10.1007/978-3-319-29926-6_2

C bond with the formation of two acceptable synthons, a neutral molecule and carbanion with an available reagent or synthetic equivalent.

Now we start with the study of retrosynthesis by the *problem-solving approach*. This approach has characteristics of seminar work promoting knowledge of organic synthesis by retrosynthetic consideration of selected *target molecules*. They are either of commercial or scientific interest, and their retrosynthetic analysis has a certain didactic value. In the next few examples we approach the disconnection of compounds with one functional group, represented by carbinols, and tertiary alcohols.

Example 2.1 Propose good disconnection for **TM 2.1**.

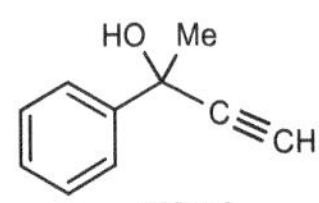

TM 2.1

Which C–C bond is preferably disconnected to the methyl, phenyl or ethynyl group depends on the stability of the carbanion, which appears as the synthon. All disconnections involve *participation of a hydroxyl group*. To evaluate the stability of the resulting carbanions, we consider the acidity of the C–H bond. For methane, pK_a amounts to 42, for benzene 40 and for acetylene 25, revealing the acetylide anion as the most stable (Table 1.2). The corresponding reagent, sodium acetylide, is easily available from acetylene and a strong base, e.g., sodium amide. The preferred disconnection leads to acetophenone **TM 2.1a** and the acetylide anion (Scheme 2.1). Participation of the OH group facilitates the disconnection of the neighboring C–C bond by the formation of the C=O bond in **TM 2.1a** with the departure of a proton.

Disconnection of acetophenone **TM 2.1a** is denoted as *retro*-Friedel–Crafts (*retro*-F.-C.) since its synthesis is completed by the Friedel–Crafts reaction. It is important to note that there is no need to generate the phenyl anion in the synthetic direction since benzene is an acceptable reagent for highly reactive acetyl chloride activated by Lewis acid.

When all groups connected to the carbinol C atom give unstable carbanions, disconnection is guided by an additional principle met in the next example.

Example 2.2 Propose possible disconnections of **TM 2.2** and explain your choice.

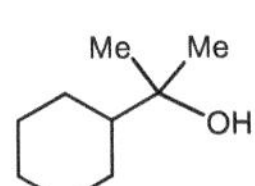

TM 2.2

Scheme 2.2 presents disconnections of the methyl and cyclohexyl group. With participation of the carbinolic hydroxyl group, disconnection affords anionic synthons, methyl and cyclohexyl carbanion. Both have proper synthetic equivalents in the corresponding Grignard reagents.

To decide on the preferred disconnection, the second principle of retrosynthetic analysis helps; *preferred disconnection leads to greater simplification of the target*

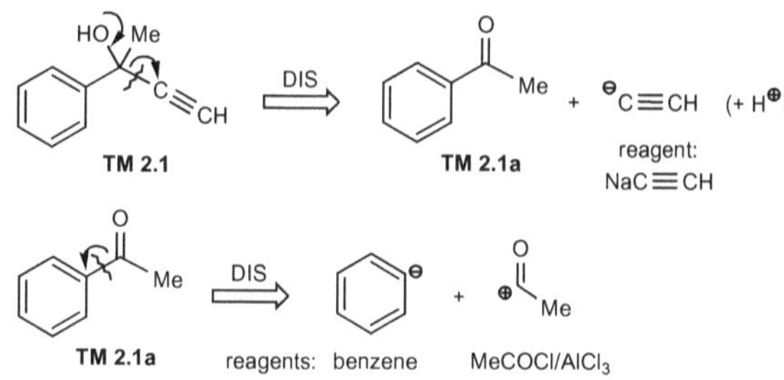

Scheme 2.1 Preferred disconnection of carbinol **TM 2.1**

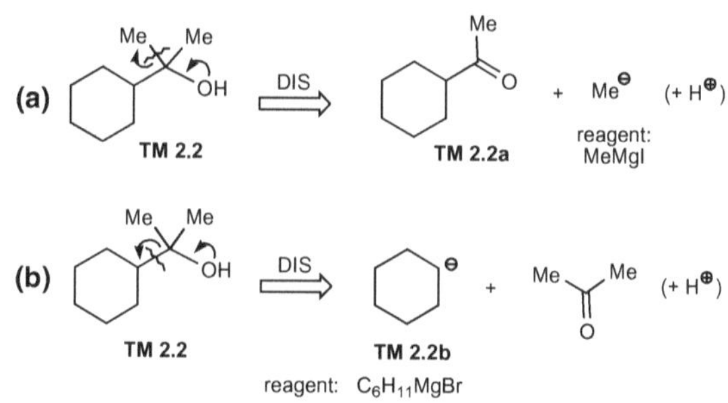

Scheme 2.2 Two possibilities for disconnection of **TM 2.2**

molecule. Maximum simplification of certain target molecules is usually achieved by disconnection, which results in two synthons of *comparable size and complexity*. Size is loosely defined by the number of C atoms in the skeleton and complexity by the number and relative position of functional groups. Greater simplification is obtained for **TM 2.2** by disconnection (b) resulting in two synthons (C_6 and C_3), closer in their dimensions then synthons obtained in disconnection (a) (C_8 and C_1). Besides, by the second disconnection easily available acetone is obtained, different from methyl-cyclohexyl ketone **2.2a**, which requires further retrosynthetic analysis.

This is the moment to consider the availability of the TMs of the second-generation **TM 2.2a** and **TM 2.2b**. Assuming a Grignard reaction in the synthetic direction, cyclohexyl-bromide is needed. On the first glance this immediate precursor of Grignard reagent **TM 2.2b** is more easily available than cyclohexyl methyl ketone **TM 2.2a** (Scheme 2.3). The two-step retrosynthetic analysis of **TM 2.2b** results in phenol, a commodity from the petrochemical industry. Its hydrogenation produces cyclohexanol, which is brominated under standard conditions.

To enter the retrosynthesis of **TM 2.2a**, we need the retrosynthetic tool presented in the following chapters. This is *functional group addition* (Sect. 1.1.2, Table 1.1). Addition of the C=C bond at the proper position in the cyclohexane ring in **TM 2.2a** offers an unexpected opportunity. This FGA leads to a cyclohexene derivative amenable to *retro*-Diels–Alder (*retro*-D.-A.) disconnection to diene and dienophile. The retrosynthetic step and mechanism of the Diels–Alder reaction are discussed in Example 2.4.

According to either of the two retrosynthetic sequences in Scheme 2.3, the synthesis of **TM 2** can be completed from easily available stating materials by

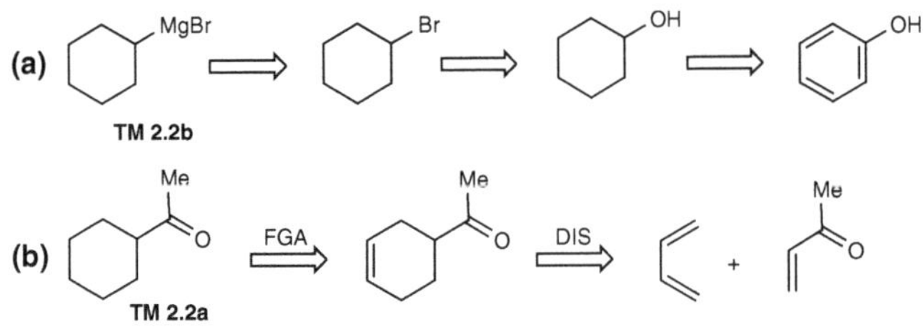

Scheme 2.3 Retrosynthetic analysis of (**a**) **TM 2.2b** and (**b**) **TM 2.2a**

well-known, conceivable reactions. Discussing this example we get acquainted with three basic principles for good disconnection:

- *disconnection should follow* the *correct mechanism*
- *disconnection should follow* the *maximal simplification of* the *target molecule*
- *by the sequence of disconnections we shall arrive at simple, easily available starting materials*

It is important that the principle of maximal simplification of the target molecule cannot be exactly defined or even quantified. Usefulness and easy understanding characterize this principle, like most rules in chemistry.

In the former examples we applied a *functional group addition* to create target molecules that are more convenient for disconnection. This concept and other interconversions of functional groups are practiced in the examples that follow.

Example 2.3 Propose retrosynthetic routes from **TM 2.3** either performing *functional group interconversion* or disconnecting the target molecule with participation of the functional group.

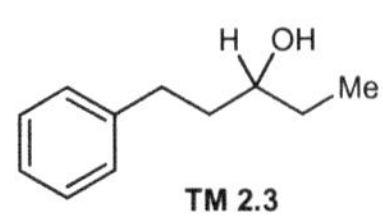

TM 2.3

Scheme 2.4 presents both retrosynthetic approaches.

FGI (a) with departure of the hydride ion and (easily removable!) proton formally results in a molecule of H_2 beside the carbonyl group. This retrosynthetic step

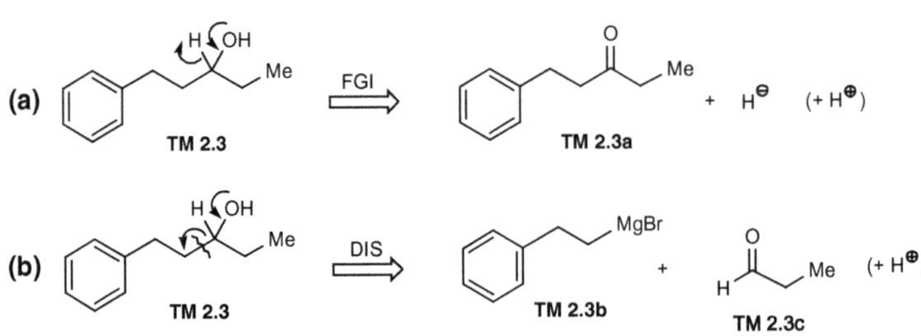

Scheme 2.4 Two retrosynthetic possibilities for **TM 2.3**

formally oxidizes the OH group with elimination of hydrogen. Disconnection (b) results in the formation of carbanion, whose synthetic equivalent is presented in the scheme. As discussed in example 1, disconnection (a) suggests the reduction of ketone **TM 2.3a** in the synthetic step. It is important to note that proposing FGI in Scheme 2.4a, we do not move far along the retrosynthetic road. Formally, any FGI with heterolytic disconnection of the C–X bond might be considered an internal redox process; the C atom is oxidized to carbocation and the heteroatom or hydrogen atom reduced to an anion. Interconversion of the functional group to a higher or lower oxidation state makes sense only when it leads to more easily available TM of the next generation. This is not the case with **TM 2.3a**, however.

The second retrosynthetic route (b) comprises disconnection of the C–C bond triggered by participation of the neighboring OH group. Grignard reagent **TM 2.3b** is presented for the carbanionic synthon, and the second one is the neutral molecule propanal **TM 2.3c**; both are available starting materials.

Example 2.4 Propose the retrosynthesis then synthesis of **TM 2.4**.

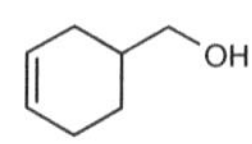

TM 2.4

This example requires more retrosynthetic imagination. We recognize the cyclohexene ring with the hydroxymethyl group in the β position of the C=C bond and consider a possible precursor for the cyclohexene ring. The benzene derivative can be excluded for good reason since partial and regioselective reduction to the substituted cyclohexene derivative represents a formidable task.

The second possibility conceives the construction of a cyclohexene ring from the proper building blocks and suggests *retro*-Diels–Alder disconnection of the cyclohexene derivative. Here we apply disconnection of two bonds resulting in two synthons, as explained in Sect. 1.1.1, Scheme 1.2, and presented for **TM 2.4** in Scheme 2.5.

The *retro*-Diels–Alder step envisages *homolytic disconnection* of two bonds in the 2,3-position related to the double bond of the cyclohexene ring. Two electrons of each σ bond and two electrons from the π bond move to the neighbor σ bonds (Scheme 2.5a). Such disconnection corresponds to the *pericyclic* or *concerted* character of the Diels–Alder reaction [1, 2].

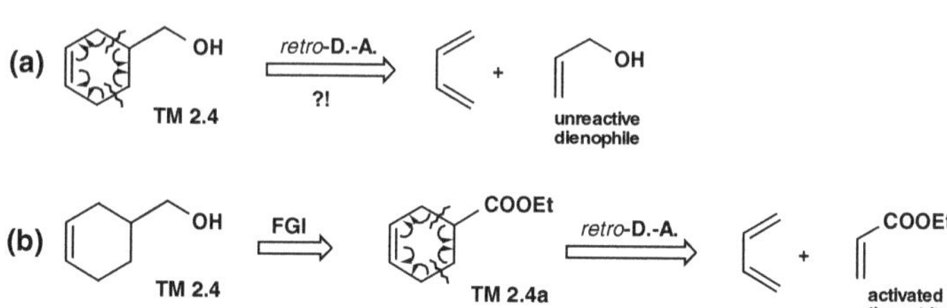

Scheme 2.5 The *retro*-Diels–Alder disconnection of **TM 2.4**

Note The concerted mechanism in the Diels–Alder reaction implies the 4π-2σ cycloaddition of two reacting partners, diene and dienophile [3–7]. To affect the productive interaction, the formation of two σ bonds, they should possess the proper electronic properties of their frontier orbitals HOMO and LUMO. The energy barrier for this reaction is lowered by the closer energies of diene's HOMO and dienophile's LUMO. This is achieved by electron-donating groups (EDG) in diene and electron-withdrawing groups (EWG) in dienophile (Scheme 2.6).

retro-Diels-Alder disconnection obeys the first rule of retrosynthesis by following the correct mechanism. Exactly for mechanistic reasons, however, the synthesis of **TM 2.4** from butadiene and allyl alcohol will fail! Disconnection (a) in Scheme 2.5 results in the unfavorable couple diene-dienophile since ethylene is substituted with the hydroxymethyl group, a weak electron-donating group rising from LUMO energy. This energetic mismatch is solved by route (b) where we first perform FGI of the hydroxymethyl group to a strong electron-withdrawing carbethoxy group in **TM 2.4a**. This retrosynthetic step leads to the reactive couple diene-dienophile.

The example of **TM 2.4** also serves to illustrate the importance of the order of retrosynthetic steps to propose a workable synthetic route (Scheme 2.7).

The last example in this section applies the former retrosynthetic concepts in a new way.

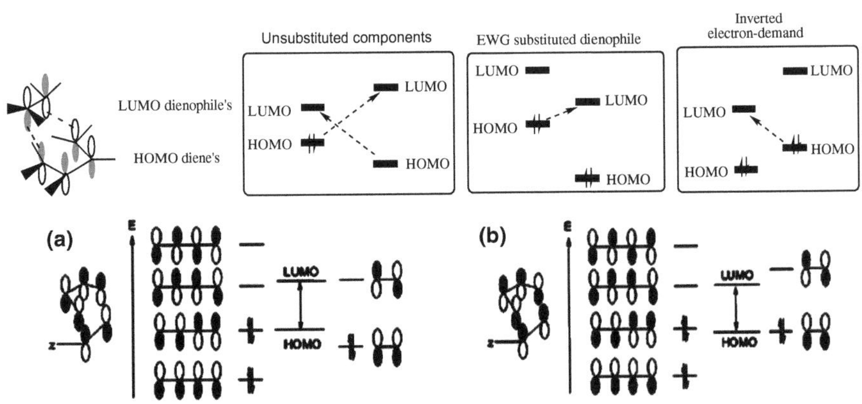

Scheme 2.6 HOMO-LUMO orbitals in butadiene and ethylene and interaction of diene and dienophile

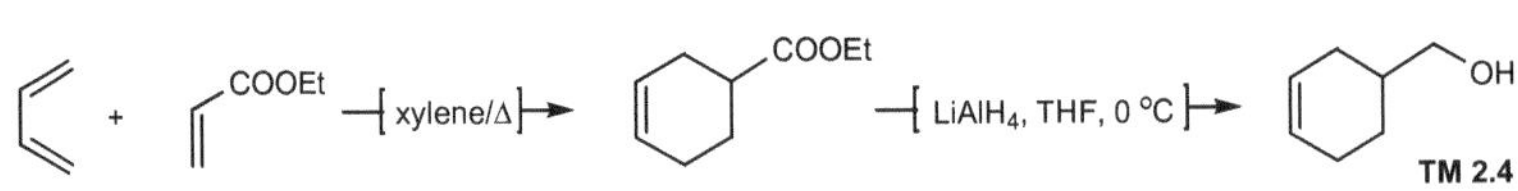

Scheme 2.7 Proposal for the synthesis of **TM 2.4**

Example 2.5 Suggest retrosynthesis then propose the synthesis of **TM 2.5**.

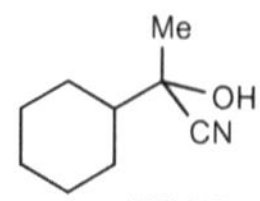

TM 2.5

Disconnection of any C–C bond from the carbinol C atom can be completed with participation of the hydroxyl group, avoiding the formation of an unstable cationic synthon. The stable cyanide anion is the preferred carbanionic synthon in the disconnection of **TM 2.5** (Scheme 2.8).

Note There are many useful reagents for the cyanide ion, from inorganic salts (NaCN, KCN) to organic cyanides (R_3SiCN, NH_4CN). **TM 2.5** contains geminal hydroxyl and cyano groups. The proposed disconnection represents the *retro*-cyanohydrin step. The *cyanohydrin reaction* is the standard route to α-hydroxy acids available on hydrolysis of cyano groups [8]. The chiral variant of this reaction is particularly important for the synthesis of biologically active compounds in an enantiomerically pure form [9, 10].

Synthon **TM 2.5a** is a neutral molecule, cyclohexyl methyl ketone. To the synthetic chemist acetophenone might spring to mind as an obvious next-generation target because of the structural relation to **TM 2.5a**. Exhaustive hydrogenation of this aromatic ketone is expected to give 1-cyclohexylethanol **TM 2.5b**, which can be oxidized to ketone by various protocols, e.g., by pyridinium chlorochromate (PCC, Scheme 2.9).

This is not a workable route, however, since the hydrogenation of acetophenone is not chemoselective and results in a mixture of products, 1-phenylethanol, ethyl benzene and ethyl cyclohexane, besides 1-cyclohexylethanol [11, 12]. The highest reported selectivity for **TM 2.5b** was 65 %, obtained with rhodium nanoparticles entrapped in boehmite nanofibers as catalyst [13].

The preferred retrosynthetic route leads over **TM 2.5a** to the next generation TM **2.5b** by the addition of a double bond (FGA) at the proper position in the cyclohexane ring followed by *retro*-D.-A. disconnection (Scheme 2.10).

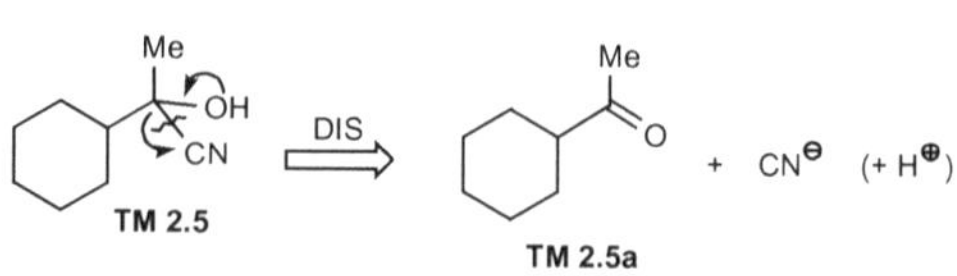

Scheme 2.8 First retrosynthetic step for **TM 2.5**

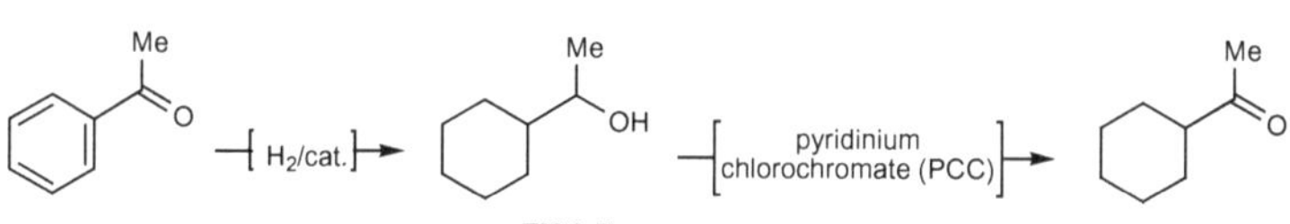

Scheme 2.9 Possible synthetic route to **TM 2.5a**

Scheme 2.10 Correct retrosynthetic analysis of **TM 2.5**

In the first step we perform *retro*-cyanohydrin disconnection, in the next step we add a double bond at the strategic position in the cyclohexane ring, and in the final step *retro*-D.-A. disconnection results in the easily available starting materials butadiene and methyl vinyl ketone.

2.3 Disconnection of Alkenes

In this section we shall closely inspect the retrosynthetic approach to simple alkenes, which includes disconnection of the C=C bond in two synthons, which have useful building blocks for synthetic equivalents. We shall not consider the formation of a double bond by β-elimination of two substituents on the vicinal carbon atoms, as for instance dehydration. More about these approaches is presented in the Sect. 4.3.1.

Double disconnection of the C=C bond can be formally presented as in Scheme 2.11.

The presented disconnection looks ugly, mechanistically incorrect and energetically unacceptable, generating synthons with double charges on the terminal C atoms. Having in mind that any disconnection is an imaginative process, we can start searching for proper reagents for such "unacceptable synthons". Surprisingly, it turns out that they exist! The reagent for "unacceptable synthon A" is benzaldehyde, where an electronegative oxygen atom donates two electron pairs to the C=O bond, compensating two formal positive charges in this synthon. To discover an acceptable reagent for "unacceptable synthon B" with a double negative charge on the terminal C atom helps another imaginative consideration. For an effective

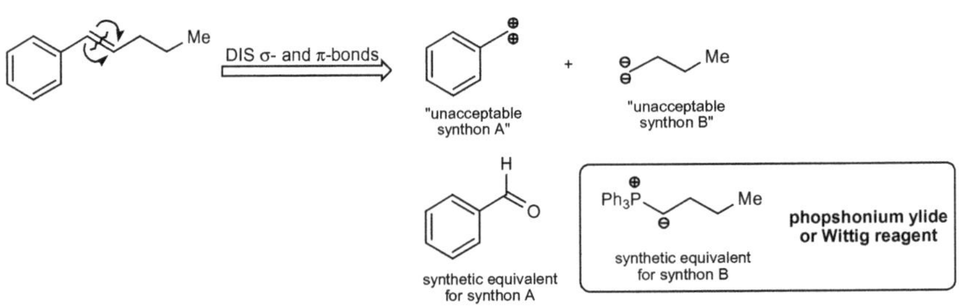

Scheme 2.11 Formal presentation of double disconnection of the C=C bond

reaction with the carbonyl group, carbanion is needed as the second reagent; so one negative charge in this synthon serves the purpose. The second negative charge is compensated by delocalization bound to an electropositive atom. The atom of choice is the P atom and as a highly effective synthetic equivalent the *phosphonium ylide* or *Wittig reagent* emerges [14–16].

Note Ylides are neutral, dipolar molecules that comprise a C atom with a formal negative charge directly bound to the positively charged heteroatom, usually nitrogen, phosphor or sulfur. They are characterized by separated charges in the *betaine structure*. Both vicinal atoms possess an octet of electrons and can be regarded as 1,2-dipolar compounds [17].

Formation of the C=C bond enables the couple carbonyl compound-ylide as a nucleophile. An electron pair of carbanions in phosphonium ylide forms one of two C=C bonds with the carbonyl C atom; the second one is formed by an electron pair "hidden" in the P–C bond. Here a general scheme is presented for the reaction of carbonyl compounds with stabilized carbanions and ylides in the preparation of alkenes (Scheme 2.12).

In the next example disconnection of an alkene reveals some typical "pitfalls" when proposing a synthetic route based on seemingly acceptable retrosynthetic analysis.

Example 2.6 Propose the retrosynthetic analysis and synthesis of **TM 2.6** without use of the Wittig reagent in the formation of the C=C bond.

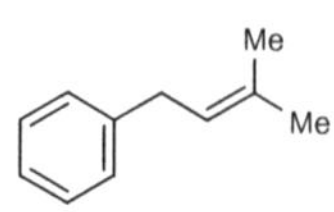

TM 2.6

Conceiving the formation of the C=C bond by the elimination of water, two possible FGIs in the first retrosynthetic step lead to two alcohols, **TM 2.6a** and **TM 2.6b**, as the targets of the second generation (Scheme 2.13).

FGI in step *a* opens a simple choice of which C–C bond in **TM 2.6a** to disconnect with the participation of the OH group. It results in the maximal simplification and two available reagents, acetone and the Grignard reagent, from 2-bromethylbenzene. FGI *b*, however, faces two possibilities for disconnection of the C–C bond in **TM 2.6b**, *b1* and *b2*. Both lead to one mole of aldehyde and one mole of Grignard reagent. Disconnection *b1* is unfavorable since Grignard reagent

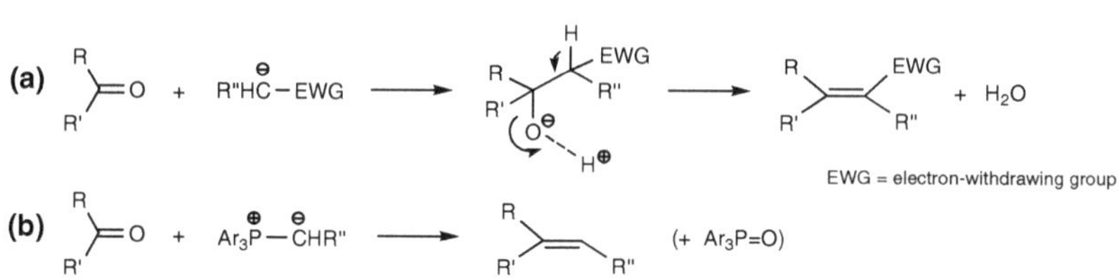

Scheme 2.12 General schemes for **a** aldol condensation and **b** Wittig reaction

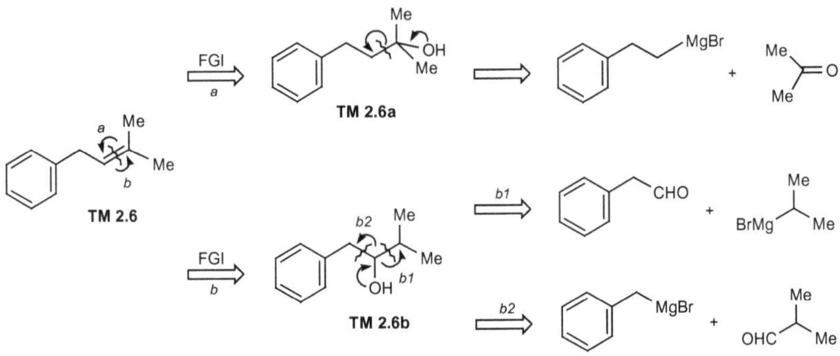

Scheme 2.13 Two possibilities in the retrosynthetic analysis of **TM 2.6**

needs to stabilize carbanion on the *sec* C atom destabilized by hyperconjugation. Disconnection *b2* leads to *iso*butyraldehyde, a large-scale industrial product available by the hydroformylation of propene catalyzed by Pt/Al_2O_3 (Sect. 3.1, Scheme 4.3). Grignard reagent derived from benzyl bromide is not a good choice because the carbanion on the benzylic C-atom is destabilized by resonance. Since the radical-anion on the benzylic C atom is less destabilized, Grignard reagent has a radical character and is prone to polymerization. Besides this argumentation against retrosynthetic consideration *b*, it should be noted that elimination of water toward the benzylic C atom competes with elimination toward the *tert*-C atom of the isopropyl group affording the structural isomer of **TM 2.6**. The origin of this competition resides in the similar C–H acidity of the two C atoms (kinetic argumentation) and comparable stabilization of the C=C bond by conjugation with the aromatic ring and by hyperconjugation with methyl groups (thermodynamic argumentation). For these reasons, the retrosynthetic route to **TM 2.6** over **TM 2.6a** is preferred.

In the frame of the next example, we consider in more detail the mechanism and steric aspects of the Wittig reaction.

Example 2.7 Suggest the retrosynthetic analysis of **TM 2.7**, and then propose a synthetic route starting from the easily available Wittig reagent.

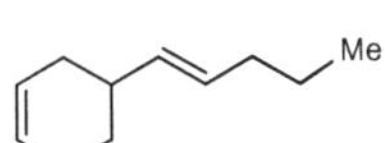

TM 2.7

The preferred double bond for disconnection is the one outside of the ring with an *E*-configuration. Disconnection of the ring will result in a formidable, branched-chain structure as a new TM! Note the position of the side chain relative to the C=C bond in the ring, suggesting the construction of cyclohexene by the Diels–Alder reaction with EWG in the side chain. *retro*-Wittig disconnection of the C=C bond in the side chain results in aldehyde **TM 2.7a** and Wittig reagent **TM 2.7b** (Scheme 2.14).

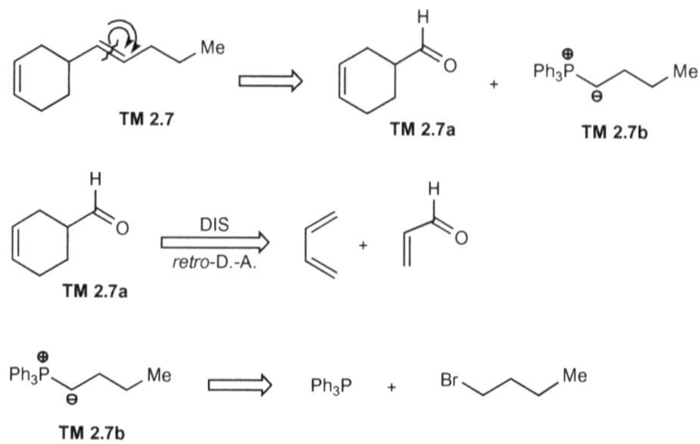

Scheme 2.14 Retrosynthetic analysis of **TM 2.7**

Target molecules of the next-generation **TM 2.7a** and **TM 2.7b** are easily available as demonstrated by the next retrosynthetic steps, the former by the Diels–Alder reaction from butadiene and acrolein (propenal) and the latter from 1-bromobutane and triphenylphosphine.

The proposed synthesis of **TM 2.7** is presented in Scheme 2.15.

On the formation of quaternary phosphonium salt by alkylation in aprotic solvent, deprotonation of C atom bound to a positively charged P atom is effected by the strong base. Phosphonium ylide reacts in the last step with aldehyde under heating in aprotic solvent to form **TM 2.7**. The most important aspect of the synthetic route in Scheme 2.15 is the stereoselective formation of the C=C bond with an *E*-configuration.

Note Let us see how this stereochemical outcome is controlled. Depending on the stability of phosphonium ylide in the Wittig reaction, formation of an *E* or *Z* isomer

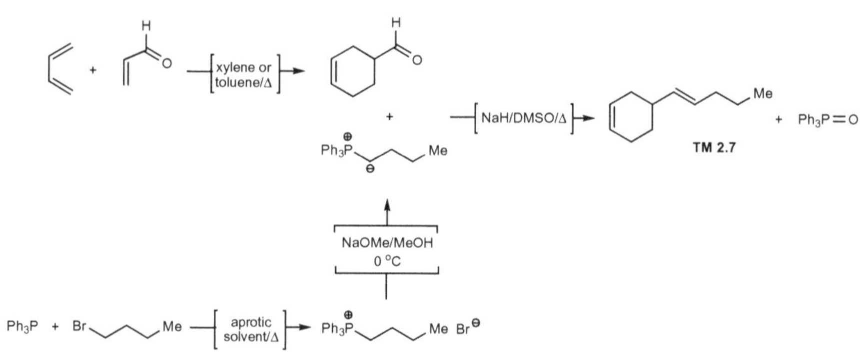

Scheme 2.15 Proposal for the synthesis of **TM 2.7**

prevails around the C=C bond. In other words, the *stability of ylides controls the stereoselective synthesis of alkenes* [18, 19].

Stabilized ylides possess EWG on an α-C atom to the phosphonium group. Unstable ylides are avoided in EWG, and the carbanion is only stabilized by the phosphonium unit. *E/Z* stereoselectivity resides in a transitory formation of a four-membered oxaphosphetane ring whose structure is dependent on the stability of ylides (Schemes 2.16 and 2.17) [18].

Unstabilized ylide forms under kinetic control of *syn*-oxaphosphetane as the consequence of the perpendicular orientation of the ylide and carbonyl group, assuring the maximal distance of larger groups. From this intermediate, Z-alkene is formed (Scheme 2.16).

When stabilized ylide reacts, equilibrium of *syn*- and *anti*-oxaphosphetanes is formed, and in the fast step the *anti*-isomer reacts, affording thermodynamically more stable *E*-alkene (Scheme 2.17).

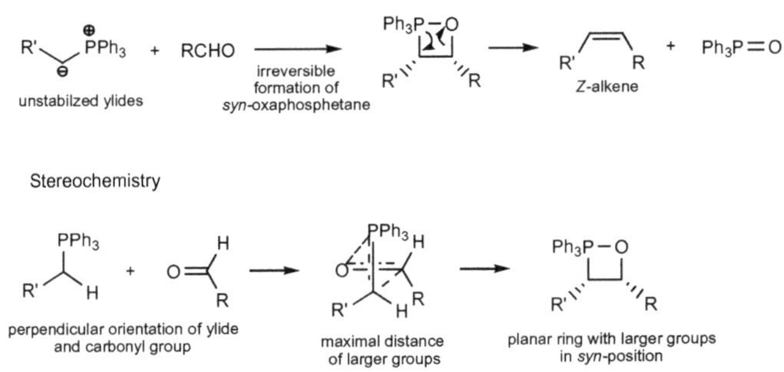

Scheme 2.16 Mechanism of the Wittig reaction with unstabilized ylide

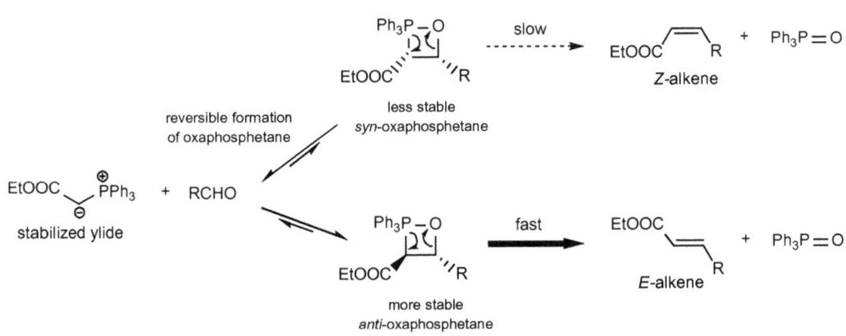

Scheme 2.17 Mechanism of the Wittig reaction with stabilized ylide

2.3.1 *Examples of the Wittig Reaction on the Industrial Scale*

The question often arises about to what extent a certain synthetic reaction is technologically feasible on the large scale and acceptable from the perspective of environmental protection. The Wittig reaction is an excellent example for consideration of the above aspects in a synthetic reaction and for the introduction of criteria for evaluation.

Besides the "atom economy" discussed in the Sect. 1.3, the feasibility of certain synthetic reactions on the large-scale is determined by other parameters, such as the price and availability of raw materials, solvents, catalysts, energy consumption and cost of equipment. Deposition of the waste and in particular the possibility for its economic recycling are decisive parameters.

Having in mind these aspects of large-scale syntheses, the question arises whether Wittig synthesis is acceptable as a large-scale production method in view of the low "atom economy," use of triphenylphosphine and formation of large quantities of triphenylphosphine oxide as a side product. To answer this question, we note that the low economy of atoms in this reaction is vastly compensated by the low-cost production of triphenylphosphine and elegant recycling of P-oxide on the large scale (Scheme 2.18).

In the production of triphenylphosphine the side products are sodium chloride and water in the catalytic recycling of phosphine oxide! These aspects greatly help to illustrate the environmental acceptability of the Wittig reaction.

In view of the different steric outcomes of the Wittig reaction with two types of ylides, we consider some examples of this reaction in the commercial production of compounds with *E* or *Z* double bonds.

Example 2.8 The unsaturated C_{18} alcohol *bombykol* is a pheromone of the silkworm female. Suggest the synthesis of **TM 2.8**, the structural analog of *bombykol*, starting from phenol as the C_6 building block.

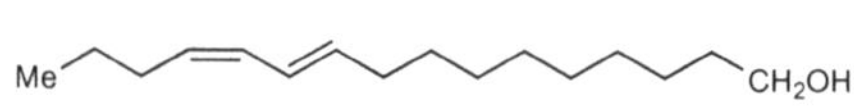

Bombykol

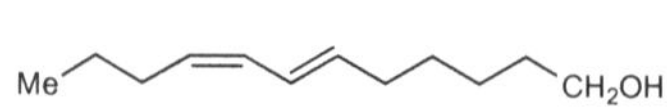

TM 2.8

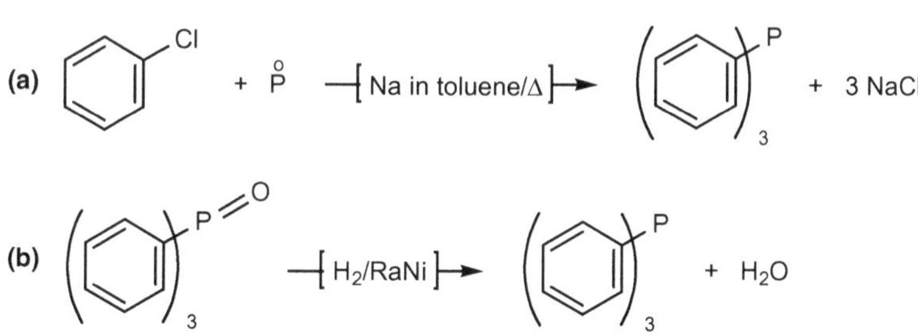

Scheme 2.18 Industrial method for the preparation and recycling of triphenylphosphine

We start retrosynthesis assuming that the terminal hydroxymethyl group in **TM 2.8** originates from phenol and so the six carbon atoms on the right side of the chain, including one in the *trans* C=C bond. This fragment obviously results in the opening of the aromatic ring. As we shall learn in Sect. 5.4 and 5.4.1, it requires two steps, Birch reduction to non-conjugated cyclohexadiene and selective ozonolysis of one C=C bond. Before reduction, the phenolic OH group is protected as methyl ether, which in the ozonolytic step becomes the COOMe group. Consequently, we can anticipate FGI of CH_2OH to COOMe in the first retrosynthetic step and propose methyl ester **TM 2.8a** as a TM of the next generation. The logical retrosynthetic assumption is also the reduction of the COOMe group in the last synthetic step. Now the dilemma arises of what double bond to disconnect first by the *retro*-Wittig, i.e., which one should be formed as the last in the synthesis. The principle of maximal simplification suggests disconnection of the C=C bond with an *E* configuration into two C_6 fragments. This disconnection results in alkene of a *Z* configuration formed from aldehyde and an *unstabilized ylide*. Now we follow a well-known principle in synthetic organic chemistry, the least stable functional group is introduced late in multistage synthesis. This principle suggests the formation of a *Z* double bond after a more stable *E* double bond, i.e., disconnection of the *E* double bond first.

These considerations suggest the retrosynthetic route outlined in Scheme 2.19.

A key point of the above retrosynthesis is the first *retro*-Wittig disconnection of **TM 2.8a** into unstabilized ylide **TM 2.8b** and conjugated aldehyde **TM 2.8c**. Inverse disconnection gives Wittig reagent with stabilized ylide on the allylic C atom, which will afford an *E* double bond instead of targeted *Z* bond. The second *retro*-Wittig disconnection of the *E* double bond in **TM 2.8c** results in stabilized ylide **TM 2.8d**. Therefore, two electron pairs are moved to C atom α- to the carbonyl group forming Wittig reagent **TM 2.8d**. The second reagent 1,6 dicarbonyl compound **TM 2.8e**, is *reconnected* into a derivate of cyclohexene **TM 2.8f**. In the synthetic direction ozonolysis of the C=C bond opens a ring in **TM 2.8f** affording 1,6-dicarbonyl structure **TM 2.8e**. On addition of the second C=C bond (FGA) to form non-conjugated cyclohexadiene **TM 2.8g** retrosynthesis brings us to the product of the Birch reduction from anisole.

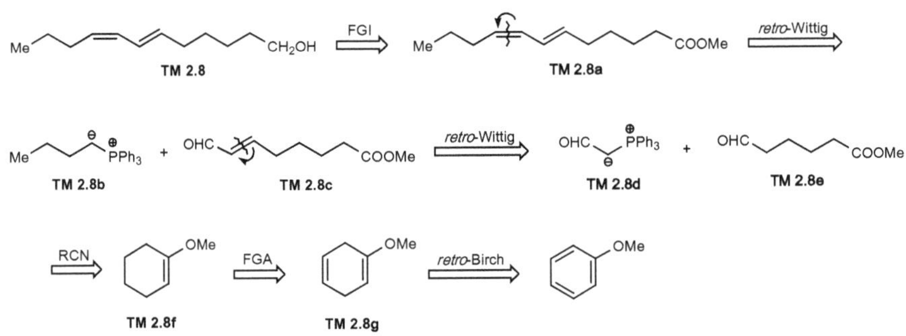

Scheme 2.19 Retrosynthetic analysis of **TM 2.8**

The proposed synthetic route to **TM 2.8** starts with anisole, which on methylation and Birch reduction affords a derivative of non-conjugated cyclohexadiene **TM 2.8g**. The importance and mechanism of this reaction are discussed in Sect.5.4. Chemoselective hydrogenation of one C=C bond is possible since the double bond in enol ether C=C–OMe is less reactive. **TM 2.8f** affords C_6 ester-aldehyde **TM 2.8e** on ozonolysis (Scheme 2.20).

The first Wittig reaction with stabilized ylide **TM 2.8d,** available from α-bromoacetaldehyde, affords conjugated *E*-alkene **TM 2.8c**. In the second Wittig reaction unstabilized ylide **TM 2.8b** forms a *Z* double bond in **TM 2.8a**. In the last step chemoselective reduction of the ester group to **TM 2.8** is completed with complex hydride.

Example 2.9 Ester with pear odor **TM 2.9** is an important product in the food and perfume industries. Complete its retrosynthesis and propose the synthesis.

TM 2.9

As in the former example, here we meet a target molecule with conjugated *E,Z* double bonds. Therefore, a similar retrosynthetic consideration is proposed (Scheme 2.21).

Unstabilized C_6 ylide **TM 2.9a** is available from 1-bromohexane, and this from *n*-hexanol (hexan-1-ol).

Note *n*-Hexanol is commodity produced catalytically by trimerization and hydration of ethylene (Scheme 2.22)

The limited quantities of oligomers formed are separated by fractional distillation of *n*-hexanol. Interestingly, this process competes with the fermentative production of *n*-hexanol from biomass, starch or cellulose. The third method for the production of *n*-hexanol is based on the hydrogenation caproic or *n*-hexanoic acid,

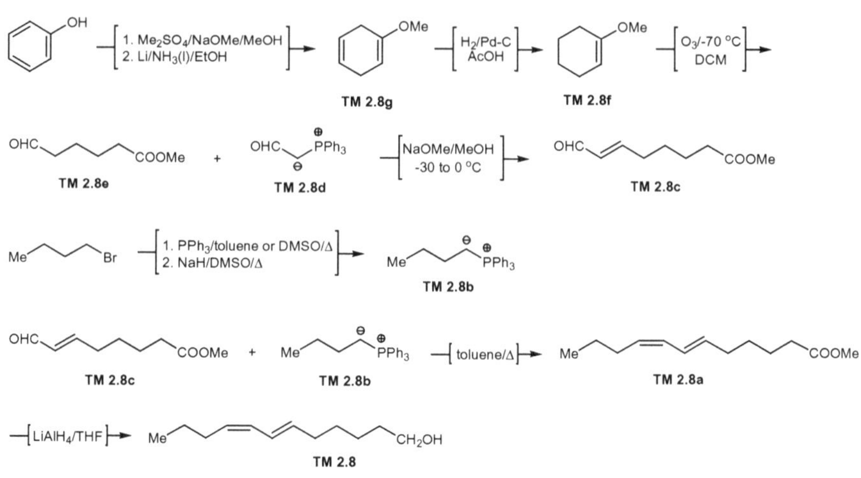

Scheme 2.20 Proposal for the synthesis of *bombykol* analog **TM 2.8**

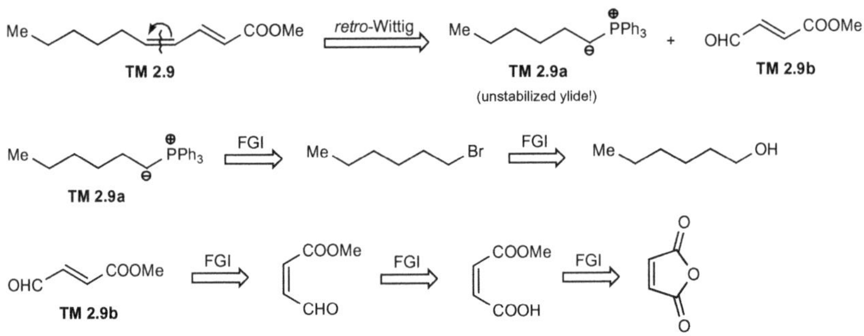

Scheme 2.21 Retrosynthetic analysis of the odor principle of pear **TM 2.9**

$$Al(C_2H_5)_3 + 6\ C_2H_4 \longrightarrow Al(C_6H_{13})_3$$

$$Al(C_6H_{13})_3 + 1/2\ O_2 + 3\ H_2O \longrightarrow 3\ C_6H_{13}OH + Al(OH)_3$$

Scheme 2.22 Industrial method for the production of *n*-hexanol

available by the fermentation of glucose. Limited quantities of this acid are produced from diethylmalonate by alkylation with 1-bromobutane, followed by hydrolysis and decarboxylation.

To complete the retrosynthesis of the C_4 target molecule **TM 2.9b**, we need to take into account easy isomerization of the more available but unstable *Z* isomer to the stable *E* isomer. Then we can correlate the structure of **TM 2.9b** with maleic anhydride as the initial building block. Two FGIs interconvert the *Z* isomer of **TM 2.9b** into maleic anhydride.

Based on this retrosynthetic analysis, a workable synthetic route is proposed (Scheme 2.23)

A key step represents the chemoselective partial reduction of acid chloride from the monoester of maleic acid. This reduction can be completed to the level of aldhyde either catalytically or by complex hydrides. In the previous example we met the reagents and reaction conditions used in the final steps.

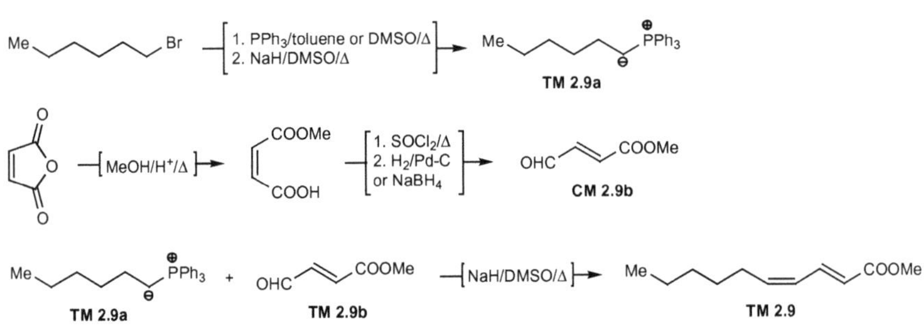

Scheme 2.23 Proposed synthesis of the odor principle of pear **TM 2.9**

2.4 Disconnection of Ketones

2.4.1 Disconnection of Dialkyl Ketones

In the introductory example (Sect. 1.3.1) we considered disconnection of **TM 1a** by *retro*-Friedel-Crafts and *retro*-Grignard. Both reactions are useful in the synthesis of aryl alkyl ketones of diverse complexity. In dialkyl ketones we consider the carbonyl group as directing the preferred C–C bond disconnection.

Example 2.10 Propose the retrosynthetic analysis for **TM 2.10** and suggest possible reagents.

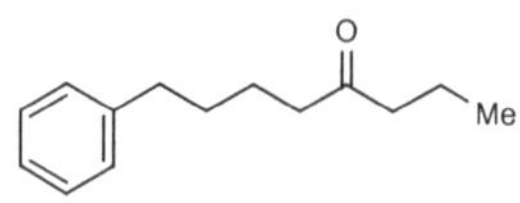

TM 2.10

Two plausible disconnections are presented in Scheme 2.24: (*a*) disconnection of the C–C bond between the α-C atom and carbonyl C atom and (*b*) disconnection of the C–C bond between the α- and β-C atoms to the carbonyl group. Both disconnections follow the principle of maximal simplification of the target molecule.

Synthons resulting from either disconnection have acceptable chemical species as reagents. We already met the disconnection of type *a* proposing retrosynthesis of **TM 1**. Disconnection *b* is based on a new concept, *introduction of an activating group*. Whereas carbocation has a logical reagent in 3-bromopropylbenezene **TM 2.10c**, carbanion cannot be selectively generated from pentan-2-one. Therefore, we conceive activating the carbethoxy group on the terminal α-C atom, enhancing its C–H acidity to pK_a ca. 11 and thus the stability of carbanion in **TM 2.10d**.

β-Keto acid **TM 2.10d** is not available; therefore, the synthetic route according to disconnection *a* is preferred. The reader should propose a synthetic route to **TM 2.10** considering the most effective approach to **TM 2.10a**.

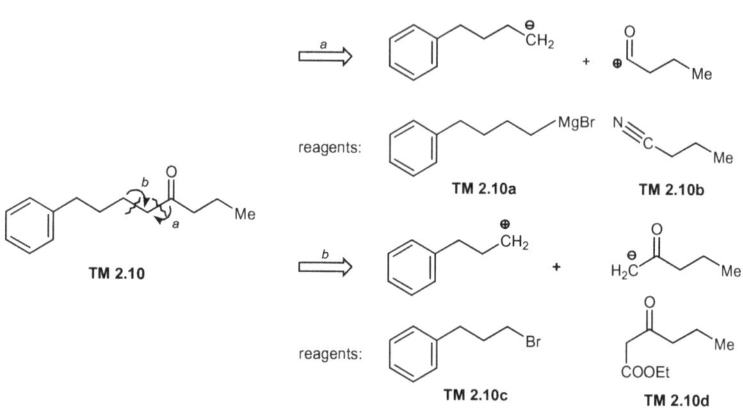

Scheme 2.24 Preferred disconnections of dialkyl ketone **TM 2.10**

Example 2.11 Complete the retrosynthesis and propose the synthesis of benzyl acetone **TM 2.11**.

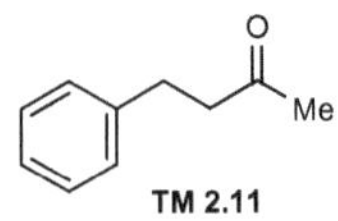

TM 2.11

In Scheme 2.25 retrosynthesis is proposed following previous argumentation for **TM 2.10**. Characteristic reaction conditions are given for the synthetic steps. Since pK_a of ethanol is 18 and pK_a of the methylenic group in acetoacetic ester is 11, deprotonation of **TM 2.10d** by sodium ethoxide is completely chemoselective. C-alkylation on the addition of benzyl bromide is followed by acidification and heating to complete the hydrolysis and decarboxylation to **TM 2.11**.

Note Activation of the C–H bond by the carbethoxy group in β-keto esters and its elimination by hydrolysis and decarboxylation, the last two steps in Scheme 2.25, deserve comment. The mechanism of decarboxylation is presented in Scheme 2.26.

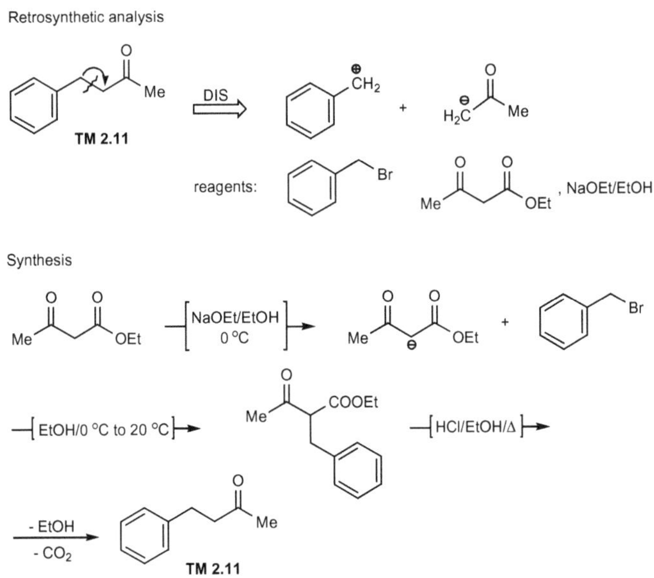

Scheme 2.25 Retrosynthetic analysis and proposed synthesis of benzyl acetone **TM 2.11**

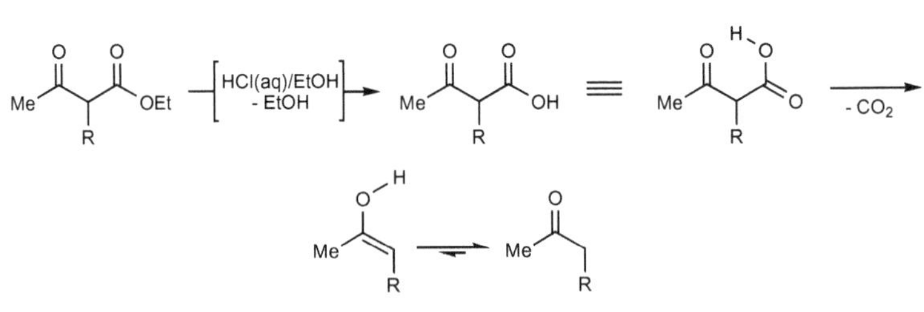

Scheme 2.26 Mechanism of elimination of the activating carbethoxy group in β-keto esters

The key step represents elimination of CO_2 in the concerted process running over the six-membered transition state and is therefore energetically favorable. Easy and selective alkylation of the methylenic group in ethyl acetoacetate, the simple decarboxylation in the last steps and availability of this starting material make it the reagent of choice for acetonide carbanion. Ethyl acetoacetate is a commodity produced by the catalytic process discussed in connection with the retrosynthetic analysis of **TM 2.13**.

Example 2.12 Propose the retrosynthetic analysis of ketone **TM 2.12** taking into account the *Z* configuration of the double bond.

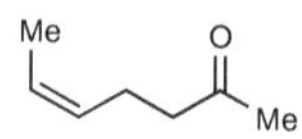

TM 2.12

This example shows how retrosynthetic consideration of the next generation target molecules is sometimes more demanding than the choice in the first step. The logical first retrosynthetic step of **TM 2.12** is the disconnection of the C–C bond in the β position to the carbonyl group generating two stable synthons, carbanion on the α-C atom to the carbonyl group and carbocation on the α-C atom to the double bond, known as the *allylic cation* (Scheme 2.27).

Since the allyl cation is stabilized by resonance with the double bond, similar to the stabilization of the benzylic cation by an aromatic ring, the reactivity of the corresponding allyl halide 1-bromobut-2-ene **TM 2.12a** is enhanced, resembling that of benzyl bromide. The anionic C_3 synthon we already met in the former example will appear in many of the disconnections that follow.

Primary bromide **TM 2.12a** is easily available by bromination of primary alcohol and this in turn by reduction of the corresponding ester, as anticipated by the first two FGIs in Scheme 2.28.

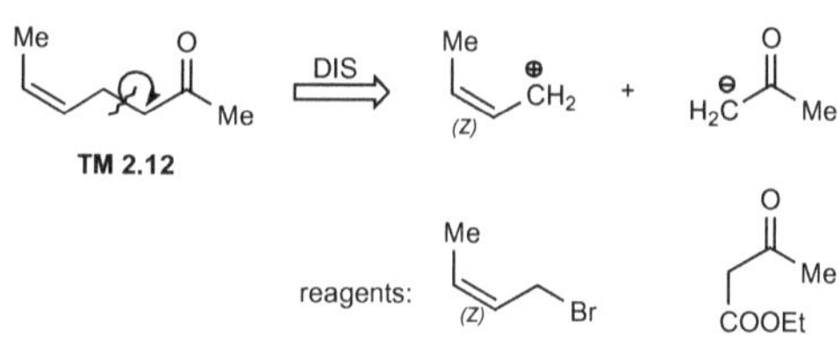

Scheme 2.27 First disconnection step in the retrosynthesis of **TM 2.12**

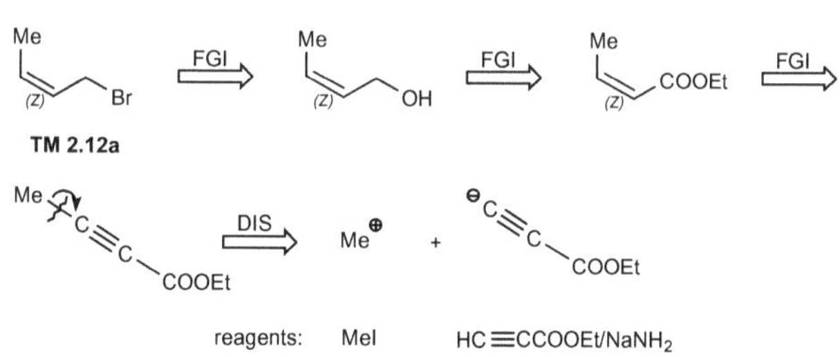

Scheme 2.28 Retrosynthetic analysis of **TM 2.12**

The main synthetic issue on the route to **TM 2.12a** represents the introduction of the C=C bond to the *Z* configuration since products of most C=C bond-forming reactions possess a thermodynamically stable *E* configuration. Here we need new knowledge; the *triple bond can be selectively reduced to a double bond with Z configuration.*

Note Partial hydrogenation of the triple bond to the *Z* double bond is possible in the presence of *Lindlar catalyst*. This is a solid, heterogeneous catalyst based on Pd deposited on calcium carbonate and doped by various morphological forms of lead [20, 21]. Lead acts as a deactivator of palladium, and Pb(II) acetate and Pb(II) oxide also serve as "catalytic poisons." Since the addition of hydrogen on the triple bond occurs *syn*-selectively, the resulting alkene possesses a *Z* configuration of the double bond.

The second approach to the *Z* C=C bond offers the Wittig reaction with unstabilized ylide as discussed in the Sect. 2.2.2, but is a less attractive method for the large scale. Anticipating *Z*-selective hydrogenation of the triple bond, we propose a third FGI to the triple bond in ethyl butynoate and its disconnection to the methyl cation and anion of ethyl acrylate.

In summary, two interconversions of ally bromide afford allylic ester, followed by FGI of the *Z* double bond to triple bond and disconnection of C_4 alkyne to methyl halide and ethyl acrylate, both available reagents.

Based on this retrosynthetic analysis, the plausible synthetic scheme for **TM 2.12a** is proposed (Scheme 2.29).

Methylation of the terminal acetylenic C atom requires deprotonation by strong bases since the pK_a of acetylene is ca. 25. *Z*-selective hydrogenation of the triple bond is followed by reduction to alcohol and bromination to **TM 12a**. In the last steps of synthesis, alkylation of ethyl acetoacetate **TM 2.12b** then decarboxylation as in Scheme 2.25 affords **TM 2.11**.

Example 2.13 Unsaturated ketone **TM 2.13** is the key intermediate in the industrial synthesis of β-carotene, a precursor of vitamin A. Start with the preferred disconnection of **TM 2.13**, continue the retrosynthesis to the reagent for the cationic

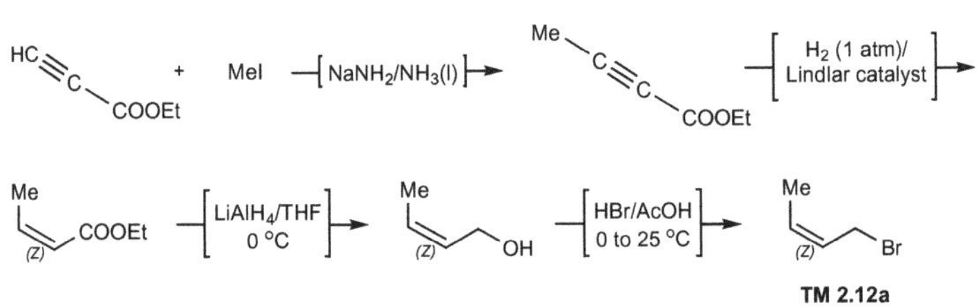

Scheme 2.29 Proposal for synthesis of **2.12a**

synthon and suggest the optimal reagent for the anionic synthon. Then propose the synthesis of **TM 2.13**.

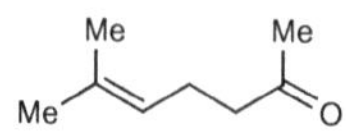
TM 2.13

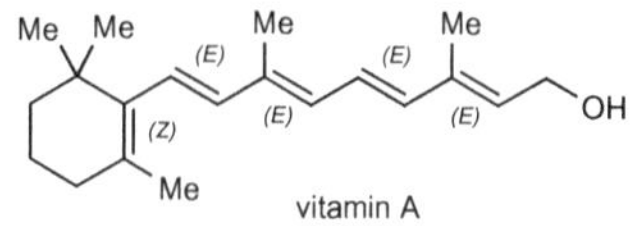

vitamin A

It is interesting to note that **TM 2.13** differs from **TM 2.12** by only one methyl group on the double bond; nevertheless, the preferred retrosynthetic analysis leads us to entirely different starting materials. As in Example 2.12, the first disconnection of the C–C bond in the α, β-position bond to the carbonyl group is preferred (Scheme 2.30).

The reagent for the allylic cation is the corresponding bromide **TM 2.13a**, a commodity with wide application in the fragrance industry. Its industrial synthesis, along with the industrial production of ethyl acetoacetate, is presented in Scheme 2.31. Both building blocks are used in the proposed synthesis of **TM 2.13**.

Starting from the petrochemicals acetylene and acetone, allylic carbinol is obtained by partial hydrogenation of the triple bond and brominated in the next step to **TM 2.13a**.

Note Here comes the important point; during bromination in the acidic medium, the intermediary allylic carbocation formed on the *tert*-C atom is in equilibrium with the more reactive *prim*-carbocation, which is brominated. This synthesis of 3,3-dimethylallyl bromide **TM 2.13a** is the basis of multi-ton industrial production since this compound is used in the agrochemical, pharmaceutical and dyestuff fields.

Two competitive methods are developed for the production of ethyl acetoacetate, both based on the dimerization of ketene (Scheme 2.31). By the first method, ketene is produced by thermal breakdown of acetone at over 300 °C [22], while the second "wet" method uses strong bases as catalysts for the elimination of hydrogen chloride from acetyl chloride [23]. Spontaneous dimerization results in a relatively stable four-membered lactone, known as diketene on the market, which on alcoholysis affords ethyl acetoacetate [24].

In the last steps of the proposed synthesis of **TM 2.13**, the alkylation of **TM 2.13b** and decarboxylation are completed according to the previously described protocol for **TM 2.11**.

Scheme 2.30 Retrosynthetic consideration of **TM 2.13**

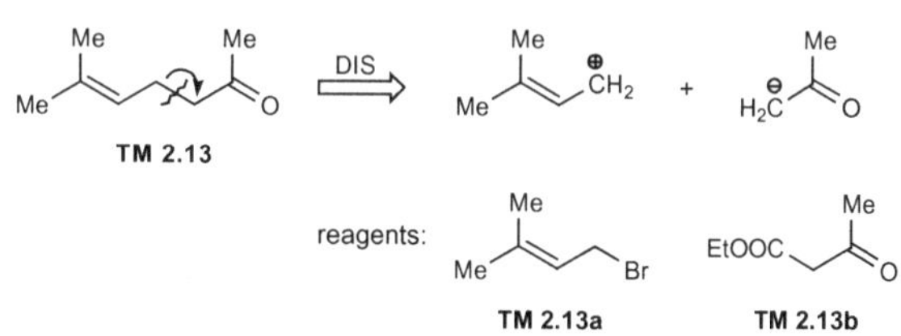

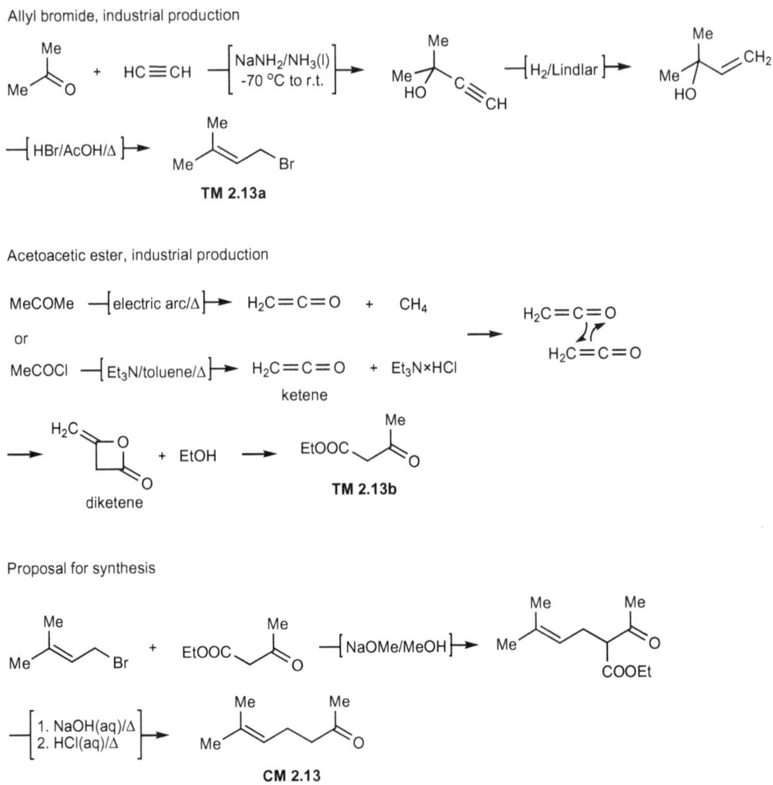

Scheme 2.31 Production of intermediates **TM 2.13a** and **TM 2.13b** and proposal for the synthesis of **TM 2.13**

2.4.2 Disconnection of Alkyl Aryl Ketones and Diaryl Ketones

As mentioned in the introduction to this chapter, *retro*-Friedel-Crafts is the basic disconnection of the Ar–CO bond in alkyl-aryl ketones. Hereafter, we will discuss two examples where problems of regioselectivity in acylation and reactivity of the aromatic ring are tackled.

Example 2.14 Propose the retrosynthesis of **TM 2.14** and indicate the problem hidden in the synthetic direction.

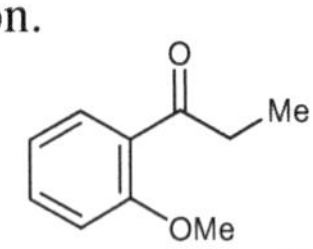

TM 2.14

The logical first step is *retro*-F.-C. disconnection to anisole (methoxybenzene), a commodity available by the methylation of phenol (Scheme 2.32).

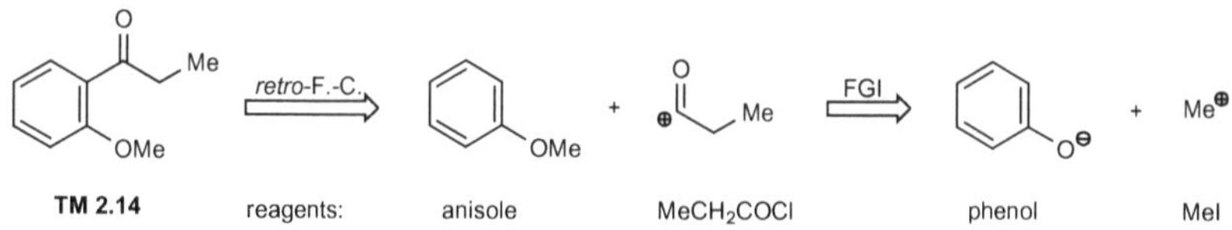

Scheme 2.32 Retrosynthetic analysis of **TM 2.14**

According to this retrosynthetic analysis, synthesis cannot be unambiguously completed. *ortho*-Acylation is sterically perturbed by the methoxy group, and an undesired *para*-isomer is prevalently formed (Scheme 2.33).

This issue can be solved by two approaches. The first is based on the separation of two structural isomers by known methods, primarily by selective crystallization or chromatographic separation on a preparative scale. The second introduces a *protecting group* to the *para*-position and eliminates it on completed acylation. In our example a good protecting group is chlorine. *para*-Chlorophenol is a commodity that is submitted to hydrogenolytic removal of chlorine on O-methylation and *ortho*-acylation using the Pt/C catalyst under controlled conditions.

Example 2.15 Propose the retrosynthesis of **TM 2.15**. Explain the preferred disconnection and solve the issue of regioselectivity in all synthetic steps.

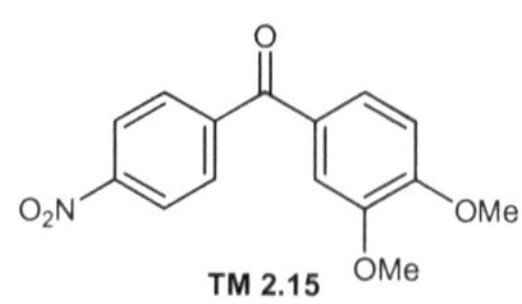

TM 2.15

An obvious retrosynthetic step is *retro*-F.-C. disconnection, and the decision between two Ar-CO bonds is unambiguous in favor of the bond to the dimethoxyphenyl unit since this substrate is activated for F.-C. acylation (Scheme 2.34). Alternative disconnection leads to nitrobenzene, an unreactive aromatic compound in F.-C. acylation. Its high inertness enables its use as the solvent for this reaction!

The prevailing formation of *meta, para*-dimethoxy isomer **TM 2.15** is directed by steric perturbation at *ortho*-positions by methoxy groups.

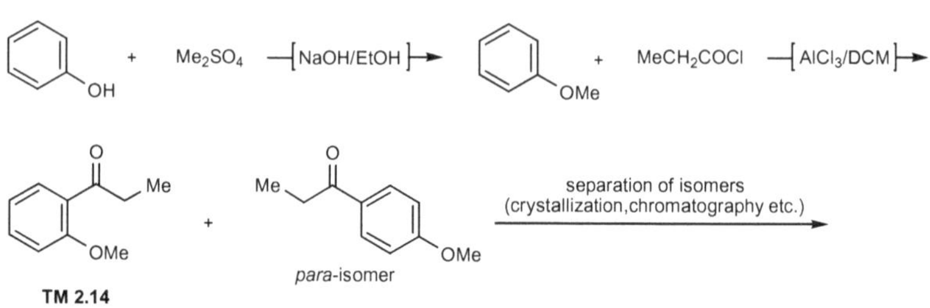

Scheme 2.33 Proposed synthesis of **TM 2.14**

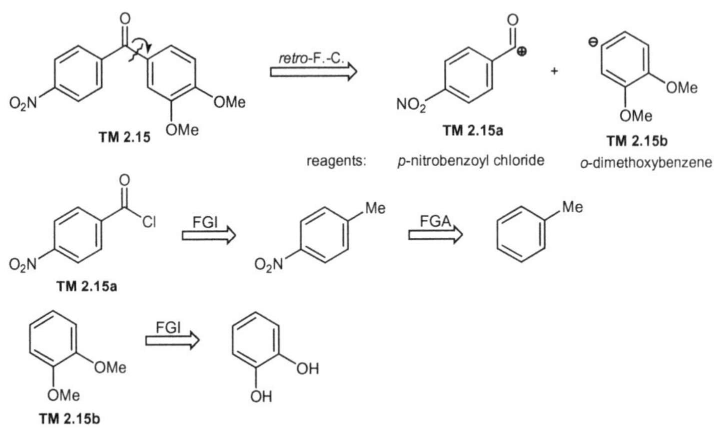

Scheme 2.34 Complete disconnection scheme for **TM 2.15**

Note Reagents for **TM 2.15a** and **2.15b** are available aromatic compounds, products of the petrochemical industry. *Para*-nitrobenzoic acid is produced by nitration of toluene to *para*-isomer as the prevailing product, followed by oxidation of methyl to the carboxylic group. *Ortho*-dimethoxybenzene is produced from *ortho*-diphenol, which in turn is available by oxidation of phenol. One technological process uses hydrogen peroxide as oxidant [25], and annual production of *ortho*-diphenol reaches 20,000 tons/year, mainly intended for the production of pesticides and perfumes.

Now we can propose the complete synthesis of **TM 2.15** (Scheme 2.35).

Note the parallel synthetic steps leading to key intermediates, which in the last step enter the F.-C. reaction affording **TM 2.15**. This is a general characteristic of *convergent syntheses* where two large building blocks are prepared separately and

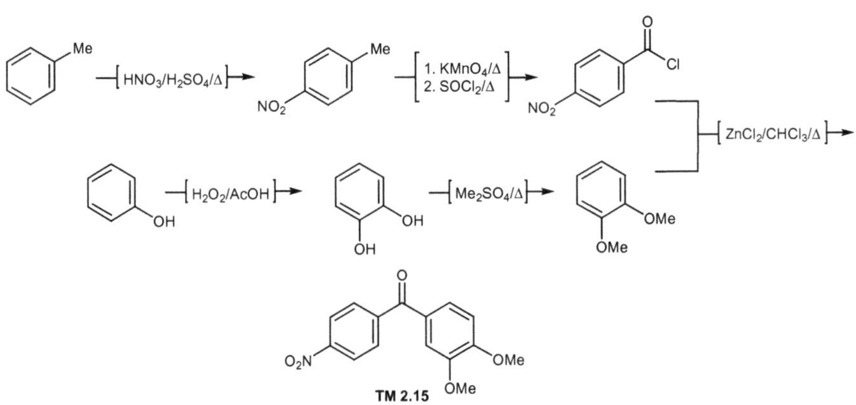

Scheme 2.35 Proposal for the complete synthesis of **TM 2.15**

in the last step coupled into a final target molecule. Convergent synthesis reflects retrosynthesis where *maximal simplification* of the target molecule is performed in the first retrosynthetic step.

2.5 Interconversion of the Nitro Group, Nitroalkanes as Building Blocks

The amino and keto groups can be introduced in the target molecule in many different ways. The amino group is usually introduced by substitution of a good leaving group with nitrogen nucleophiles or by reduction of the imino or nitro group. The second approach is preferred since ammonia and amines are weak nitrogen nucleophiles. The keto group is most frequently introduced by oxidation of *sec* alcohols or by Grignard reaction as discussed in Sect. 1.3.1, Schemes 1.7 and 1.9.

With the next examples we introduce a new synthetic strategy for target molecules with oxygen and nitrogen functional groups. It is based on the use of building blocks with nitro groups available on the industrial scale by nitration of hydrocarbons. Reduction to the amino group or oxygenation to the keto group completes the approach to these functionalities.

Example 2.16 Consider the retrosynthesis of the anti-appetizer *chlorphentermine* **TM 2.16** and propose its synthesis.

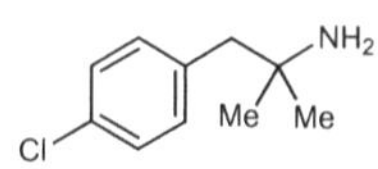

TM 2.16
Chlorphentermine

The presence of the Me_2CNH_2 group suggests the use of the Me_2CHNO_2 building block. Hence, in the first retrosynthetic step we propose FGI to the nitro group in **TM 2.16a** (Scheme 2.36).

This is an appealing solution since disconnection of the central C–C bond in **TM 2.16a** results in stable synthons, benzyl cation and α-carbanion of 2-nitropropane. In the above scheme, the corresponding reagents are immediately presented and an unambiguous and short synthesis can be proposed (Scheme 2.37).

Importantly, in the scheme the generation of carbanion before the addition of benzyl chloride is indicated since the methoxy anion behaves as a competitive nucleophile. On completed alkylation, **TM 2.16a** is reduced chemoselectively

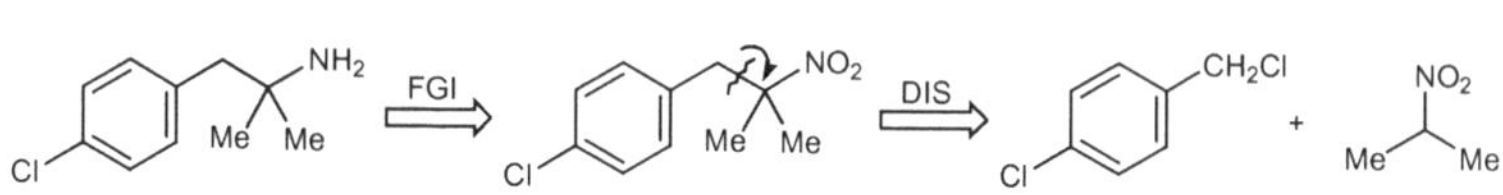

Scheme 2.36 Retrosynthesis of *chlorphentermine* **TM 2.16**

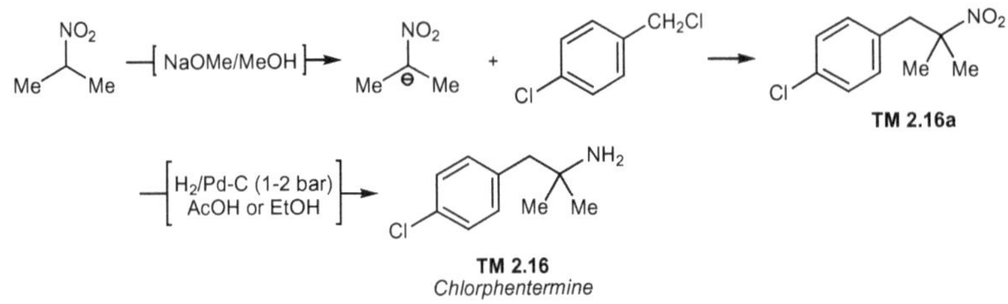

Scheme 2.37 Proposal for the synthesis of *chlorphentermine* **TM 2.16**

under controlled conditions affording **TM 2.16**. Higher hydrogen pressure or a larger ratio of the catalyst might cause the hydrogenolysis of chlorine.

Example 2.17 Consider the retrosynthesis of diamine **TM 2.17**, a structural congener of **TM 2.16**, and propose a short synthesis.

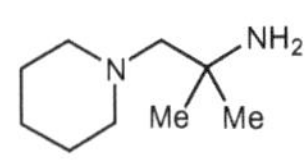

TM 2.17

Again we have an amino group on the *tert*-C atom where it cannot be introduced by direct nucleophilic substitution. Therefore, the FGI of the amino to nitro group springs to mind (Scheme 2.38).

Disconnection of **TM 2.17a** requires new knowledge. This structure corresponds to the Mannich base since the *tert*-amino group is present in the β-position to the strong electron-withdrawing nitro group. The *retro*-Mannich type disconnection of two bonds leads to simple starting materials, piperidine, formaldehyde and 2-nitropropane. In the same scheme are proposed reaction conditions for the synthesis of **TM 2.17**. More details on the *Mannich reaction* are presented in Sects. 4.4.2 and 6.1. Here it suffices to mention that this *three-component reaction* affords β-amino carbonyl compounds known as *Mannich bases*.

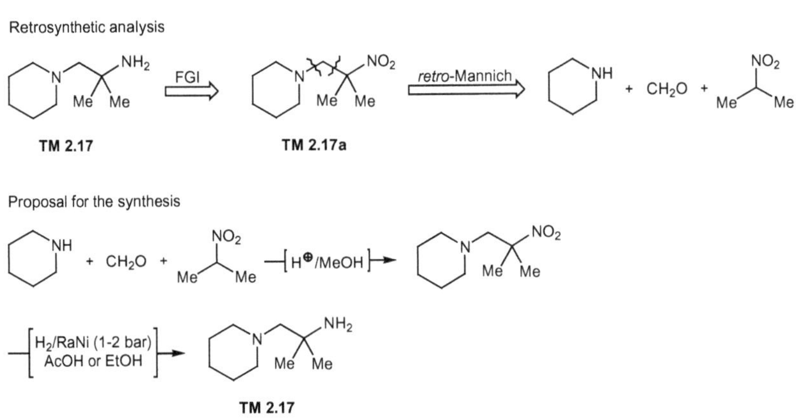

Scheme 2.38 Retrosynthetic analysis and proposal for the synthesis of **TM 2.17**

Example 2.18 Complete the retrosynthetic analysis of C_8 diamine **TM 2.18**, an important monomer for the industrial production of polyamides, and suggest its three-step synthesis starting from easy available commodities, nitromethane, acrylonitrile and isobutyraldehyde.

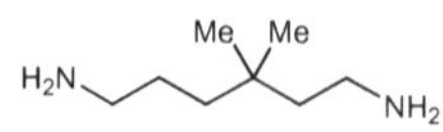

TM 2.18

Obviously, both primary amino groups originate from the CN and NO_2 group in two building blocks. The next consideration requires more retrosynthetic skill. Since the CN group is incorporated as acrylonitrile, this C_3 building block is the origin of the amino group and three C atoms left from the quaternary one. Disconnection of the C–C(Me_2) bond requires the formation of carbanion on the *tert*-C atom of the Me_2C group and acrylonitrile as an electrophile. This reaction is known as the *Michael addition* of stable carbanions to enones as electrophiles and is discussed in more detail in Sect. 4.4. The amino group on the right side and terminal C atom obviously originate from nitromethane. The remaining C_4 unit at the branching point then belongs to isobutyraldehyde.

Having identified all building blocks in **TM 2.18,** we can propose retrosynthetic Scheme 2.39.

The first two FGIs interconvert two amino groups into their precursors in **TM 2.18a**; next FGA introduces the double C=C bond in **TM 2.18b**, enabling *retro*-aldol disconnection of nitromethane and **TM 2.18c**. Cyano-aldehyde **TM 2.18c** affords acrylonitrile and stabilized carbanion of isobutyraldehyde on *retro*-Michael disconnection.

According to this retrosynthetic consideration, the short synthesis of **TM 2.18** is proposed (Scheme 2.40).

The particular convenience of this three-step synthesis represents the *contemporaneous reduction* of the C=C bond and two nitrogen functionalities in the last step.

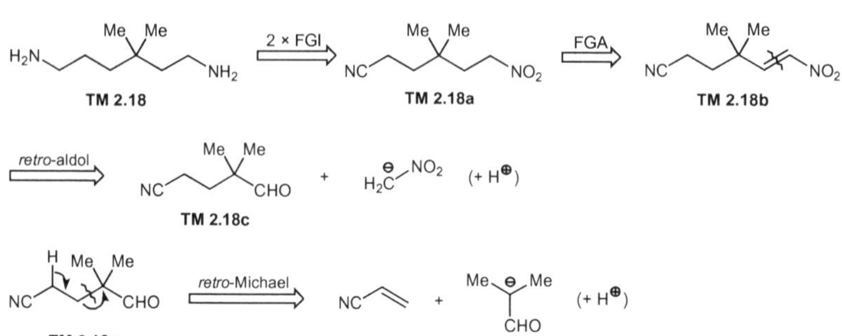

Scheme 2.39 Retrosynthetic analysis of **TM 2.18**

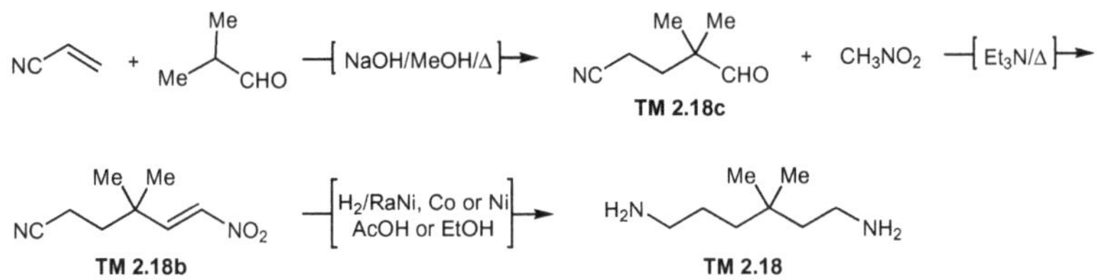

Scheme 2.40 Proposal for the synthesis 1,6-diamine **TM 2.18**

Example 2.19 Consider the synthesis of 1,3-disubstituted cyclohexane **TM 2.19** using C_5 diene as a strategic building block.

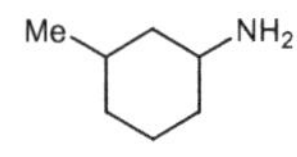

TM 2.19

This information eliminates the cyclohexane derivative or its aromatic precursor as stating material. It also indicates the C_2 unit is the second building block and suggests the construction of a carbon skeleton by Diels–Alder reaction. Diene C_5 with a Me group is present in isoprene (2-methylbuta-1,3-diene). C_2 dienophile should be activated by EWG, and the nitro group serves this purpose best. This analysis suggests amino-nitro FGI to **TM 2.19a** as the target molecule of the next generation. By addition of the C=C bond in a strategic position on the cyclohexane ring, we arrive at key intermediate **TM 2.19b**. This cyclohexene derivative is now prone to r*etro*-D.-A. disconnection to isoprene and nitroethylene **TM 2.19c** (Scheme 2.41).

In spite of its relative instability nitroethylene (pure compound readily decomposes at r.t., but is stable in benzene solution over months!) is conveniently produced in kilo-quantities by dehydration of 2-nitroethanol according to the method outlined in the proposed synthesis of **TM 2.19** (Scheme 2.42) [26].

Note that **TM 2.19a** does not appear as an intermediate in the synthetic direction since both unsaturated functionalities in the product of the Diels–Alder reaction are contemporaneously reduced.

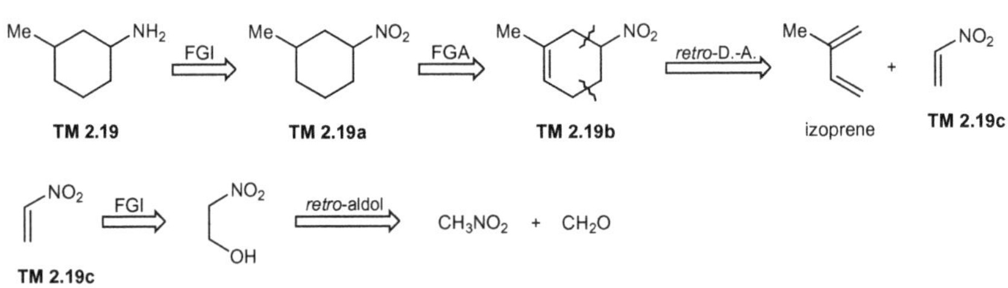

Scheme 2.41 Retrosynthetic analysis of **TM 2.19**

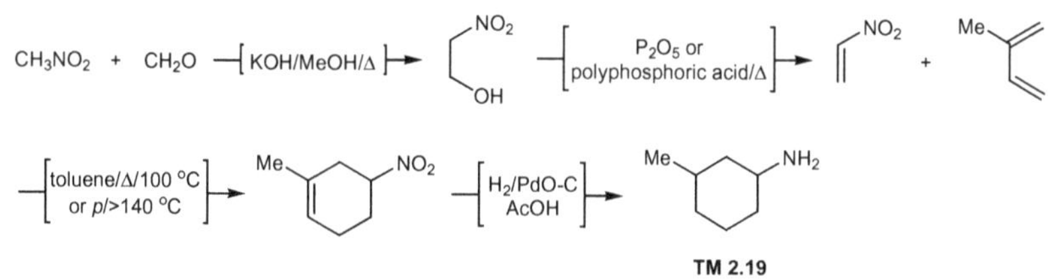

Scheme 2.42 Proposal for the synthesis of **TM 2.19**

Example 2.20 Propose the retrosynthetic analysis of non-conjugated cyclic enone **TM 2.20** and suggest the three-step synthesis.

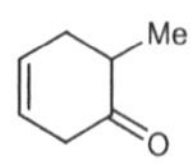

TM 2.20

The presence of a double bond and keto group in the cyclohexane ring indicates that the carbon framework is not available by partial hydrogenation of the benzene derivative. To solve the retrosynthetic puzzle, we need new knowledge. This rests in the possibility to transform a nitro to keto group by the oxygenation of the enolic form of the nitro group. The usual protocols use strong protic acids, the Nef reaction [27] or Lewis acids, preferably $TiCl_3$ [28]. Retrosynthetic analysis needs interconversion of the keto to nitro group in **TM 2.20a** in the first step, an imaginative FGI based on the above reaction (Scheme 2.43).

retro-Diels-Alder disconnection of **TM 2.20a** results in butadiene and 1-nitropropene as dienophile. *retro*-Aldol disconnection of **TM 2.20b** results in nitromethane and acetaldehyde. A proposal for the synthesis is given in the scheme, denoting the reactive form of the nitro group in the course of oxygenation.

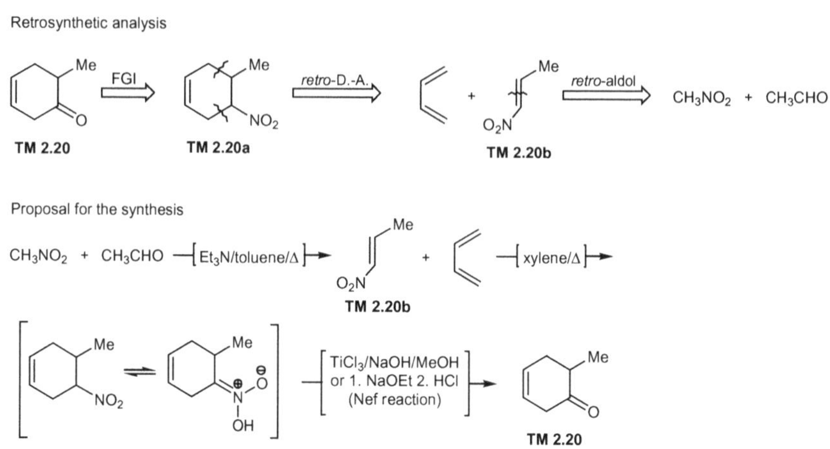

Scheme 2.43 Retrosynthetic analysis and proposal for the synthesis of **TM 2.20**

References

1. Diels O, Alder K (1928) Liebigs Ann Chem 460:98–122
2. Diels O, Alder K (1929) Liebigs Ann Chem 470:62–72
3. Carey FA, Sundberg RJ (2007) Advanced organic chemistry. Part B. Reactions and synthesis, 5th edn. Springer, Berlin, pp 475–481
4. Smith JG (2011) Organic chemistry, the Diels-Alder reaction, 3rd edn. McGraw-Hill, New York, pp 588–599
5. Nicolaou KC, Snyder SA, Montagnon T, Vassilikogiannakis G (2002) Angew Chem Int Ed Engl 41:1668–1698
6. Spino C, Pesant M, Dory Y (1998) Angew Chem Int Ed Engl 37:3262–3267
7. Yoon T, Danishefsky SJ, de Cala S (1994) Angew Chem Int Ed Engl 33:853–855
8. Brown WH, Foote CS, Iverson BI, Anslyn EV, Novak BM (2010) Organic chemistry, 6th edn. Thomson Brooks&Cole, pp 592–594
9. Effenberger F (1994) Angew Chem Int Ed Engl 33:155–1564
10. Brunel J-M, Holmes J (2004) Angew Chem Int Ed Engl 43:2752–2778
11. Červený L, Zuzana Dobrovolná Z, Zdeněk Bělohlav Z, Kluson P (1996) Collect Czech Chem Commun 61:764–769
12. Lenarda M, Casagrande M, Moretti E, Storaro L, Frattini R, Polizzi S (2007) Catal Lett 114:79–84
13. Park IS, Kwon MS, Kim N, Lee JS, Kang KY, Park J (2005) Chem Commun 5667–5669
14. Wittig G (1980) Science 210:600–604
15. Wittig G (1980) Angew Chem 72:671–675
16. Cadogan JIG (ed) (1979) Organophosphorus reagents in organic synthesis. Academic Press, London
17. McMurry J (2008) Organic chemistry, 7th edn. Thomson Brooks&Cole, pp 720–722
18. Maryanoff BE, Reitz AB, Mutter SM, Inners RR, Almond Jr. HR, Whittle RR, Olofson RA (1986) J Am Chem Soc 108:7664–7678
19. Robiette R, Richardson J, Aggarwal VK, Harvey JN (2006) J Am Chem Soc 128:2394–2409
20. Lindlar H, Dubuis R (1973) Org Synth Coll 5:880–884
21. Molnár A, Sárkány A, Varga M (2001) J Mol Catal A 173:185–221
22. Stutzenegger A (1954) US 2802872
23. Gómez-Bombarelli R, González-Pérez M, Pérez-Prior MT, Manso JA, Calle E, Casado J (2008) Chem Res Toxicol 21:1964–1969
24. Yube K, Furuta M, Mae K (2007) Catal Today 125:56–63
25. Fiegel H, Voges H-W, Hamamoto T, Umemura S, Iwata T, Miki H, Fujita Y, Buysch H-J, Garbe D, Paulus W (2002) Phenol derivatives, in Ullmann's encyclopedia of industrial chemistry. Wiley-VCH, Weinheim
26. Ranganathan D, Rao CB, Ranganathan S, Mehrotra AK (1980) J Org Chem 45:1185–1189
27. Noland WE (1955) Chem Rev 55:137–155
28. Ballini R, Petrini M (2004) Tetrahedron 60:1017–1047

Chapter 3
Stereoisomers and Stereoselective Reactions—"Departure into Third Dimension"

Abstract The basics of stereoselective reactions and reaction stereochemistry—the relation of stereoselectivity to the topology of tetrahedral and planar units in organic molecules—are discussed. The kinetic control of enantioselective reactions and characteristics of enantioselective and diastereoslective reactions is presented. Asymmetric syntheses are exemplified by the hydrogenation of C=O and C=NR bonds in prochiral substrates catalyzed by organometallic complexes with chiral phosphine ligands. The mechanism of asymmetric alkylation of stabilized carbanions in specifically designed chiral substrates and the practicability of this method in the preparation of optically pure α-alkyl carboxylic acids are discussed. The synthetic approach to chiral auxiliaries and importance of recycling are presented.

3.1 Introduction

In this book, along with the retrosynthesis and synthesis of the target molecule "in two-dimensional space," examples of the asymmetric or chiral variant of the synthetic reaction are presented. High cognitive and methodological barriers exist to approaching where the "third dimension" is introduced along with standard retrosynthetic analysis. Introduction of the "third dimension" in organic synthesis requires consideration of the *relative and absolute configuration of stereogenic or chiral center(s)* formed in the *stereoselective reaction*. In the retrosynthetic direction this usually means transformation of the relative or absolute configuration on the chiral center to the *prochiral unit or group* in the target molecule of the next generation.

The progress of modern synthetic organic chemistry is largely related to the discovery of new *asymmetric or enantioselective reactions*, particularly those catalyzed by chiral catalysts. In this chapter, basic knowledge on stereoisomerism and reaction stereochemistry is correlated with selected examples of asymmetric reactions to facilitate the discussion of stereoselective and asymmetric reactions in the next chapters.

V. Šunjić and V. Petrović Peroković, *Organic Chemistry from Retrosynthesis to Asymmetric Synthesis*, DOI 10.1007/978-3-319-29926-6_3

3.2 Retrosynthesis and Stereochemical Aspects of Synthetic Reactions

From an achiral molecule, one enantiomer of the chiral target molecule is formed in an *asymmetric reaction*, also known as *enantioselective reaction*. Retrosynthetic analysis considers chiral target molecules as racemates and at this stage does not consider asymmetric synthesis to one enantiomer. When the second stereogenic center in the chiral molecule is introduced in a stereoselective manner, one of two possible diastereomers is formed. Such a synthetic reaction is called *diastereoselective*. For the sake of simplicity, during retrosynthetic analysis the formation of diastereomers is not related to the optical purity of the target molecule. This means that starting chiral molecules are considered racemates, which in a diastereoselective step affords one of the two possible racemic products.

Before consideration of stereoselective reactions in this chapter, we had the opportunity to discuss one example of stereoselectivity "in two-dimensional space." This refers to reactions where *E*/*Z* isomers are formed around a double bond, specifically in target molecules with an RCH=CHR′ unit (Wittig reaction, Sect. 2.2).

Retrosynthetic analysis primarily considers the electronic properties of synthons, i.e., the stabilizing effect of neighboring groups on the *charged species in the ground state*. This approach is not amenable to analysis of asymmetric syntheses since their outcome is determined by the *stereoelectronic properties of the transition state*. *Stereoselective reactions are kinetically controlled*, and the rational approach to analysis of the stereoelectronic properties of the *energetically preferred transition state* is difficult and usually supported by specialized computer programs.

Different stereoelectronic properties of similar groups in reacting molecules result in *chemoselectivity*, with higher reactivity of one group as compared to others. This difference is relatively easy to anticipate in the course of retrosynthetic analysis. The outcome of an asymmetric reaction is much more difficult to forecast in view of the difficulty in designing a transition-state structure on the route to the preferred enantiomer.

In order to present the principles of various kinds of stereoselective reactions, we start with classification of the possible isomers formed in these reactions. To this aim we designed a set of isomers with the common molecular formula $C_{11}H_{12}O$ (Fig. 3.1).

In order to distinguish between *constitutional or structural isomers* and *stereoisomers* we need to observe that *structural isomers do not possess the same connectivity*. Stereoisomers, in turn, possess the same connectivity, but they are related or not as an object and its mirror image, i.e., they are *enantiomers* or *diastereomers*.

Figure 3.1 shows the *absolute configurations* of stereogenic centers of diastereomeric and enantiomeric molecules. The *configuration* is defined as the relative position of atoms in the space that characterize a single stereoisomer. The concept of an absolute configuration is particularly important in stereochemistry. It

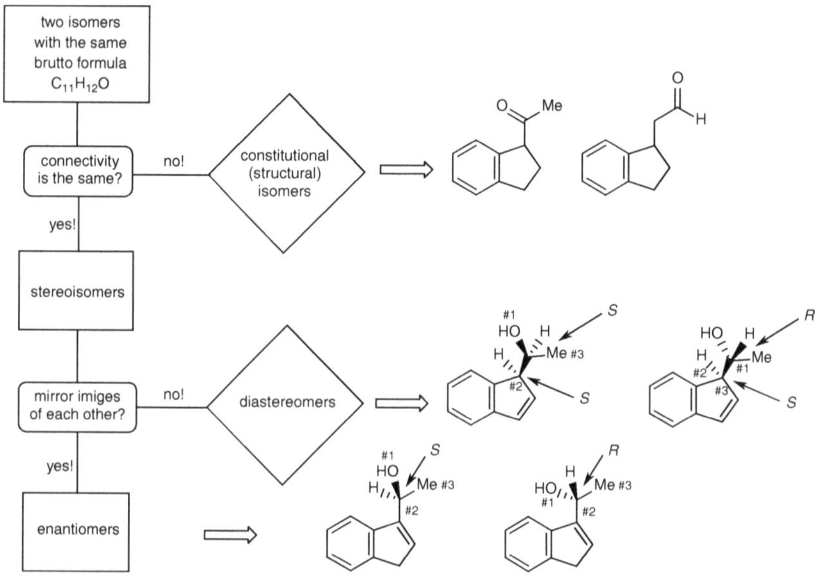

Fig. 3.1 Examples of the classification of stereoisomers

is assigned by the descriptors *R* or *S* according to the *Cahn-Ingold-Prelog* (CIP) *convention*. According to the CIP convention, priority is determined for all atoms and groups bound to the stereogenic center. The atom or group of the highest priority gets the number 1 and that of the lowest priority number 4 (Fig. 3.1).

For more detailed study of organic stereochemistry, an explanation of the *R*/*S* and *E*/*Z* systems around the tetrahedral and planar (trigonal) atoms, the traditional use of the D and L descriptors (based on the correlation of the configuration to D- and L-glyceraldehyde), *erythro*- and *threo*- assignation of the configuration on two vicinal atoms and the helical descriptors, *M* and *P*, consultation of the known textbooks of stereochemistry [1–3] as well as the chapter on stereochemistry in the Anslyn/Dougerthy textbook on physical chemistry is suggested [4].

3.3 Basics of Stereoselective Reactions

While presenting principles of retrosynthetic analysis, we do not devote attention to the formation of stereoisomers. There is a good reason for "departure into the third dimension" after the introductory chapters where the stereochemical aspects of retrosynthesis and the stereoselectivity of the corresponding synthetic steps are not considered.

Examples of enantio- and diastereoselective reactions are presented in Scheme 3.1 for the set of structures as in Fig. 3.1.

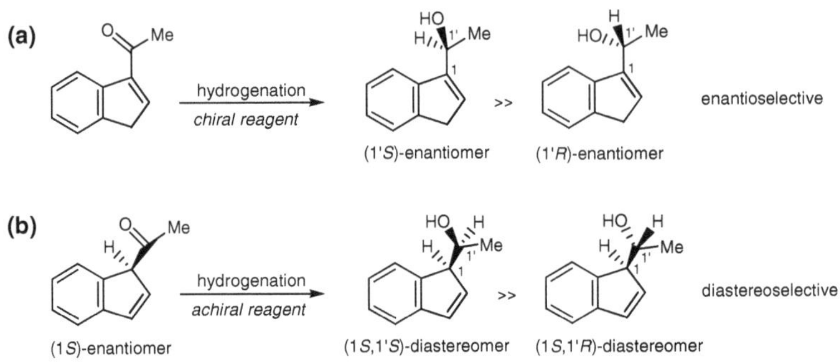

Scheme 3.1 Two types of stereoselective hydrogenation of the carbonyl group

In both examples the ketone carbonyl group is reduced to the *sec* hydroxyl group in either (a) an achiral molecule or (b) a chiral molecule. On hydrogenation of achiral ketone one enantiomer might be formed preferentially [the (1*S*)-configuration is deliberately selected in the scheme]. To perform this reaction in an enantioselective manner, (a) the presence of a *chiral reagent or catalyst* is a prerequisite. In example (b) the preferred formation of one diastereomer [the (1*S*, 1′*S*) configuration is deliberately selected] results from the chiral, optically pure (1*S*)-enantiomer. There is no need for a chiral reagent or catalyst in the diastereoselective reaction of the chiral substrate. This type of reaction was traditionally, but not conveniently named *asymmetric induction*.

3.4 Topological Relation and Stereoselectivity

Topological relations between ligands on the tetrahedral C atom and between faces around the plane of the trigonal C atom are of prime importance for understanding stereoselectivity. The terms *topological, topic* and *topicity* originate from the Greek word *topos*, the place. Figure 3.2 presents molecules that possess protons on the tetrahedral carbon with various topological characteristics. Figure 3.3 presents molecules that possess a planar carbonyl group or trigonal carbon atom and various topological characteristics of the faces around the C=O double bond. Molecules **1** and **2** are achiral; molecules **3** and **4** are chiral with one stereogenic center. The symmetry relation for the hydrogen atoms Ha and Hb on denoted methylenic groups is different. In achiral molecules **1** and **2** there is a symmetry plane that interconverts Ha into Hb. Here we introduce the concept of *chemical equivalence*. In the first two molecules the atoms Ha and Hb are *chemically equivalent* since they can be interconverted by a certain symmetry operation.

Substitution of one H atom by the new group results in one of the enantiomers. Therefore, these H atoms in achiral molecules **1** and **2** are properly denoted as *enantiotopic*. *Enantiotopicity* is destroyed under any *chiral influence*, in particular in the presence of a chiral center in the same molecule. Hydrogen atoms Ha and Hb

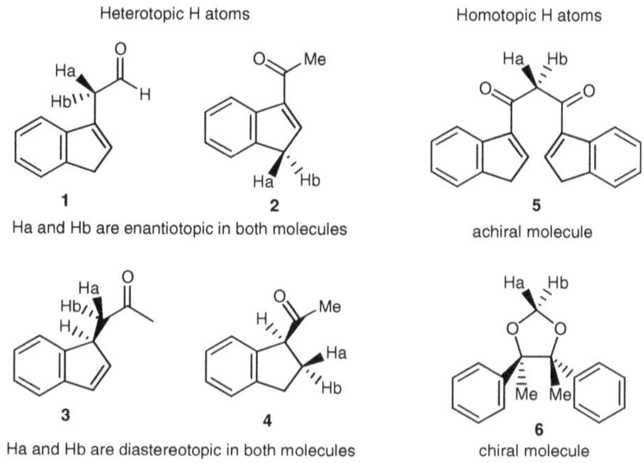

Fig. 3.2 Examples of the topological characteristics of H atoms on tetrahedral C atoms

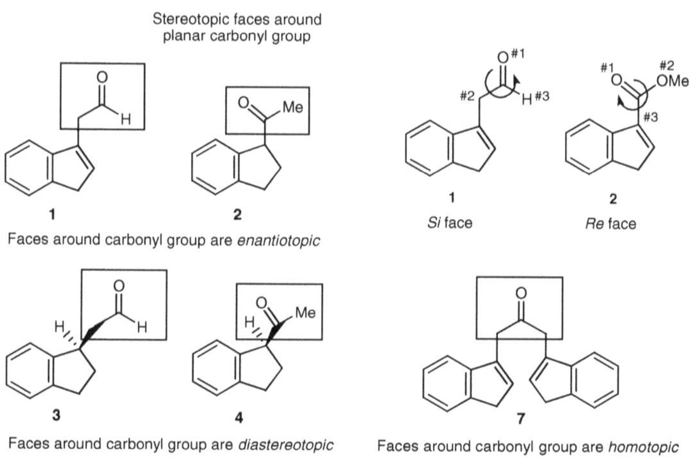

Fig. 3.3 Molecules with topologically different faces around the planar carbonyl group

in chiral molecules **3** and **4** are therefore *diastereotopic*. They are different in all chemical aspects, exhibiting different chemical shifts in the ^{1}H-NMR spectra, different C–H acidity and C–H bond length and therefore different reactivity. Exchange and substitution of any of them results in *diastereomers* that differ in all physicochemical properties.

Finally, let us consider the atoms Ha and Hb in achiral molecule **5** and chiral molecule **6**. They are *homotopic*, chemically and symmetrically equivalent. The central C atom in achiral molecule **5** bears two times two of the same ligands and can be described by the general formula CH_2L_2. Chiral molecule **6** is not a *meso*-form, and there is no mirror plane bisecting this molecule since two stereogenic centers possess the same absolute configuration.

Now we can use a similar argumentation to analyze the symmetry properties of a *planar unit* in the selected molecule (Fig. 3.3).

In all the molecules the planar carbonyl group is presented in the drawing plane. In molecules **1**, **2** and **7** the drawing plane is at the same time the *symmetry plane* or *mirror plane*. Addition of certain reagent H-L to the carbonyl group of **1** and **2** results in enantiomers, depending on the face from which H-L (nucleophile) approaches the planar C=O group. For compound **7**, independently on the face from which the nucleophile approaches the carbonyl group, only one achiral molecule is formed.

Enantiotopic faces around the planar group in a certain molecule are denoted by descriptors *Re* and *Si*, similar to the CIP convention for the description of an absolute configuration of the tetrahedral stereogenic center. This convention is presented for molecules **1** and **2** in Fig. 3.3. The opposite descriptors around the carbonyl group in **1** and **2** are the consequence of the changed priority order of the groups connected to the carbonyl C atom. In the first case, the anti-clockwise rotation connects groups 1, 2 and 3; in the second it is clockwise rotation. It is important to note that the addition of reagent H-L, for instance, from the *Re* face does not unavoidably result in an *R* absolute configuration on the new stereogenic center. It depends on the relative priority of the incoming ligand L relative to the other two groups.

3.5 Stereoselective Processes and Kinetic Control

The chemical reaction is *kinetically controlled* when its rate is determined by the energy of the transition state. When two reactions are competing, the faster is the one with a lower transition state energy. A practical consequence of this kinetic relation is the larger quantity of stereoisomer formed via the reaction with the lower transition state. To illustrate the control of enantioselectivity by the kinetic parameters, we use the relation of their Gibbs activation energies (Fig. 3.4).

In this scheme the effect of the chiral catalyst on the transition state energy of the reaction of the prochiral substrate to enantiomers is presented. A stronger interaction of the chiral catalyst, presented as a scalene triangle (the chiral object in two-dimensional space!) with a prochiral substrate along the reaction coordinate, is indicated by sidewise interaction and on the route to the minor enantiomer by a point-wise interaction. In the absence of the chiral catalyst, the transition state energies will be equal on the route to either enantiomer resulting in the racemate.

The ratio between two enantiomers is expressed as the *enantiomeric excess* (e.e. %) by a simple equation:

$$\text{e.e.}\% = (R - S)/(R + S) \times 100$$

As already stated, the *direction of enantioselectivity* is difficult to anticipate. The main reason resides in the subtle difference between structures of the two transition

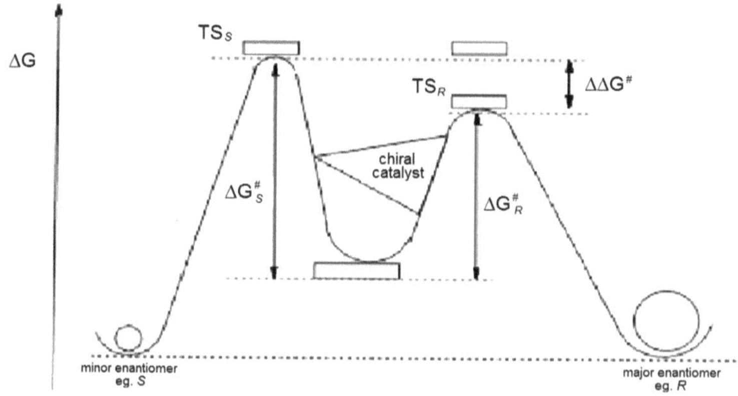

Fig. 3.4 Schematic presentation of the effect of the chiral catalyst on the transition states in the enantioselective reaction

states, TS_R and TS_S, and consequently in a very low difference in their respective energies of activation $\Delta\Delta G^{\#}$.

The enantiomeric excess is exponentially correlated with $\Delta\Delta G^{\#}$, and it is worth illustrating the practical consequence of this exponential correlation. Thus, ca. 10 % e.e is available at the $\Delta\Delta G^{\#}$ of ca. 1.5 kJ/mol, ca. 70 % e.e., when the difference in TS energies is above ca. 5 kJ/mol and approaches 100 % e.e. at the $\Delta\Delta G^{\#}$ of ca. 8–9 kJ/mol. We can understand how small this energy difference is by comparison with the energy barrier for the rotation around the C–C bond in ethane of ca. 8 kJ/mol—this process is defined as "free rotation"!

3.6 Reaction Stereochemistry, More About Enantio- and Diastereoselectivity

Before discussion of the selected examples of enantioselective reactions in Sect. 3.7, we shall consider some general principles of *reaction stereochemistry*. Generally, the influence of the absolute configuration at the stereogenic center in the reacting molecule on the *direction of diastereoselectivity* is easier to forecast than the effect of the chiral catalyst on the *direction of enantioselectivity*. In other words, the absolute configuration on the second or any other stereogenic center in the chiral molecule is more easily predictable and controllable than the configuration of the first stereogenic center from an achiral molecule.

Examples in Fig. 3.5 serve to illustrate how different symmetry properties of ketones **8–10** define the stereochemistry of the products formed on the reduction of the carbonyl group.

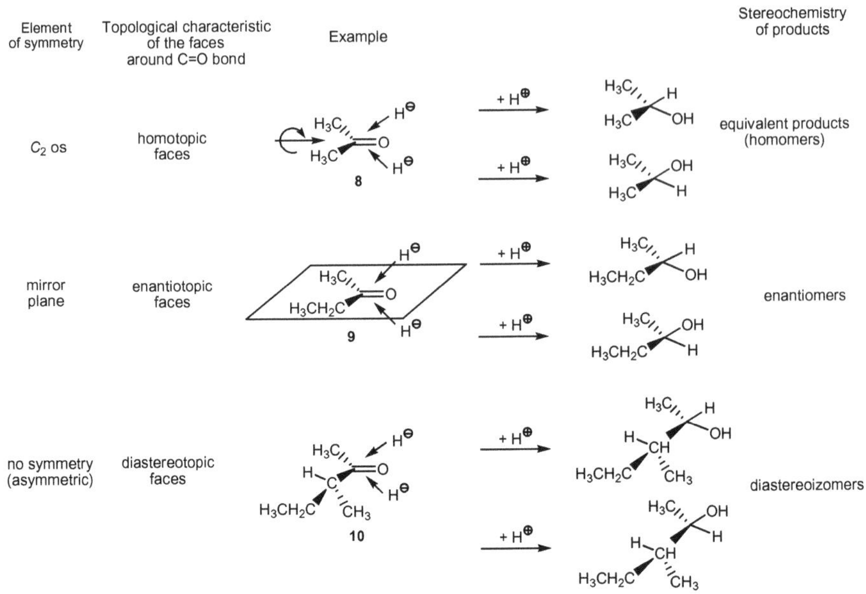

Fig. 3.5 Stereochemical outcome of the reduction of ketones **8–10**

When the hydride ion approaches the carbonyl group in acetone **8** from the opposite faces, achiral 2-propanol is formed. When this reagent approaches the same group in ketone **9**, two enantiomers are formed and two diastereomers in the same reaction of **10**. Now we consider the products of reduction of in more detail carbonyl group in compounds **8–10** depending on the *symmetry of the reagent.*

Achiral reagents, such as, e.g., $LiAlH_4$, as well as *chiral reagents*, such as, e.g., isopinocampheylborane, $[(Ipc)_2BCl$, more about the structure and mechanism of hydrogenation is presented in Sect. 1.3.2, Scheme 1.12] on the hydrogenation of acetone will give one product only, independent of the direction of the approach of the hydride ion to the carbonyl group.

Achiral reagents will react with prochiral 2-butanone **9**, giving racemic butan-2-ol.

Chiral reagents will approach butanone from the *preferred enantioface*, affording predominantly one enantiomer.

When the *achiral* or *chiral* reagent approaches chiral ketone **10** from *preferred diastereoface* one, diastereomer is the prevailing product. *Double asymmetric induction* takes place when chiral ketone **10** is reduced by the chiral reagent or catalyst. As a result we can expect enhanced or lowered diastereoselectivity in comparison to the reaction with the achiral reagent. This is a consequence of the match or mismatch between the two chiral effects.

The stereochemical relation between the products in Fig. 3.5 reflects the symmetry relation of the transition states in the reduction. Let us first consider the case when the *achiral hydride reagent* is used.

- For acetone **8** only one transition state exists, and propan-2-ol is formed as the only product
- For butanone **9** transition states for the approach of the hydride ion to the carbonyl group "from the top" or "from the bottom" are *enantiomeric* and therefore have the same activation energy (Fig. 3.5). Both enantiomers will be formed in equal quantities, hence the racemate
- Chiral ketone **10** is approached by hydride to the transition states, which are in *diastereomeric* relation, and hence possess different activation energies. This results in the prevailing formation of one diastereomer.

Now we analyze what happens when *optically pure chiral hydride* reagent is used.

- In reaction with acetone **8** only one achiral product is formed.
- In prochiral butanone **9** the chiral reagent distinguishes enantiotopic faces. This results in enantioselectivity as assumed on the simple symmetry argumentation. An open question is how large an enantiomeric ratio is reached, as illustrated in the diagram in Fig. 3.4.
- In the reduction of enantiomerically pure chiral ketone **10**, the reducing agent prefers an approach from the less perturbed diastereotopic face of the C=O bond. An achiral reagent will already produce two diastereomeric alcohols in a ratio differing from 50:50 (asymmetric induction). The chiral reducing agent can affect the diastereomeric ratio to make it larger or smaller than that obtained with an achiral reagent. Moreover, reduction with this reagent can result in a prevailing diastereomer of the opposite configuration than obtained with an achiral reagent, i.e., with *inversion of the diastereomeric bias*.

In conclusion,

- *homotopic faces* cannot be differentiated by chiral reagents
- *enantiotopic faces* can be differentiated by chiral reagents
- *diastereotopic faces* can be differentiated by chiral and achiral reagents.

3.7 Examples of Asymmetric Syntheses

3.7.1 *Hydrogenation of the C=O Bond Catalyzed by Chiral Organometallic Complexes*

Example 3.1 Enantioselective reduction of ketones to *sec* alcohols **11–14** is completed by Rh(I) catalytic complexes with three chiral ligands, one phosphine oxide and two bidentate phosphines (Scheme 3.2) [5–7].

The results indicate that non-systematic variation of the structure of chiral ligands leads to a casual degree of enantioselectivity. This statement is valid for all enantioselective catalytic processes. Systematic study of chiral catalysts and reaction conditions is required to reach high enantioselectivity.

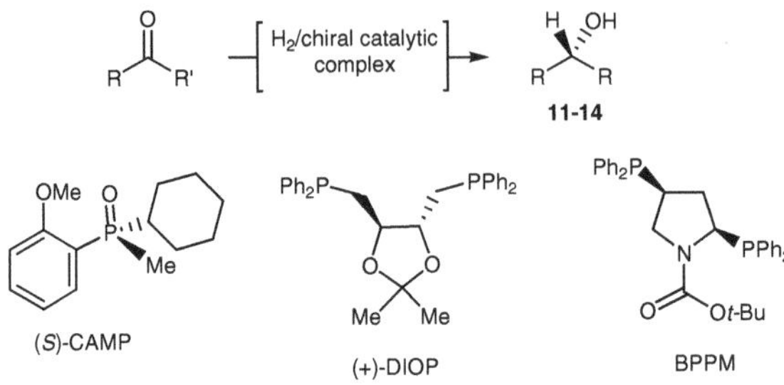

product	R	R'	e.e./%	abs. config.	ligand	ref.
11	Me	Et	12	(*R*)	(*S*)-CAMP	1
12	Me	napht-1-yl	39	(*R*)	(+)-DIOP	2
13	Me	CO_2Pr	76	(*R*)	BPPM	1
14	Ph	Me	80	(*R*)	(+)-DIOP	3

Scheme 3.2 Enantioselective reduction of ketones to *sec* alcohols **11–14** by Rh(I) complexes with chiral ligands

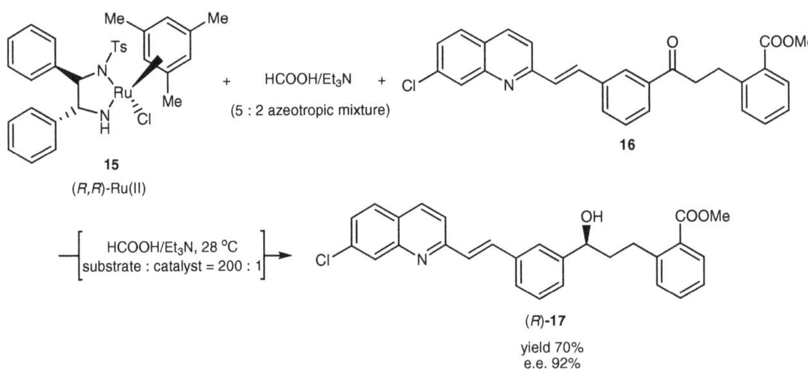

Scheme 3.3 Reduction of ketone **16** by catalytic complex **15**

Example 3.2 High enantioselectivity in the reduction of ketone **16** to alcohol **17** is obtained by catalytic complex **15** (Scheme 3.3).

Alcohol **17** is a key intermediate in the synthesis of *Montelukast*, a broadly used drug in the therapy of asthma. The biological activity of (*R*)-enantiomer is much higher than that of (*S*)-enantiomer; therefore, the direction of enantioselectivity and optical purity of the product are crucial for the technological feasibility of this synthetic step. For drugs used in human therapy in the optically pure form, an optical purity of 98–100 % e.e. is required.

The reduction is characterized by the application of formic acid as the source of hydrogen. Affected by pre-catalytic complex **15**, in the presence of the *tert* amine, this reagent decomposes, forming catalytic complex **15a** and CO_2 (Scheme 3.4).

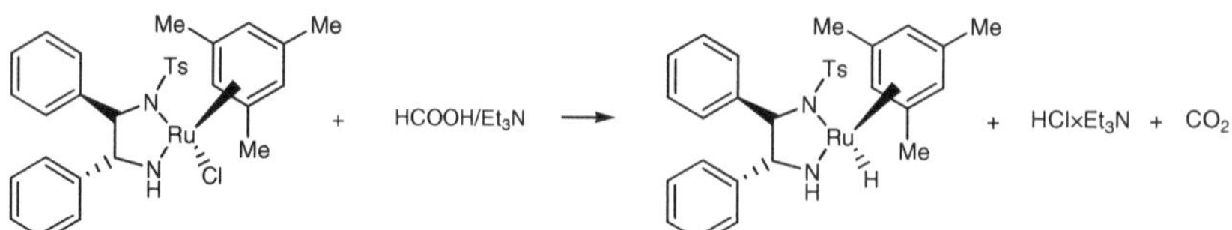

15
(R,R)-Ru(II)

15a

Scheme 3.4 Formation of catalytic complex **15a**

The hydride ion coordinated in complex **15a** is transferred from one enantiotopic face to ketone **16**, affording the (*R*)-enantiomer of alcohol **17**. Catalytic complex **15** is the result of a long-term project of the structure optimization combining the large stereocontrolling unit 1,2-diphenylethylenedimane-*N*-tosylate, strong π-electron-donating ligand 1,3,5-trimetylbenzene and chlorine as an easily exchangeable ligand for the hydride ion [8, 9]. Moreover, this "robust" complex does not need the inert atmosphere of nitrogen or argon in the reactor, an important condition for the industrial application of organometallic complexes as catalysts.

3.7.2 *Hydrogenation of the C=N Bond Catalyzed by Chiral Organometallic Complexes*

Synthesis of α- and β-amino acids in the enantiomerically pure form is often completed by asymmetric reduction of the corresponding imino precursors. Since imines are considerably less stable than parent aldehydes and ketones, their asymmetric hydrogenation has been less explored. Here follow two examples.

Example 3.3 *N*-acylhydrazones are more stable than imines, and their enantioselective reduction to *N*-acylhydrazines is successfully completed by the catalytic Rh(I) complex of chiral bidentate phosphine ligand **22** (Scheme 3.5) [10].

In the last step of this process, the catalytic hydrogenolysis of the N–N bond affords *sec* amines and α-amino acids **18–21** with high optical purity [11].

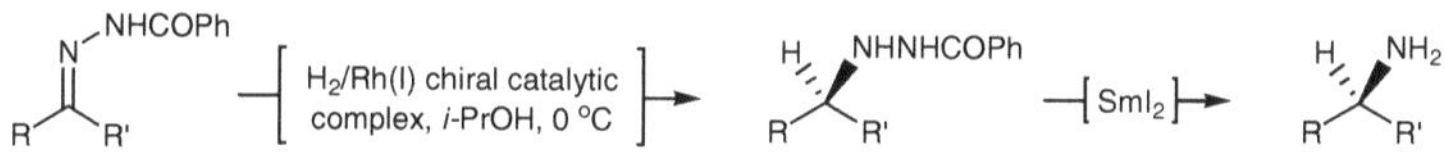

18-21

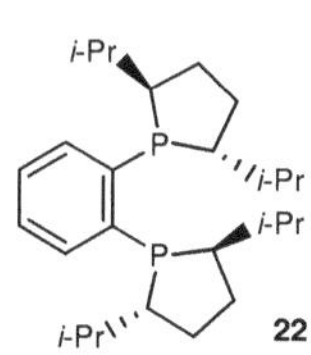

22

product	R	R'	e.e./%	abs. config.
18	Ph	Me	92	(*R*)
19	CO_2Et	Me	89	(*R*)
20	CO_2Et	Ph	91	(*R*)
21	cyclohexyl	Me	72	(*R*)

Scheme 3.5 Enantioselective reduction of acylhydrazones

Example 3.4 (−)-(*R*)-*Sitagliptin* **24** is an oral hypoglycemic used in the therapy of diabetes mellitus type II. This complex molecule incorporates the amide of β-amino acid with an (*R*) configuration. After extensive research enantioselective hydrogenation of its precursor **23** was successfully solved (Scheme 3.6) [11, 12].

The same study revealed the existence of imino-enamino equilibrium in **23** dependent on the pH and polarity of the reaction medium. The more stable enamine form is presented in Scheme 3.6. However, **23** is bound within catalytic complex **23a** in the imino form as the N,O-bidentate ligand affording a reduction with deuterium *sitagliptin* selectively deuterated on the chiral C-3 atom (Scheme 3.7) [13].

Numerous complexes for the hydrogenation of imines have been screened in this study, among them DIOP (2,3-*O*-isopropylidene-2,3-dihydoxy-1,4-bis-(diphenylphosphino)butane), BINAP (2,2'-bis(diphenylphosphino)-1,1'-binaphtyl), DuPHOS (1,2-bis(2,5-diisopropylphospholano)benzene) and many others. All of them afforded (−)-(*R*)-*sitagliptin* **24** with substantially lower e.e., usually below 30 %. This outcome nicely illustrates how difficult the rational design of the structure of the chiral ligand for the catalytic complex is, which is highly enantioselective with the specific substrate.

Example 3.5 Suggest the asymmetric reaction of *tert*butyl 4-(trifluorophenyl) but-2-enoate **23b** with the chiral counterpart to obtain the enantiomer of the

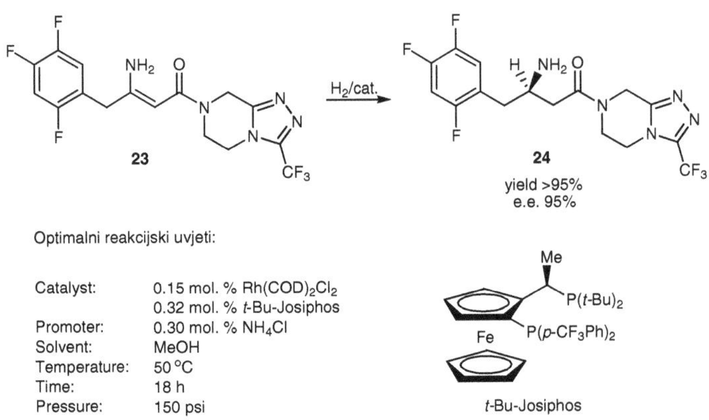

Scheme 3.6 Enantioselective reduction of **23**, a precursor to (−)-(*R*)-*sitagliptin* **24**

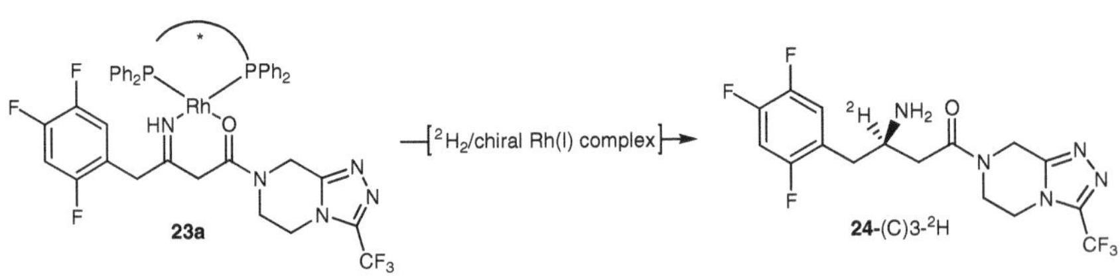

Scheme 3.7 Asymmetric deuteration at C-3 in complex **23a**

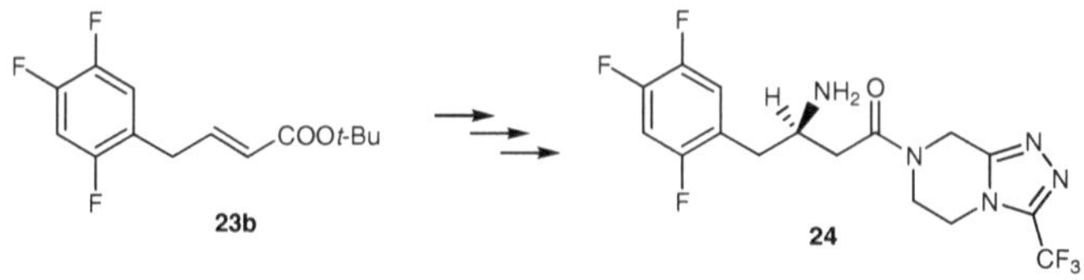

Scheme 3.8 Outline of the symmetric synthesis of (−)-(*R*)-*sitagliptin* from enone **23b**

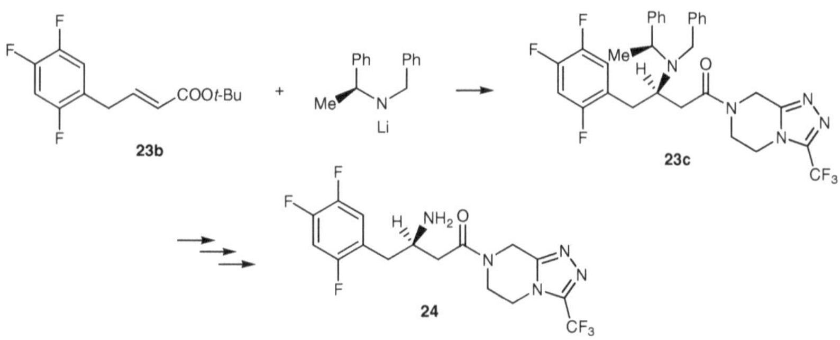

Scheme 3.9 Key step in the asymmetric synthesis of (−)-(*R*)-*sitagliptin*

immediate precursor of *sitagliptin* (the configuration of the chiral reagent does not need to be defined). This is a chiral variant of the well-known addition reaction and the first step in the synthesis of (*R*)-*sitagliptin* (Scheme 3.8).

The answer suggests the *Michael-type asymmetric addition* of an enantiopure amine or its more reactive anion to enone to **23b**. The authors used a lithium amide reagent for the addition to obtain the key chiral intermediate **23c** (Scheme 3.9) [14]. The selected (*S*)-configuration of the phenyl ethylamino unit in the chiral amide anion induces the (*R*)-configuration in the precursor of (−)-(*R*)-*sitagliptin.*

The reported asymmetric synthesis of (−)-(*R*)-*sitagliptin* was completed in seven steps from commercially available starting materials. Acid-catalyzed hydrolysis of β-dalkylamino *tert*butyl ester **23c**, coupling with the triazolopyrazine building block to amide, reduction of amide to *tert* amine and hydrogenolysis of the N-benzyl and N-phenylethyl unit, afforded (−)-(*R*)-*sitagliptin* in 43 % overall yield and 42 % e.e.

3.7.3 Asymmetric Alkylation of Stabilized Carbanion

The classical example of alkylation of stabilized carbanion is the alkylation of diethyl malonate. Double alkylation followed by hydrolysis and decarboxylation affords α-substituted carboxylic acids. The same target molecules are available by α-alkylation of carboxylic esters, which requires much stronger bases since the pK_a of the α-CH_2 group is about 25.

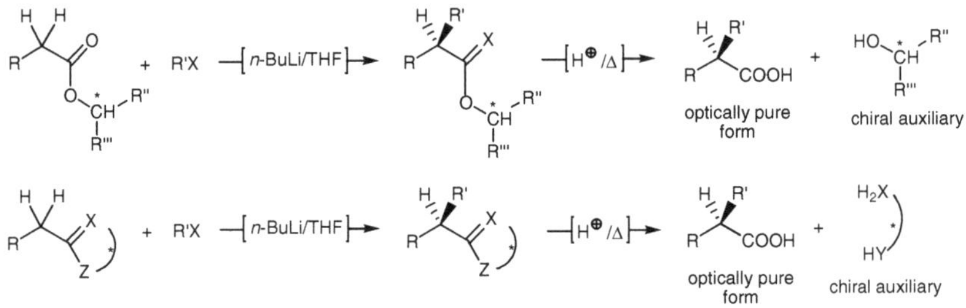

Scheme 3.10 Chiral variants for alkylation of carbanion α to the carboxylic group

Traditional non-catalytic syntheses of α-substituted carboxylic acids in the optically pure form by alkylation use chiral information incorporated into the carboxylic group. This is usually achieved with esters of chiral alcohols or by incorporation of the chiral heterocyclic unit, as shown in Scheme 3.10 where the absolute configuration of the products, α-substituted carboxylic acids, is arbitrarily selected.

In the following example we consider stereoselective α-alkylation of selected chiral carboxylic acid derivatives.

Example 3.6 Mayers et al. prepared chiral oxazoline **27** from propionic acid and easily available optically pure alcohol (1*S*,2*S*)-1-phenyl-2-aminopropan-1,3-diole **25** as a convenient substrate for stereoselective α-alkylation (Scheme 3.11) [15, 16]. It is interesting to note the selective cyclization of diol **25** to oxazoline **26** by the more nucleophilic benzylic OH group in the presence of the primary OH group.

Optically pure aminoalcohol **25** is waste in the production of the antibiotic *Chloramphenicol* by the process that uses its (1*R*, 2*R*)-enantiomer; hence, it is a cheap raw material. Protection of the primary OH group in **26** by methylation to **27** was regarded by the authors as a "necessary harm" when using cheap starting material **25** [16].

Deprotonation of **27**, a template for stereoselective alkylation with a strong base like *n*-BuLi or LDA, affords a carbanion in the aza-allylic position, stabilized by resonance (Scheme 3.12).

In aza-enolate **27a** the methyl group is present in the *trans*-position to the N atom, as confirmed by ^{13}C-NMR spectroscopy. This method reveals a constant *Z/E* ratio of 95:5 over a large temperature range enabling the chelation of the Li cation to N,O atoms of the chiral bidentate ligand. The alkylating agent R-Hal approaches the α-C atom "from the bottom" of the double C=C bond directed by the dipole interaction of halogen with the chelated Li cation. The approach of R-Hal "from the top side" is perturbed by the strong steric hindrance of the phenyl group. Both effects act synergistically and control the diastereoselective C alkylation forming a new stereogenic center in the (*S*) configuration with up to 90 % diastereoselectivity. Hydrolysis of the oxazoline ring is accompanied by the partial loss of optical purity affording (*S*)-α-alkyl-propionic acids with 75–80 % e.e.

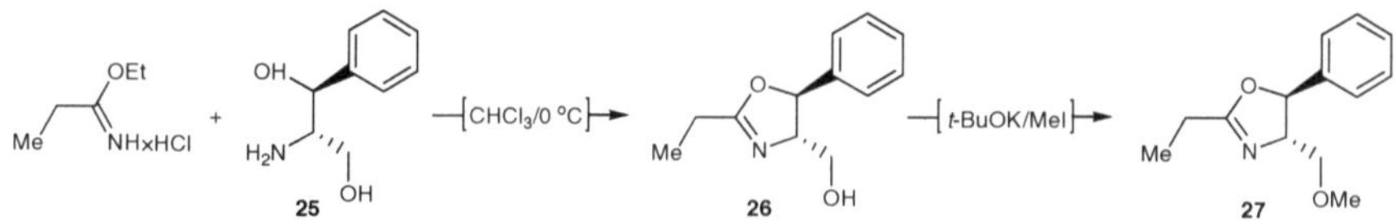

Scheme 3.11 Preparation of chiral oxazoline **27**

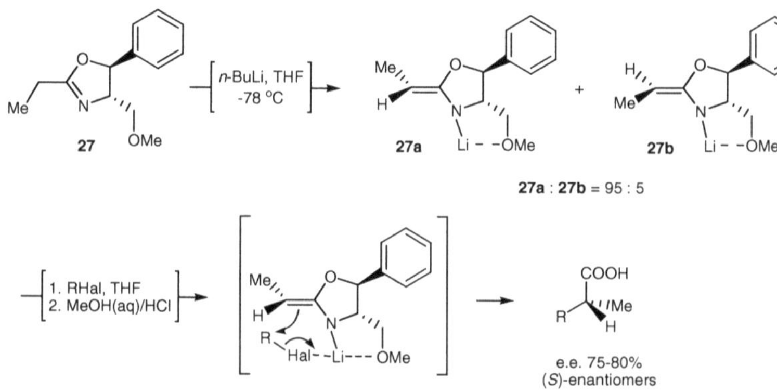

Scheme 3.12 Mechanism of stereoselective alkylation of chiral oxazoline **27**

Chiral oxazolone with Me instead of the CH_2OMe group on the C-4 atom in **27** was prepared and alkylated, and hydrolytic degradation was performed according to Scheme 3.13. Not quite surprisingly, final α-alkyl propionic acids with e.e. below 20 % were obtained! This result confirmed the key role of chelation of the Li cation to the OMe group and the polar interaction with halogen departing from the alkylating agent. It turns out that the OMe group in **27**, regarded as a "necessary harm," is of paramount importance for high stereoselectivity in the alkylation step and for obtaining final products with high optical purity!

One can logically conclude that enantiomeric auxiliary **25** with the opposite configuration at both chiral centers will reorient the stereoselectivity of the alkylation to obtain enantiomeric (*R*)-α-alkyl-propionic acids. This is generally valid for a single asymmetric step but need not be the case for two consecutive reactions. Starting from 2-methyloxazoline derivative **28** instead of 2-ethyl derivative **27**, the double alkylation creates a new stereogenic center (Scheme 3.13). Reversal of alkylation by alkylating agents R-Hal and R′-Hal leads to the opposite configurations at a newly formed chiral center and consequently to enantiomers of α-alkyl carboxylic acids.

The table in Scheme 3.13 demonstrates the dependence of the *direction and degree* of asymmetric alkylation on the order of addition of alkylating agents. When the first alkylation is completed with the smaller alkylating agent and the second with the sterically more demanding one, the stereoselectivity and consequently the enantiomeric purity of the final products are lower.

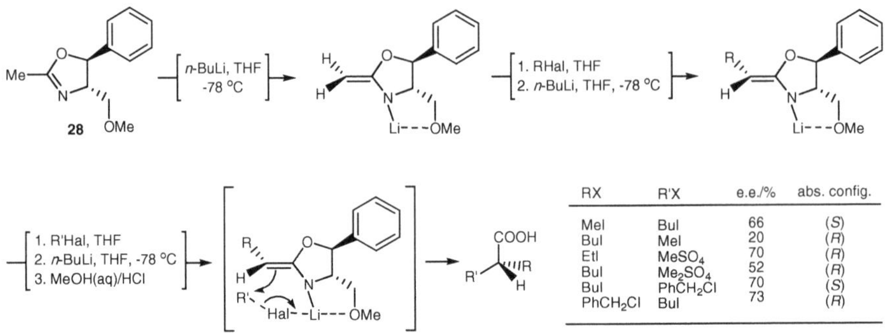

RX	R'X	e.e./%	abs. config.
MeI	BuI	66	(S)
BuI	MeI	20	(R)
EtI	$MeSO_4$	70	(R)
BuI	Me_2SO_4	52	(R)
BuI	$PhCH_2Cl$	70	(S)
$PhCH_2Cl$	BuI	73	(R)

Scheme 3.13 The order of alkylation of **28** leads to opposite enantiomers

In conclusion, the asymmetric alkylation of chiral enolates and enamines can be completed with high stereoselectivity affording final products with high optical purity [17]. The chiral economy of this and other noncatalytic methods that use chiral auxiliary agents in a stoichiometric quantity depends on their availability and effective recycling in the process.

References

1. Nogradi M (1981) Stereochemistry, basic concepts and applications. Akademia Kiado, Budapest
2. Eliel E, Wilen SH (1994) Stereochemistry of organic compounds. Wiley Interscience, New York
3. Juaristi E (1991) Introduction to stereochemistry and conformational analysis. Wiley, New York
4. Anslyn EV, Dougherty DA (2005) Modern physical organic chemistry, chapter 6, Stereochemistry. University Science Books, Herndon, VA
5. Solodar J (1975) Chem Tèch 421–423
6. Hasaguchi S, Kanehira K, Kumada M (1981) Tetrahedron Lett 22:4417–4420
7. Landis CR, Halpern J (1987) J Am Chem Soc 109:1746–1754
8. Hashiguchi S, Fujii A, Takehara J, Ikariya T, Noyori R (1995) J Am Chem Soc 117:7562–7563
9. Fujii A, Hashiguchi S, Uematsu N, Ikaraya T, Noyori R (1996) J Am Chem Soc 118:2521–2522
10. Burk MJ (1991) J Am Chem Soc 113:8518–8519
11. Shultz CS, Kirska SW (2007) Acc Chem Res 40:1320–1326
12. Hsiao Y, Riviera NR, Rosner T, Kirska SW, Njolito E, Wang F, Sun Y, Armstrong III JD, E. Grabowski EJJ, Tillyer RD, Spindler F, Malan C (2004) J Am Chem Soc 126:9918–9919
13. Noyori R, Kitamura M, Ohkuma T (2004) Proc Nac Acad Sci USA 101:5356–5362
14. Davies SG, Fletcher AM, Lv L, Roberts PM, Thomson JE (2012) Tetrahedron Lett 52:3052–3055
15. Lutomski KA, Mayers AI (1984) Asymmetric synthesis (Morrison D, ed), vol 3. Academic Press, New York, pp 231–274
16. Meyers AI, Snyder ES, Ackerman JJH (1978) J Am Chem Soc 100:8186–8189
17. Procter G (1996) Asymmetric synthesis. Oxford Science Publications, Oxford, pp 41–68

Chapter 4
Disconnection with Participation of Two Functional Groups

Abstract The concept of charge alteration in compounds with one functional group and the resulting match or mismatch of partial charges in compounds with two functional groups is introduced. The favorable relation of oxygen functionalities in a 1,3- and 1,5-dioxygenated pattern supporting logical disconnections is demonstrated. The Mannich reaction, Michael addition and Robinson annelation are presented as synthetic approaches to this target molecule based on illogical disconnections. Diastereo- and enantioselective (asymmetric) aldol reactions are presented and the mechanism for the preferred formation of *syn/anti* products discussed. Examples of non-catalytic and catalytic asymmetric aldol and Mannich reactions are given and their mechanistic aspects and experimental protocols presented. Asymmetric syntheses of optically pure compounds with a 1,3-CO pattern (α-alkyl-β-hydroxy carboxylic acids, β-hydroxy ketones, dihydropyranones) and asymmetric Mannich reactions on the route to the optically pure herbicide (*S*)-*fenpropimorph*, (*S*)-phenylglycine and (2*S*, 3*R*)-α-amino-γ-keto acids are discussed.

4.1 Introduction

Until now, we have considered the disconnection of C–C bonds by participation of one functional group. In this chapter, we discuss examples where two functional groups control selection of the preferred C–C bond for disconnection by their electronic effects. Most heteroatoms (O, N S, Hal) in organic molecules are more electronegative, whereas P, Si and all metals are more electropositive than carbon. The polarization C-heteroatom bond dictates the direction of the imaginative flow of the electrons in the disconnected C–C bond at a distance from the functional group.

For the following discussion it is important to recall that any functional group in organic molecules rises the charge separation along the carbon chain. The through-bond distance of two functional groups in bifunctional molecules defines

V. Šunjić and V. Petrović Peroković, *Organic Chemistry from Retrosynthesis to Asymmetric Synthesis*, DOI 10.1007/978-3-319-29926-6_4

the position of the C–C bond and electron flow in disconnection. In this context, the terms *match* and *mismatch* of charges in the target molecules are presented, and the principal modes of logical disconnections of the 1,3- and 1,5-dioxygenated pattern in target molecules are discussed. Examples are given for asymmetric synthesis of chiral target molecules with these two deoxygenated patterns.

In the next chapter, specific issues with illogical disconnections of target molecules with an even number of atoms separating two oxygen functionalities are addressed.

4.2 Match and Mismatch of Charges in Bifunctional Molecules

To understand the electronic effects of two functional groups at a certain distance, we start by considering the effect of one functional group on the saturated and conjugated carbon chain. As presented in Fig. 4.1 for the C_5 chain, polarization of the C–X bond where X is a more electronegative atom alternatively changes the charges and space of the atomic orbital on the C atoms along the carbon framework.

For the saturated structure (a) the largest charge polarization is within the polar C–X bond, which then diminishes monotonously, and charges of the opposite sign appear in alternation. Conjugated double bonds along the chain substantially alter this picture of the charge distribution; positive and negative charges of approximately the same intensity appear alternatively along the conjugated chain.

The concepts of resonance and charge alternation are nearly synonyms and useful for the description of electron distribution of conjugated π systems bound to heteroatoms [1, 2]. Alternation of the charge along a saturated chain was definitively confirmed by the semiempirical CNDO/2 method [3].

The principle of *charge alternation* in both saturated and unsaturated conjugated systems is a useful mnemonic tool in retrosynthetic analysis. This principle was introduced in retrosynthetic analysis by E.J. Corey, along with the concept of logical and illogical disconnection [4], whereas T.L. Ho has broadened the application of this principle [5]. The concept of "logical" or "illogical" heterolytic disconnection of the C–C bond resulting in two charged synthons corresponds *to the polarity of real reagents*. This concept correlates the alternation of polarity in the target molecule with polarity in its precursor molecule.

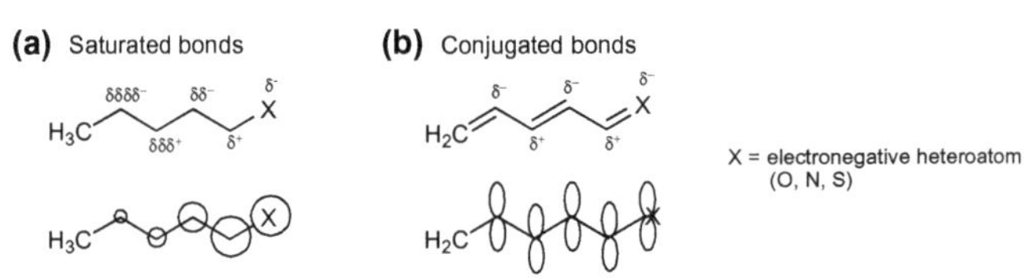

Fig. 4.1 Charge distribution and atomic orbitals in the chains consisting of five C atoms

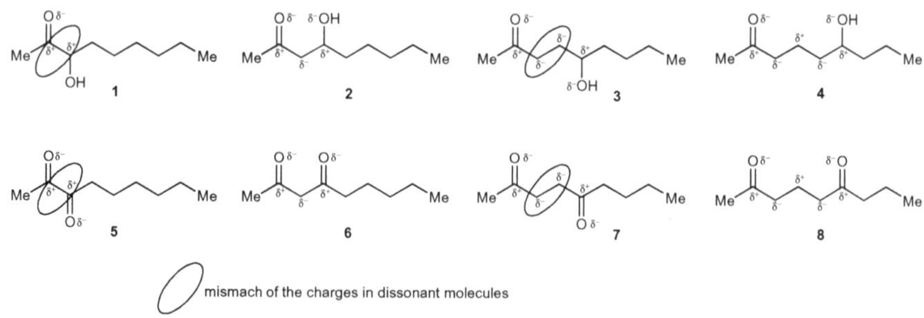

Fig. 4.2 Structures with match and mismatch of the alternating charges

When two functional groups with an electronegative heteroatom are present in an organic molecule, a charge alternation can result in their *match* or *mismatch* as presented for two oxygen functionalities in Fig. 4.2.

In this text, we use the terms *match* and *mismatch* of charges in the target molecules. These structures were also denoted as *consonant* or *dissonant* by D. A. Evans [6] or as *normal* or *umpolung* fragments by D. Seebach [7]. Examples in Fig. 4.2 are from a 1,2- to 1,5-*dioxygenated pattern* and represent molecules with two functional groups that comprise oxygen atoms separated by two to five C-atoms or one to four C–C bonds. This concept is valid for all organic molecules disubstituted with electronegative heteroatoms. In retrosynthesis we neglect their distance through space, an important factor in the consideration of stereoselective reactions.

The distance between two functional groups through bonds determines the optimal bond for heterolytic disconnection *according to the correct mechanism* resulting in two synthons of the opposite charge and defined as *disconnection with participation of two functional groups*. Preferred or logical disconnections are possible when the number of C atoms between two functional groups is uneven and not preferred or illogical when this number is even. The origin of this relation resides in charge alternation, presented in Fig. 4.2.

4.3 1,3-Dioxygenated Pattern (1,3-CO)

4.3.1 1,3-Hydroxycarbonyl Compounds

In this and the next chapter we consider disconnection with participation of two oxygen functionalities, called the *dioxygenated pattern*. The rationale behind this selection rests in the fact that oxygen functionalities can be conveniently transformed into functional groups with other heteroatoms. Schematic structures of the 1,3-dioxygenated pattern are presented as **TM I–III** (Fig. 4.3).

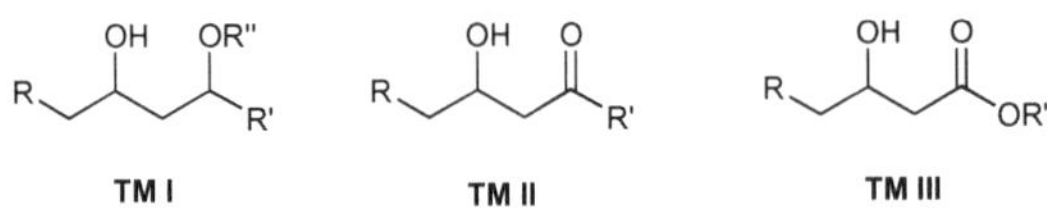

Fig. 4.3 Structures with a 1,3-dioxygenated pattern

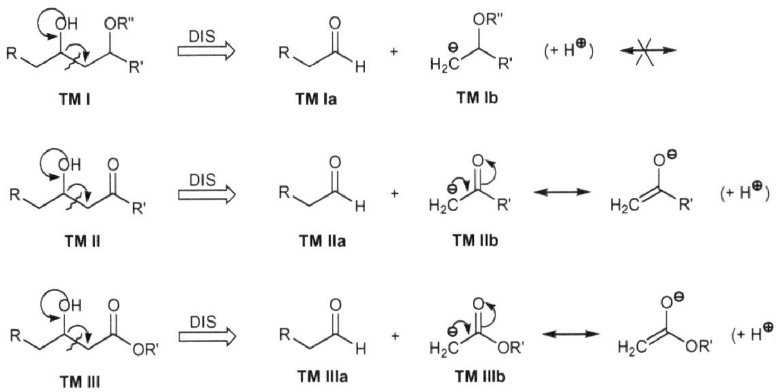

Scheme 4.1 Disconnections of a 1,3-dioxygenated pattern in **TM I–III**

Maximal simplification of any of the three target molecules will be achieved by disconnection of one of two *central or internal* bonds (Scheme 4.1).

It is important to note that the oxidation level of the oxygenated functionality determines the position of disconnection. Changing the oxidation state of the oxygenated functionality, we can control the position of the C–C disconnection. As already seen, the OH group facilitates the disconnection of the C–C bond to the carbinolic C atom (Sect. 1.3.1, Scheme 1.10). The C–C bond to the OH group is preferably disconnected, and the disconnection in **TM I** results in a carbonyl compound and unstabilized carbanion (plus proton). In **TM II** and **TM III**, instead, the emerging carbanion is stabilized by resonance with a second oxygen functionality. Disconnection is therefore directed by the *participation of two functional groups*. The carbonyl group in **TM II** and carboxyl group in **TM III** stabilize the carbanion by resonance, which is not possible for the OR" group in **TM I**.

The synthetic reaction in which two carbonyl components are connected to a more complex molecule with a 1,3-hydroxycarbonyl pattern as in **TM II** is the well-known *aldol reaction*, which takes place between the enolate and carbonyl form of aldehydes or ketones. The mechanism of this reaction and its stereochemical course will be discussed later in this chapter. The examples that follow serve to suggest the decision concerning which internal C–C bond in 1,3-dioxygenated compounds to disconnect.

Example 4.1 Propose the preferred disconnection for **TM 4.1** and then the synthesis from easily available starting materials. Suggest how to use **TM 4.1** for the

Scheme 4.2 Preferred disconnection of **TM 4.1**

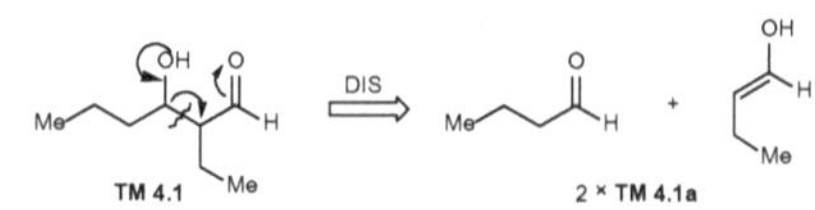

synthesis of 2-ethylhexanol **TM 4.2**, a commodity produced in the amount of 2.5 million tons/year.

The 1,3-hydroxycarbonyl unit in **TM 4.1** corresponds to the pattern in **TM II**. Participation of both functional groups suggests disconnection of the internal C–C bond as shown in Scheme 4.2. For simplicity in this scheme the enolic form of the carbonyl compound is presented as the second component of disconnection instead of its enolate anion plus proton, as shown in Scheme 4.1 for **TM II**.

The result of disconnection is a pleasant surprise; **TM 4.1a** repeats in two molecules of *n*-butanal.

Note This C_4 aldehyde is a commodity produced at 6 million tons/year by the catalytic formulation of propene. Depending on the catalyst, isobutanal is obtained by the same route (Scheme 4.3).

Some technological aspects of the process are worth comment. Formylation (a) is catalyzed by the Rh(I) complex of tris(3-sulfophenyl)phosphine trisodium salt (TPPTS). Ligand solubilizes the Rh(I) complex in water because of the presence of three sulfonyl sodium groups, and its high stability enables high turnover of the catalytic cycle. *n*-Butanal is insoluble in water and separates as the upper layer, while the aqueous layer that contains catalysts can be reused in the process.

To answer the question on the use of **TM 4.1** for production of **TM 4.2**, we observe that in **TM 4.2** the *sec* hydroxyl group is eliminated and the carbonyl group reduced. Whereas reduction of the carbonyl group in **TM 4.1** can be completed by numerous reducing agents, hydrogenolysis of the hydroxyl group is possible only after activation, e.g., as methanesulfonate (mesylate) or *para*-toluenesulfonate (tosylate). Much more convenient is the two-step approach, the base-catalyzed water elimination in **TM 4.1** and reduction of both unsaturated functional groups in the last step (Scheme 4.4).

Exceptionally, here we entered the retrosynthetic analysis of **TM 4.2** only after the proposed synthesis of this target molecule. The retrosynthetic scheme suggests

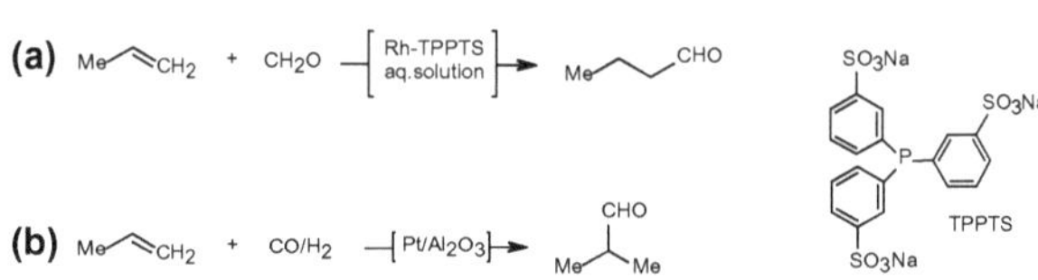

Scheme 4.3 Schemes of catalytic production of *n*-butanal and isobutanal

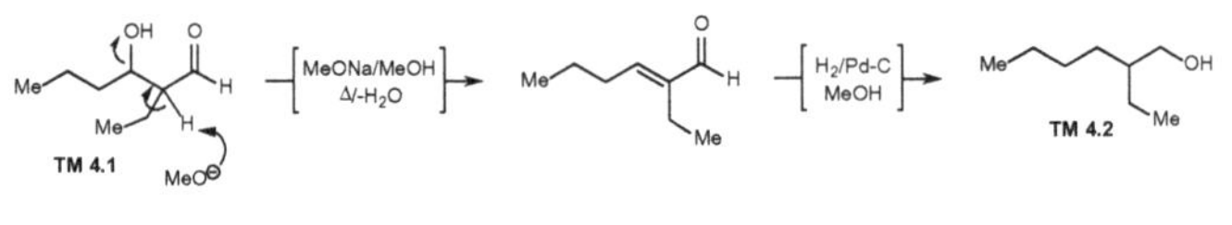

Scheme 4.4 Two-step synthesis of **TM 4.2**

Scheme 4.5 Retrosynthetic steps from **TM 4.2** to **TM 4.1**

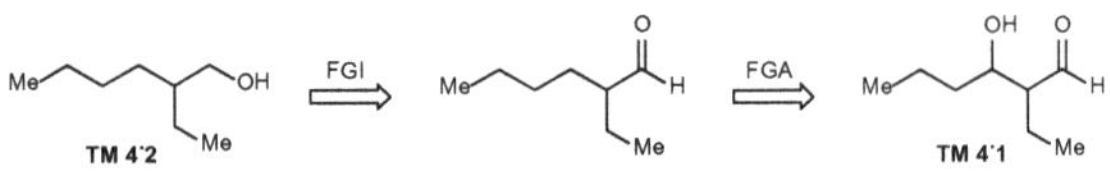

the interconversion and addition of a functional group, FGI and FGA in a sequence (Scheme 4.5).

Example 4.2 Structural isomers **TM 4.3** and **TM 4.4** are equally acceptable for certain physico-chemical studies. Which one do you prefer for an easier synthetic approach?

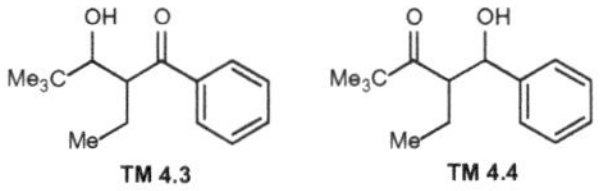

Inspecting **TM 4.3** and **TM 4.4**, one observes the 1,3-hydroxycarbonyl pattern, and disconnection of one internal C–C bond in the *retro*-aldol direction comes to mind (Scheme 4.6).

Disconnections result in one molecule of aldehyde **TM 4.3a** or **TM 4.4b** and ketone **TM 4.3b** or **TM 4.4a**, presented in its enolic form according to the mechanism of disconnection.

Note Both aldehydes are products of industry commodities. Benzaldehyde is produced by oxidation of toluene in liquid phase by MnO_2/H_2SO_4 (Scheme 4.7a) or

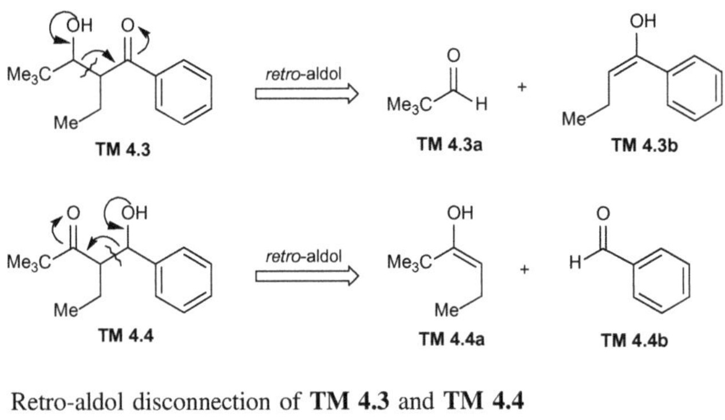

Scheme 4.6 Retro-aldol disconnection of **TM 4.3** and **TM 4.4**

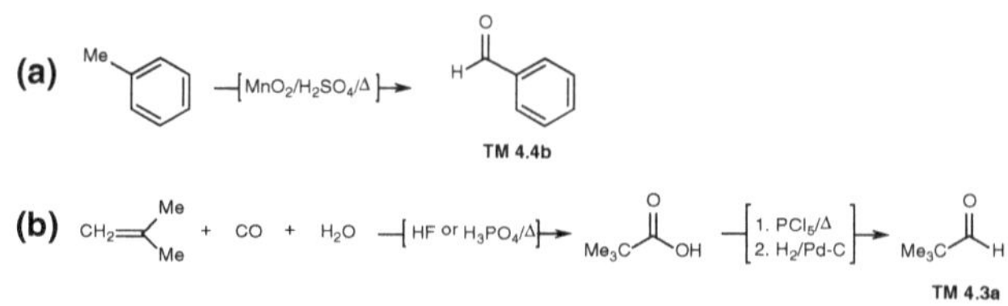

Scheme 4.7 Technological methods for production of benzaldehyde and pivaloyl aldehyde

in gas phase at 400 °C by V_2O_5/K_2SO_4. Pivaloyl aldehyde (*tert*-butyl aldehyde) is available by reduction of pivaloyl chloride, while pivalic acid is produced on the large-scale by hydroxycarbonylation of isobutene (Scheme 4.7b).

Phenyl propyl ketone (butyrophenone) **TM 4.3b** is available by common Friedel-Crafts acylation. On the other hand, *tert*butyl propyl ketone **TM 4.4a** is not an easily available C_8 building block. Butanoic acid is produced by either fermentation of starch using the *Bacillus subtilis* strain or oxidation of *n*-butanal, available by formylation of propene (Scheme 4.3).

More easily available **TM 4.3** is therefore the preferred model compound for study of the physico-chemical properties of 1,3-hydroxycarbonyl unit.

4.3.1.1 Departure into Third Dimension; Enantioselective Aldol Reaction

retro-Aldol disconnection is an example of the disconnection of the 1,3-hydroxycarbonyl pattern. In the synthetic direction, formation of the C–C bond requires *alkylation of α-carbanion* or its enolate by carbonylic reagent. An aldol reaction characterizes the nucleophilic attack of an electronegative α-C atom of enol or enolate anion, depending on the pH of the medium, on the carbonyl C atom of the second component (Scheme 4.8).

This reaction is of great importance in its *chiral variant* as an *asymmetric aldol reaction*. There are many practical methods for stereochemical control in either the non-catalytic or catalytic variant of this reaction. Two stereogenic centers are formed in a single reaction step, with an exception when the terminal methyl group reacts as an enol component (Scheme 4.8) or symmetrical ketone and formaldehyde as a carbonyl component (Scheme 4.9). Aldehydes are much more convenient than ketones as an electrophilic carbonyl counterpart and are preferably used. In Scheme 4.9 pairs of *syn*- and *anti*-diastereomers formed in an asymmetric aldol reaction are presented.

Scheme 4.8 Alternative aldol reactions

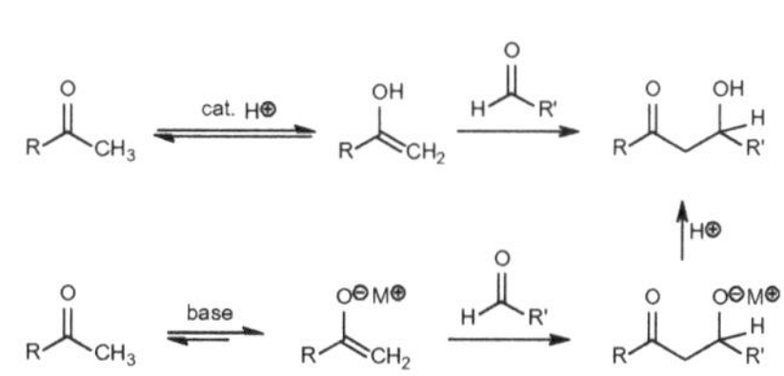

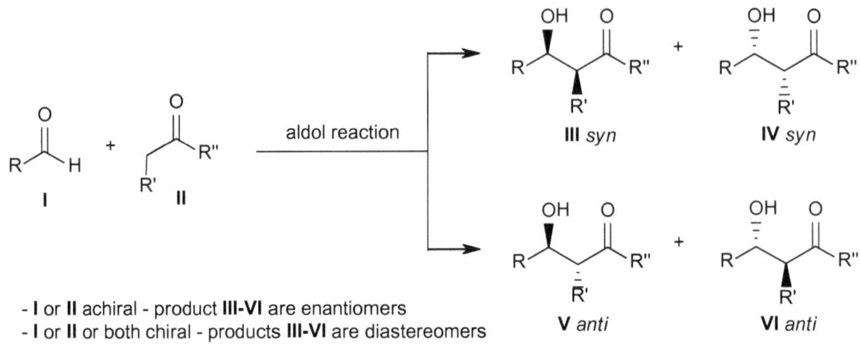

Scheme 4.9 Stereoisomeric products in the aldol reaction

Synthetic chemists are primarily interested in preparing one of diastereomeric products **III–VI** in the *optically pure form*. To design an asymmetric aldol reaction we need to understand the mechanism of diastereoselective formation of *syn*- and *anti*-racemates.

The nucleophilic component reacts in the aldol reaction in its enolic form, usually chelated to a metallic cation as *E*- or *Z*-enolate. The stereochemistry of metal chelates plays a primary role in the stereoselectivity of the aldol reaction by controlled formation of *syn*- or *anti*-diastereomers. As a rule *E*-isomers of enolates lead to *anti*-aldol products and *Z*-isomers to *syn* products. This outcome is explained by the Zimmerman-Traxler mechanism, which invokes a six-membered transition state (Scheme 4.10) [8, 9].

For the sake of clarity in the above scheme, the absolute configuration for one of the two possible enantiomeric *syn*- and *anti*-products is presented. Note that on the route to an *anti*-isomer both larger groups R and R' are found in a *pseudo-equatorial* position in the cyclic transition state, present in the more stable chair conformation. In the transition state leading to the *syn*-isomer group R is found in the less preferred, crowded *pseudo-axial* position. Here we follow the suggestion to use the prefixes *axial/equatorial* only for substituents on the cyclohexane ring in the

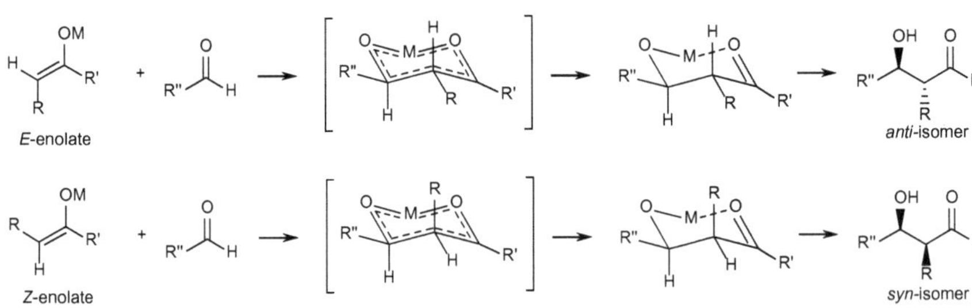

Scheme 4.10 Presentation of Zimmerman-Traxler transition states

chair conformation. In cases where a substituent is present in the six-membered heterocyclic or chelate ring in the chair conformation, the prefix *pseudo* is added [9].

Depending on the absolute configuration of the chiral component in the aldol reaction, the prevailing enantiomer will be formed as *a syn*- or *anti*-product, depending on the structure of the enolate. The chiral component can be either one of the reactants or catalyst. In the first case we have *non-catalytic asymmetric aldol reaction*, in the second a *catalytic asymmetric aldol reaction*. The configuration at the chiral center in the chiral component determines the direction of enantioselectivity. Both cases are discussed in the following sections.

4.3.1.2 Non-catalytic Enantioselective Aldol Reaction

An aldol reaction can be completed enantioselectively with an α-carbanion of the chiral carbonyl compound reacting as chiral enolate. Since the α-C atom in chiral enolate is planar, alkylation is diastereoselective because of the recognition of diastereotopic faces by the carbonyl counterpart. Alkylation of the configurationally stable tetrahedral carbanion on the α-C atom is a relatively rare case and is known as the *memory of chirality*.

The topological relation in the chiral substrate changes along the reaction coordinate; *diastereotopy of protons* on the α-methylene group transforms into *diastereotopy of faces* around the planar C=C bond in enol. A similar consideration is valid for the asymmetric catalytic aldol reaction with prochiral substrates; enantiotopicity of hydrogen atoms transforms into enantiotopicity of faces around the planar C=C bond in enolate.

In the next two examples, synthesis of α-alkyl-β-hydroxy carboxylic acids with high diastereo- and enantioselectivity is presented. We shall see how transformation of the reaction conditions, especially addition of an *achiral auxiliary agent*, inverts the direction of diastereoselectivity. Both examples treat rather complex organometallic reactions; hence, their study might be proper after consulting chapters on organometallic reactions in organic chemistry textbooks.

Example 4.3 This example demonstrates a notable difference in stereoselectivity between alkylation and aldol reactions of enolates derived from chiral oxazolidinones. Lithium enolates of oxazolines **27** and **28** proved exceptionally efficient in the control of the stereoselectivity of alkylation (Sect. 3.7.3, Schemes 3.12 and 3.13) but react with low stereoselectivity in aldol reactions. Instead, high stereochemical control of aldol reactions is achieved with boronic enolates of chiral oxazolidinones **9–13** (Scheme 4.11).

Two impressive results are achieved with the aldehyde general formulae R'CHO: first, nearly exclusive formation of *syn*-isomers **14a–18a** and second high optical purity of the prevailing diastereomers, as confirmed by 99–100 % e.e. of isolated α-substituted β-hydroxycarboxylic acids. Products **14a–18a** and **14b–18b** are in a diastereomeric relationship since both series retain the same configuration

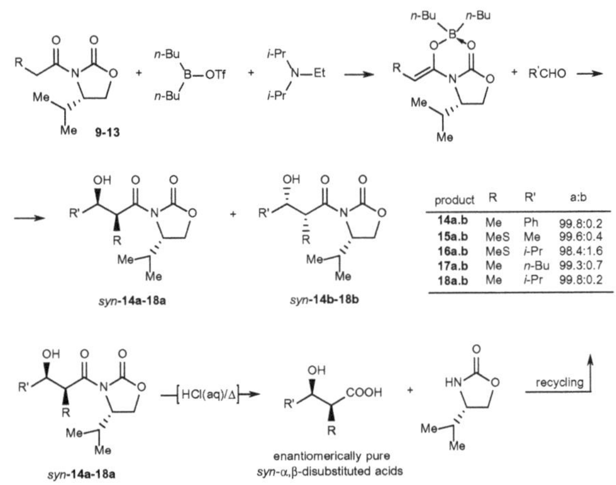

product	R	R'	a:b
14a.b	Me	Ph	99.8:0.2
15a.b	MeS	Me	99.6:0.4
16a.b	MeS	i-Pr	98.4:1.6
17a.b	Me	n-Bu	99.3:0.7
18a.b	Me	i-Pr	99.8:0.2

Scheme 4.11 Stereoselective aldol reaction of boronates of *N*-acyloxazolidinones **9–13**

of the inducing stereogenic center in the oxazolidinone ring. Elimination and recycling of the chiral auxiliary are completed by mild hydrolysis of imides **14–18**.

Chiral oxazolidinones possess nearly all properties required for economic application on the large scale; both enantiomers are available in two steps from D- or L-amino acids; they are configurationally stable and easy to recycle. "Tailing to measure" of chiral oxazolidinone derivatives is possible, as shown in the syntheses of some biological products [10, 11].

Example 4.4 Formation of *Z*-boronates derived from chiral oxazolinones **9–13** is preferred because of reduced non-bonding through-space interactions between the R and *iso*-propyl group on the chiral center (Scheme 4.11). It is therefore understandable that most aldol reactions with this substrate run in the *syn*-selective mode. Prevailing absolute configurations (1*S*, 2*R*) on the new chiral centers in **14–18** (shown in the Scheme!) are determined by the *absolute stereochemistry* of chiral oxazolidinone. *Absolute stereochemistry* comprises a set of configurational and conformational properties of certain chiral molecules. It dictates "facial selectivity" in chiral oxazolidinones; the direction of approach by the carbonyl group of R'-CHO is as an electrophile to the double C=C bond of *Z*-enole.

Reacting systems are discovered wherein boronic enolates surprisingly afford *anti*-products with varying diastereomeric excesses [12]. Coordination of aldehyde to Lewis acid forms a stable complex under reaction conditions, which coordinates to boronic enolates, resulting in a transition state that is no longer of the Zimmerman-Traxler type (Scheme 4.12).

The approach of coordinated aldehyde to the enolic C=C bond minimizing steric perturbations leads to *anti*-isomers of aldols. Diastereomeric excess of *anti*-isomer with an absolute configuration presented in Scheme 4.12 varies; still, this method has a significantly enriched application in chiral oxazolidinones.

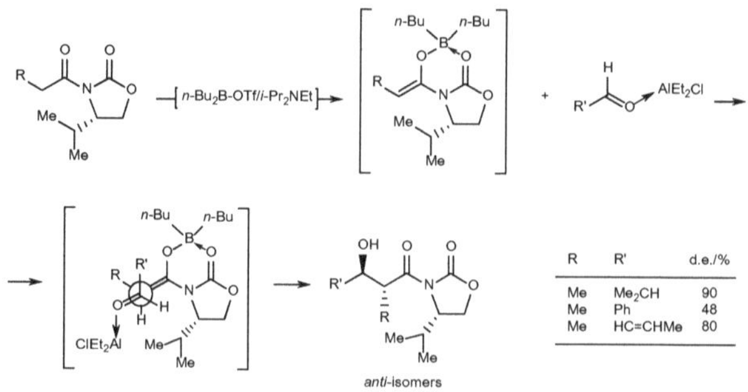

R	R'	d.e./%
Me	Me_2CH	90
Me	Ph	48
Me	HC=CHMe	80

Scheme 4.12 *anti*-Directed aldol reaction of boronic complexes of chiral oxazolidinones

4.3.1.3 Catalytic Enantioselective Aldol Reaction

The high value of catalytically performed reactions as compared to non-catalytic variants is particularly evidenced in the field of enantioselective reactions. Chemists cannot complete enantioselective reactions without certain chiral information in the reacting system. This information is regularly derived from the chiral compounds present in nature, collectively named the *chiral pool of the nature*. Their availability is often limited, which is not an issue when they are used as catalysts, but causes significant costs of non-catalytic reactions when they are needed in equimolar quantities. The practical value of the catalytic approach to enantioselective processes cannot be overestimated. *Asymmetric catalysis* characterizes the *amplification of chirality*; one chiral molecule of the catalyst generates an enormous number of chiral molecules of the product in the optically pure form. This results with high *chiral economy* of catalytically performed enantioselective syntheses.

Arguments in favor of catalytic asymmetric reactions are supported by the following examples.

Example 4.5 Borate complex **VIII**, derived from the monoester of tartaric acid **VII**, proved to be a highly efficient catalyst for the aldol reaction of silyl enolates and aldehydes (Scheme 4.13).

The high practical value of this reaction rests in the fact that the relative and absolute configurations of products **22–25** are independent of the *E/Z* geometry of silyl enol ether. *E*-and *Z*-isomers of substrate **20** give the same *syn*-product **24** with >95 % e.e. This outcome is explained by the intermediary formation of an "open transition state" since the dominant interaction between groups R and R" precludes the route to the *anti*-isomer via the cyclic transition state [13].

Example 4.6 In the previous example catalytic complex **VIII** is a derivative of tartaric acid. More easily accessible is boronate **X**, derived from *N*-tosyl-L-tyrosine **IX**. This catalyst proved highly effective in the synthesis of (3*S*)-α-hydroxy ketones **26–30** starting from silyl enolates and aldehydes (Scheme 4.14) [14].

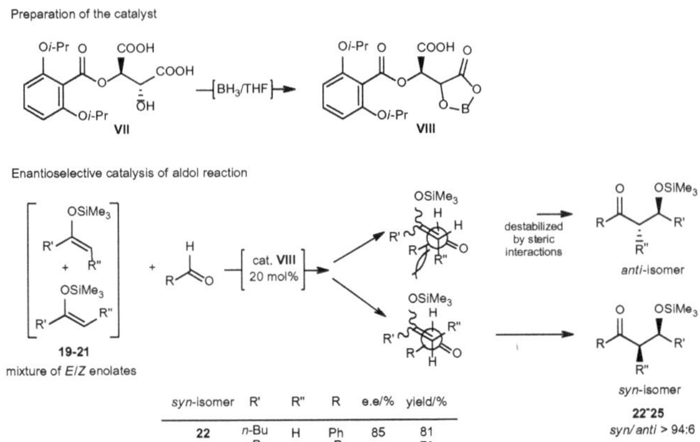

syn-isomer	R'	R"	R	e.e/%	yield/%
22	*n*-Bu	H	Ph	85	81
23	*n*-Bu	H	*n*-Bu	80	70
24[a]	Et	Me	Ph	96	96
24[b]	Et	Me	Ph	94	97
25	$(CH_2)_4$		Ph	>95	57

[a] from *E*-isomer **20**
[b] from *Z*-isomer **20**

Scheme 4.13 Mechanism and stereoselectivity of an aldol reaction catalyzed by borate complex **VIII**

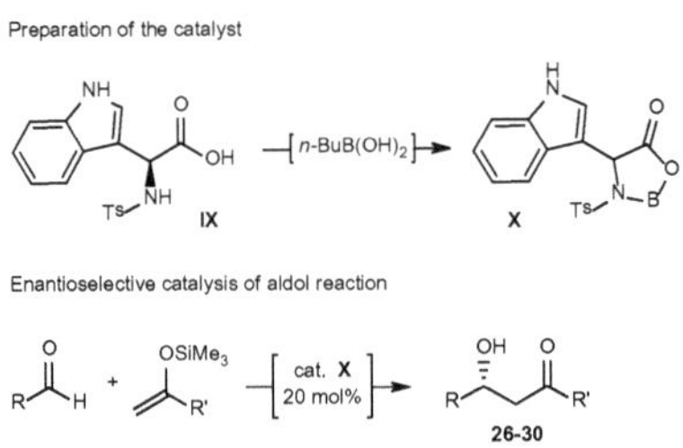

aldol	R	R'	e.e./%	yield/%
26	Ph	Ph	89	82
27	cyclohexyl	Ph	93	67
28	*i*-Pr	Ph	89	94
29	2-furyl	Ph	92	100
30	Ph	*n*-Bu	90	100

Scheme 4.14 Enantioselective aldol reactions catalyzed by boronate **X**

The complexity of the catalytic systems for the preparation of 3-hydroxyketones in their enantiomerically pure form is notable. Nevertheless, an asymmetric aldol reaction in its catalytic variant represents a workable synthetic method for the preparation of compounds **26–30** with high optical purity.

Example 4.7 The asymmetric synthesis of optically pure aldol products **31–34** that on cyclization in acidic medium afford dihydropyranones **35–38**, important chiral intermediates in the synthesis of analogs of monosaccharides and other biologically active compounds, is presented in Scheme 4.15.

The chiral catalyst is boronate **X**, described in the previous example. Note that absolute configuration in **31–34** is maintained in the cyclic products **35–38**. On protonation a vinylic methoxy group became the better leaving group, and cyclization proceeded with retention of the configuration on the chiral center.

Scheme 4.15 Preparation of optically pure dihydropyranones **35–38**

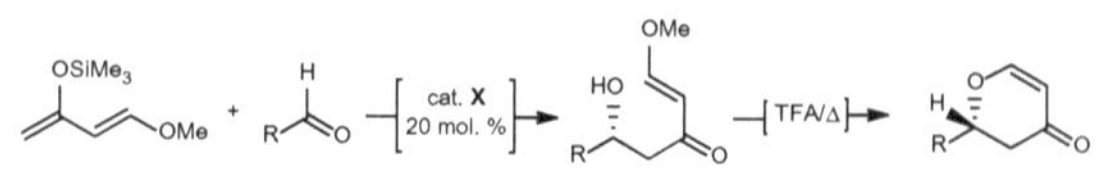

31-34

35-38

pyranone	R	e.e./%	yield/%
35	Ph	82	100
36	2-furyl	67	83
37	cyclohexyl	76	80
38	$PhCH_2CH_2$	69	57

4.3.2 1,3-Dicarbonyl Compounds

The second class of 1,3-dixygenated compounds is characterized by two carbonyl groups in the 1,3-position. Depending on the structure of group R", they belong to 1,3- or β-diketones, 1,3-ketoaldehydes and 1,3-ketocarboxylic acids and their derivatives. Disconnection of either internal C–C bond in **TM IV** can be completed by participation of one carbonyl group (Scheme 4.16).

Differently from the 1,3-hydroxycarbonyl pattern, disconnection of the 1,3-dicarbonyl pattern results in two logical synthons, an α-carbanion stabilized by conjugation and an acyl cation. The acyl cation has a number of acceptable reagents in reactive derivatives of carboxylic acids. Reactive enough with carbanions are carboxylic acid esters affording 1,3-dicarbonyl compounds in the well-known *ester condensation*, also known as *Claisen condensation*. The next example serves as an introduction to the retrosynthetic analysis of this pattern.

Example 4.8 Complete the retrosynthetic analysis of **TM 4.5** and suggest its synthesis.

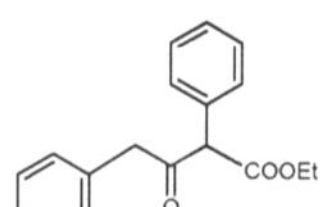

TM 4.5

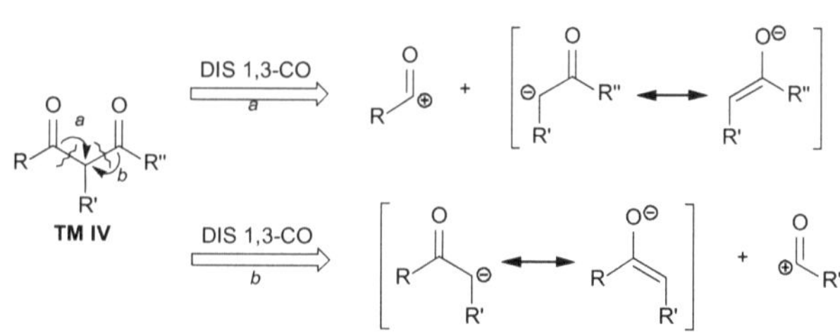

Scheme 4.16 Two possible disconnections of 1,3-dicarbonyl pattern in **TM IV**

Scheme 4.17 Two possible disconnections of **TM 4.5**

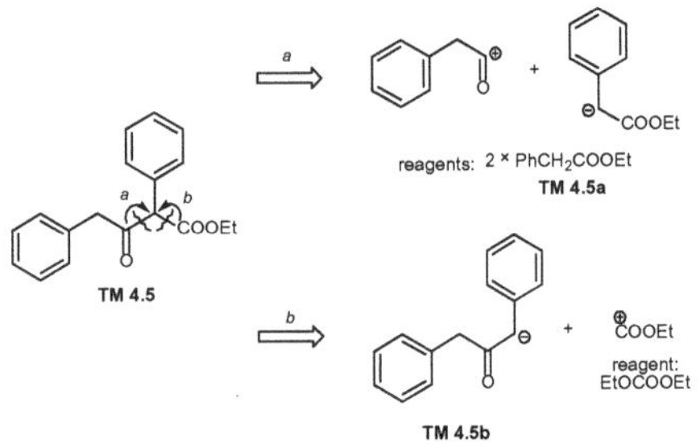

The 1,3-dicarbonyl pattern in **TM 4.5** suggests obvious disconnection *a* leading to the maximal simplification of the molecule. Two C_7 synthons have the same reagent, an ester of phenylacetic acid (Scheme 4.17).

One economic industrial production of phenylacetic acid is the oxidation of ethylbenzene by an aqueous solution of potassium bichromate in an autoclave at elevated temperature (Scheme 4.18).

It is surprising that the contradicting principle of maximal simplification disconnection *b* competes with disconnection *a* (Scheme 4.17). The carbanion of 1,3-diphenylacetone and carbethoxy cation appear as synthons. 1,3-Diphenylacetone is a commercial product, available from phenylacetic acid over the intermediary formed, β-keto acid, which under reaction conditions decarboxylates (Scheme 4.19).

On the other hand, the carbethoxy cation, a seemingly illogical synthon, has an available synthetic equivalent in diethyl carbonate. Diethyl carbonate is produced by oxidative carbonylation of ethanol, promoted by various heterogeneous catalysts; one of the most effective is the mixed catalyst $CuCl_2/PdCl_2$ deposited on charcoal (Scheme 4.20).

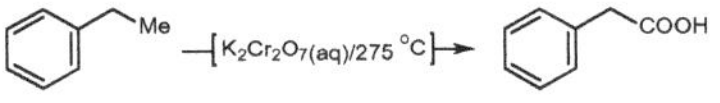

Scheme 4.18 Industrial method for production of phenylacetic acid

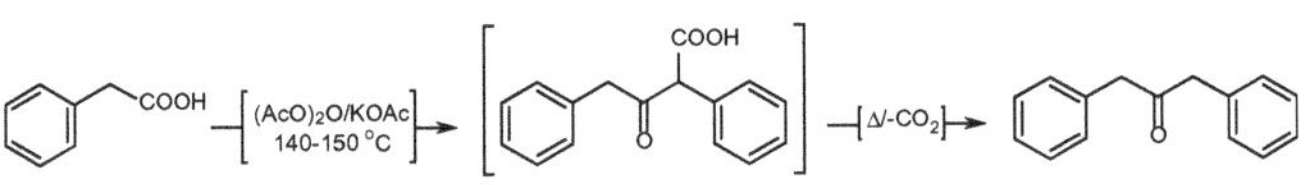

Scheme 4.19 Industrial production of diphenyl acetone

Scheme 4.20 Method for production of diethyl carbonate

Scheme 4.21 Synthetic routes to **TM 4.5**

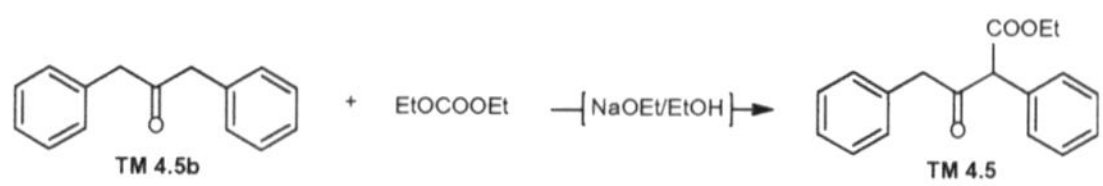

The conclusion emerges that both disconnections are acceptable in view of the availability of starting materials. In spite of the "pure simplification" in the retrosynthesis of **TM 4.5**, synthesis according to Scheme 4.21 is preferred.

4.3.3 *Concept of Hard and Soft Acids and Bases (HSAB)*

The practicability of synthesis **TM 4.5** according to the disconnection that includes "poor simplification" is confirmed by the next example, which also serves as an introduction to the important concept of *hard and soft nucleophiles and electrophiles.*

Example 4.9 Propose a synthetic route to **TM 4.6** and explain why the obvious disconnection according to maximal simplification of this target molecule is hampered by non-workable synthesis.

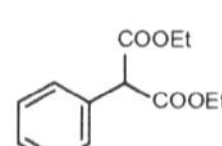

TM 4.6

On first glance we recognize the 1,3-dicarbonyl pattern of diethyl malonate, which suggests the disconnection of the C–C bond to the aromatic ring. In view of the easy alkylation of diethyl malonate, one would expect a workable arylation of the diethylmalonate anion by aryl halides. However, this reaction represents a difficult synthetic task! Here we meet the important principle of reactivity in organic chemistry. The high stability of the malonate carbanion is due to its resonance stabilization by two geminal carbethoxy groups. Delocalization of the negative charge by resonance makes this carbanion a very *soft nucleophile* and therefore not reactive with *hard electrophiles* such as aryl halides, specifically bromobenzene **TM 4.6a** (Scheme 4.22).

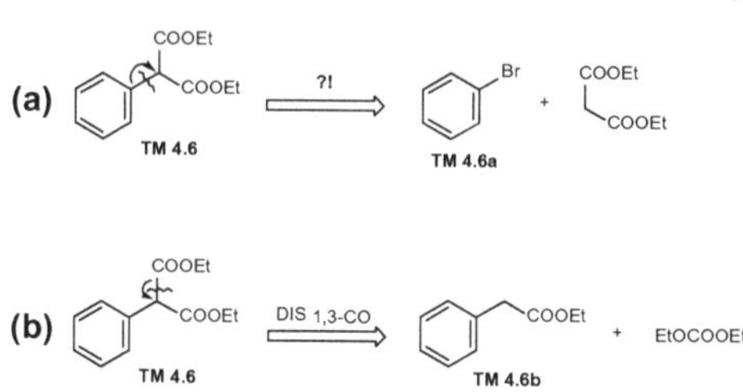

Scheme 4.22 Two possible disconnections of **TM 4.6**

The concept of *hard and soft acids and bases* (HSAB) is usually related to the physico-chemical properties of organic compounds and is not much recognized in synthetic organic chemistry [15, 16]. There is also the less often used concept of *hard and soft nucleophiles* and *electrophiles* [17]. The fundamental distinction between these two concepts, i.e., between basicity and nucleophilicity on the one hand and acidity and electrophilicity on the other, lies in that fact that basicity and acidity are *thermodynamic properties*, whereas nucleophilicity and electrophilicity are *kinetic terms*.

As a rule, soft electrophiles react faster with soft nucleophiles and hard electrophiles with hard nucleophiles, but there are some exceptions. These synthetically important exceptions are the consequence of the prevalent control of reactivity of soft nucleophiles by the (high) energy of their HOMO, whereas hard nucleophiles are characterized by a low energy of HOMO but high localization of negative charge. Therefore, interaction between soft nucleophiles and electrophiles is mainly controlled by the energy of their *frontier orbitals*, whereas interaction between hard nucleophiles and electrophiles controls the *charge localization* [17]. Therefore, we can summarize:

- hard-hard interaction is fast because of the strong Coulombic interaction
- soft-soft interaction is fast because of the strong HOMO-LUMO interaction

To illustrate this principle, let us consider the reactivity of acyl and alkyl chlorides with amines and thiols [18]. Amines are hard nucleophiles and react quickly with acyl chlorides as hard electrophiles. The reaction of thiols as soft nucleophiles with acyl chlorides is surprisingly slow. The reactivity of amines and thiols with soft alkyl chlorides is inversed. All these patterns of reactivity are explainable by the *hard–soft principle*.

Now we can continue with the retrosynthetic analysis of **TM 4.6**. The *hardness* of aryl halides with similar electronic properties of the C atom in the benzene ring corresponds to the *hardness* of acyl chlorides with chlorine bound to the sp^2 C-atom of the C=O group. The *hardness* of bromobenzene **TM 4.6a** explains the inertness against the *soft* malonate anion, as indicated by questionable disconnection (a) in Scheme 4.22.

As suggested by the discussion of HSAB, we invoke alternative disconnection (b) of **TM 4.6**. On the first glance, an *illogical* disconnection leads to the available staring material, ethyl phenylacetate and diethyl carbonate. In the synthetic direction, ester condensation is a well-know and workable process.

Example 4.10 Suggest the disconnection of **TM 4.7**.

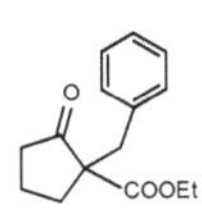

TM 4.7

The first retrosynthetic step represents the logical disconnection of the exocyclic C–C bond with the formation of two large synthetic blocks (Scheme 4.23). The

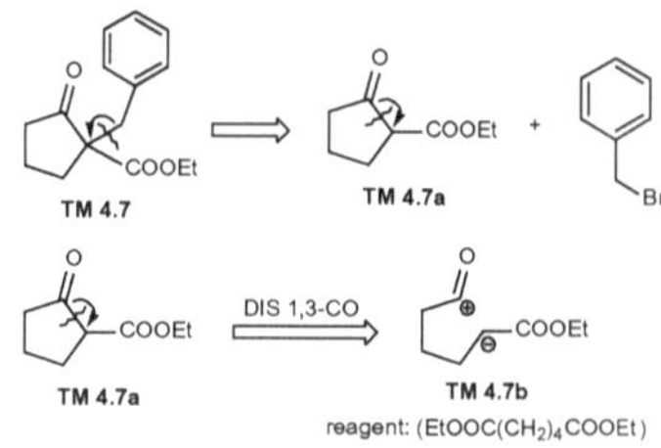

Scheme 4.23 *retro*-Dieckmann disconnection of **TM 4.7**

next target molecule, **TM 4.7a**, comprises a 1,3-dicarbonyl pattern with a keto group inside the cyclopentane ring and an exocyclic carboxylic group. Disconnection of the C–C bond in the ring is preferred, although we mentioned that ring-opening disconnections often result in more complex target molecules! *retro*-Ester condensation in **TM 4.7** results in easily available diethyl ester of adipic acid **TM 4.7b**.

The intramolecular variant of ester condensation is known as the *Dieckmann reaction* and preferred for construction of thermodynamically favored medium rings without steric strain, in particular cyclopentanone and cyclohexanone derivatives.

Note Benzyl bromide is produced by the bromination of toluene at room temperature in air using MnO_2 as catalyst. Adipic acid is produced by the industrial methods presented in Scheme 4.24.

Method (a) uses nitric acid for the oxidation of the mixture of cyclohexanol/cyclohexanon available by the hydrogenation of phenol; process (b) is based on hydroxycarbonylation of 1,3-butadien and process (c) on a catalytic "green chemistry" reaction with water as the only side product. In this process, cyclohexene is oxidized by hydrogen peroxide in the presence of tungsten-based catalyst under *phase-transfer catalysis* (PTC).

Phase-transfer catalysts promote the transport of reactants from one phase to the other where the reaction takes place. This means that the reaction medium is biphasic, usually consisting of two immiscible liquids, and the transfer is completed between the liquid/gas or solid/liquid phase [19, 20]. PTC processes have other advantages: enhanced conversion, fewer side products, emanation of expensive solvents needed

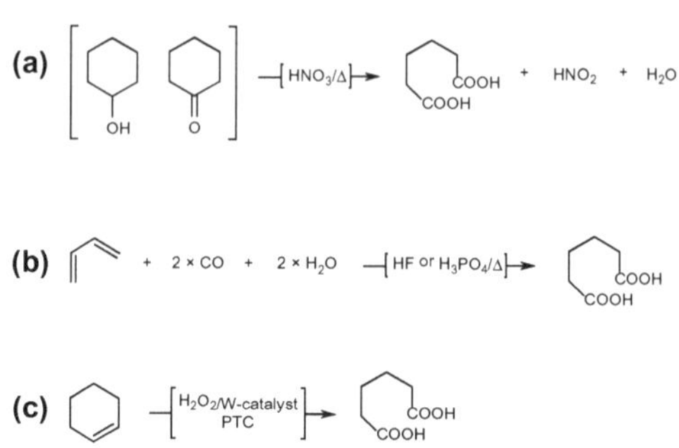

Scheme 4.24 Industrial methods for production of adipic acid

for dissolution of all reactants in a single phase, elimination of expensive starting materials and limited quantities of waste. PTC satisfies the "green chemistry" criteria (Sect. 3.1) using water as one of the reaction media and diminishing the consumption of organic solvents [21].

4.4 1,5-Dicarbonyl Pattern (1,5-CO)

Two carbonyl groups in the target molecule **TM V** are separated by five C atoms, and two possibilities emerge for the disconnection of internal bonds. The first one is disconnection of the C–C bond next to one of two carbonyl groups, and the second one is disconnection of one central C–C bond (Scheme 4.25).

Disconnection (a) results with stable acyl carbocation **TM Vb**, for which we introduced a series of reagents (Sect. 1.4.1, Scheme 1.15). The second synthon is unstable carbanion **TM Va**, for which we do not have an acceptable reagent besides organometallic species. Problems with the use of organometallics in the preparation carbonyl compounds were already discussed in Chap. 2.

Disconnection (b) of the central C–C bond emerges as unexpectedly convenient and involves a mechanism that we will discuss in more detail. Since the C–H acidity of the α-C atom in **TM V** is high enough, we can include C–H bond α- in the second carbonyl group in the disconnection of the central C–C bond in the α,β position to a neighboring carbonyl group. The net result of *disconnection with participation of* the *C–H bond* is the generation of two logical synthons of the second generation, conjugated enone **TM Vc** and stable carbanion **TM Vd**. Removal of the proton in disconnection (b) is used in similar schemes to complete the stoichiometry of the disconnection process. In the synthetic direction, this means deprotonation of the methylenic C–H bond with a strong base.

Convenient disconnection of the 1,5-dicarbonyl pattern results in the α,β unsaturated compound, enone, with a highly reactive C=C bond. Partners in the synthesis are neutral enone and α-carbanion of the second carbonylic reagent to complete the well-known *Michael addition* [22, 23]. Hence, the disconnection of the central bond in the 1,5-dicarbonyl pattern is denoted as the *retro*-Michael.

Michael addition is a *vinylogous analog* of the aldol reaction, and most considerations of the aldol reaction in Sect. 4.3 apply here. The reaction is catalyzed by

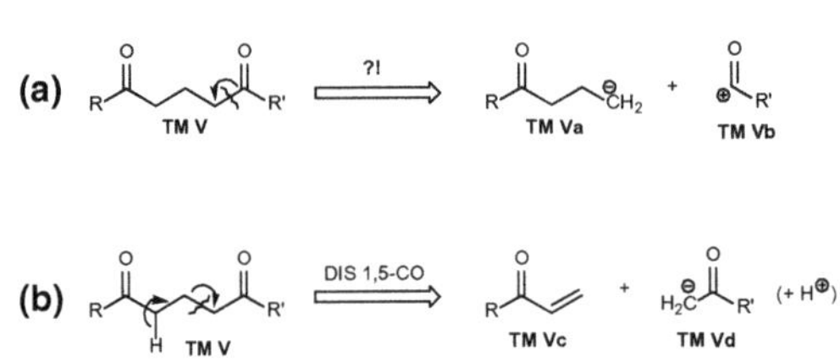

Scheme 4.25 Disconnections of internal C–C bonds in **TM V**

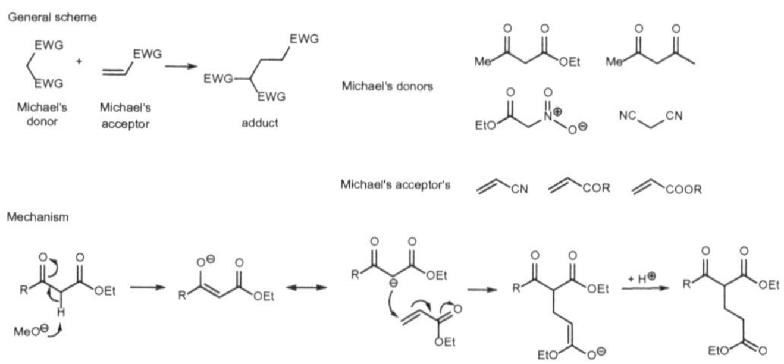

Scheme 4.26 General scheme and mechanism of Michael addition

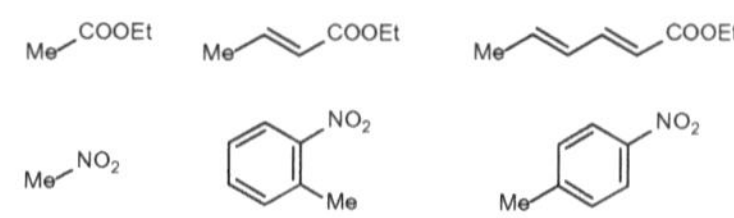

Fig. 4.4 Examples of molecules where the vinylogy principle gives similar C–H acidity to methyl groups

both acids and bases, but more efficiently by bases according to *the* mechanism presented in Scheme 4.26 [24].

Note The vinylogy principle is important in retrosynthetic analysis. It states that two groups separated by one or more conjugated bonds exhibit the same reactivity as if they were directly connected. This is a consequence of the effect of the first group through conjugated double bonds. Figure 4.4 presents few molecules where the *vinylogous position* of the terminal methyl groups results in similar C–H acidity.

Selection of the preferred central C–C bond for disconnection in **TM V** in Scheme 4.25 depends on the electronic properties of the R and R' groups and should result in the most stable synthons. On heterolytic disconnection, the electron pair from the σ bond moves to the α-C atom in **TM Vd** where the carbanion is stabilized by the neighboring electron-withdrawing group. To avoid the formation of an unstable carbocation on the β-C atom, concerted deprotonation of the acid C–H bond in the α-C atom implies participation of the second C=O group in the 1,5-dicarbonyl unit affording strongly electrophilic enone **TM Vc**.

Example 4.11 Propose the retrosynthesis and then suggest the conditions for the synthesis of **TM 4.8**.

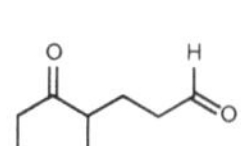

TM 4.8

Two carbonyl groups in **TM 4.8** are present in the 1,5-position; one belongs to the terminal aldehyde and the other to cyclic ketone. Generally, the C–H acidity of the α-C–H bond in aldehydes is somewhat higher (pK_a 17) than in ketones (pK_a 20). The carbonyl C atom in aldehydes is a stronger electrophile than in ketones. Ketone enolates have more nucleophilic α-C atoms than aldehyde enolates. The former

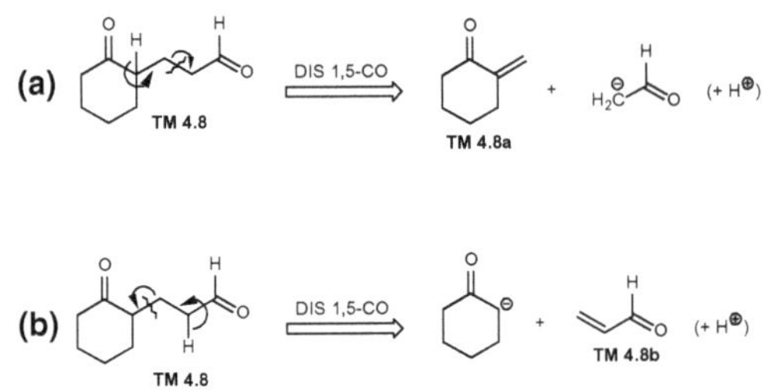

Scheme 4.27 Alternative 1,5-CO disconnections of **TM 48**

Scheme 4.28 Proposed synthesis of **TM 4.8**

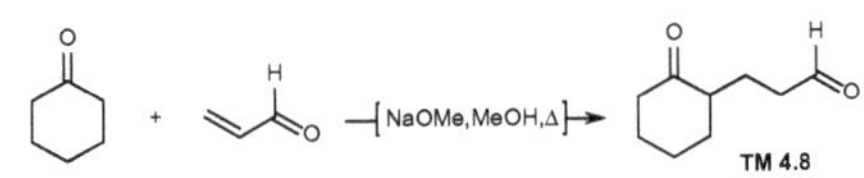

argumentation suggests disconnection (a) of **TM 4.8** and the latter disconnection (b) (Scheme 4.27).

Disconnection (a) results in the α-methylenic derivative of cyclohexanone, an enone whose synthesis will be shown in Sect. 4.4.2, Scheme 4.43, and acetaldehyde. Disconnection (b) leads to the easily available raw materials cyclohexanone and acrolein. Acrolein is an industrial commodity produced by thermal oxygenation of propene with oxygen at 250 °C. Cyclohexanone is commercially produced by co-catalyzed oxidation of cyclohexane or by controlled hydrogenation of phenol [25]. The availability of both starting materials suggests a one-step synthesis of **TM 4.8** workable on the large scale (Scheme 4.28). The molar ratio of reactants is controlled to avoid double α,α′-alkylation of cyclohexanone.

In the next example we shall see that the preferred reagents in the synthetic steps depend on the correct order of retrosynthetic steps when alternative disconnections are conceivable.

Example 4.12 Propose convenient synthetic routes to **TM 4.9** considering two alternative disconnections.

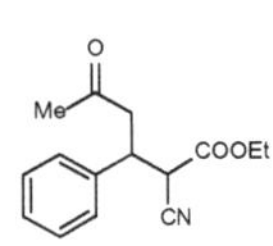

TM 4.9

First we observe that **TM 4.9** comprises a 1,5-dicarbonyl pattern and select two central C–C bonds for disconnection (Scheme 4.29). According to route (a), the first retrosynthetic step, *retro*-Michael disconnection, affords an acetonide anion and target molecule of the second generation, building block **TM 4.9a**. In the second step, *retro*-aldol disconnection of **TM 4.9a** affords two easily available reagents, benzaldehyde and ethyl cyanoacetate.

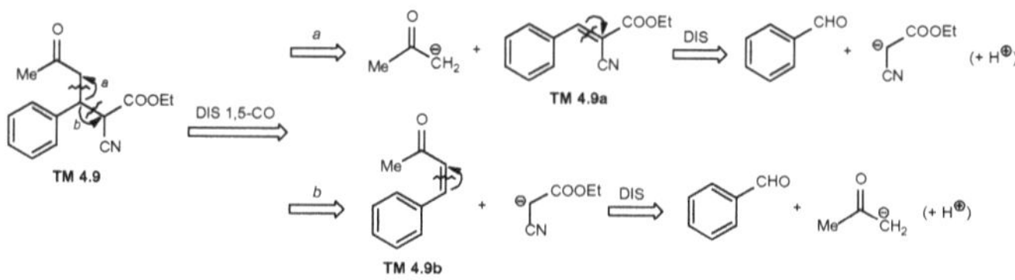

Scheme 4.29 Alternative 1,5-CO disconnections of **TM 4.9**

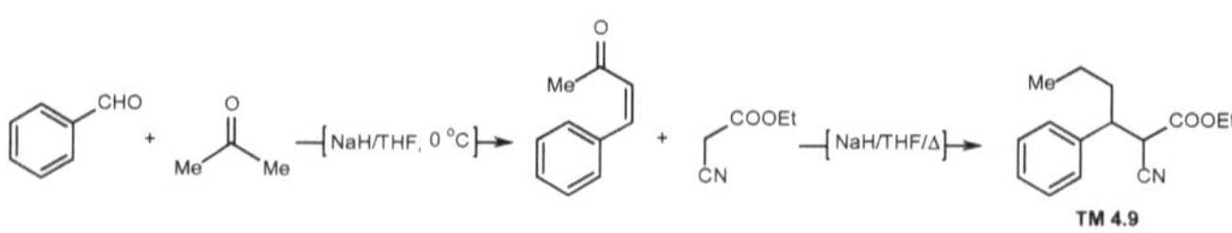

Scheme 4.30 Preferred synthetic route to **TM 4.9**

retro-Michael disconnection in the first step (b) results in enone **TM 4.9b** and one molecule of ethyl cyanoacetate. In the second step, disconnection of enone leads to benzaldehyde and acetonide anion.

Surprisingly, both disconnection routes lead to the same starting materials, benzaldehyde, ethyl cyanoacetate and acetone. We described ethyl acetoacetate as the preferred synthetic equivalent for acetone in Sect. 2.3.1, Example 2.11. For the synthesis of **TM 4.9b**, a workable method is the condensation of acetone with an excess of benzaldehyde, a non-enolizable component to prevent polymerization of acetonide anion.

To decide whether disconnection (a) or (b) suggests a synthetically more convenient route, we consider the relative reactivities of single reactants. Benzaldehyde is more reactive with the acetonide anion than with enone **TM 4.9b**. Therefore, in the first step, aldol condensation is preferred, followed by Michael addition of enone to the carbanion of ethyl cyanoacetate. Consequently, retrosynthesis (b) suggests the preferred synthetic route to **TM 4.9** (Scheme 4.30).

4.4.1 From Retrosynthesis to Robinson Annulation

In the previous chapter, we described Michael addition in the route to 1,5-dicarbonyl compounds where the R or R' group is often methyl. The methyl group in the 1,6-position to the distant carbonyl group enters *intramolecular aldol condensation* under the same reaction conditions, affording the cyclohexanone derivative (Scheme 4.31).

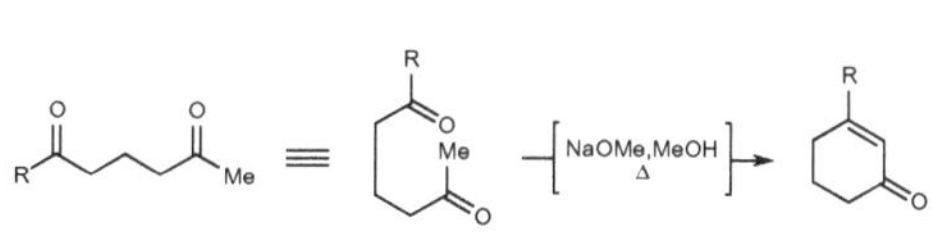

Scheme 4.31 Intramolecular condensation of 1,5-dicarbonyl compounds

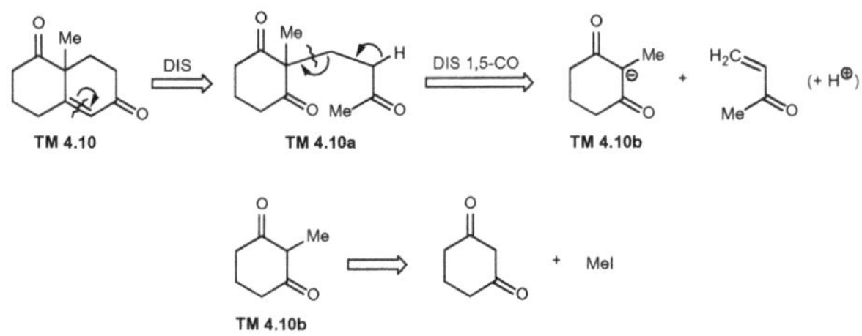

Scheme 4.32 Retrosynthetic analysis of **TM 4.10**

Note Michael addition-aldol condensation with cyclization is an example of a *tandem reaction,* discovered by Nobel Prize laureate R. Robinson and named the *Robinson annulation* [26]. *Tandem reactions* are two or more reactions that occur in a defined order without the isolation of intermediates [27]. Robinson annulation affords mono-, bi- and tricyclic derivatives of cyclohexanone, important intermediates in many syntheses of natural products, in particular steroids.

In the next examples, we shall practice retrosynthetic analysis of these structures.

Example 4.13 Propose the retrosynthetic analysis and then synthesis of **TM 4.10.**

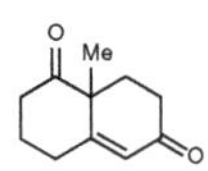

TM 4.10

In the bicycle target molecule we recognize a 1,5-CO pattern and enone unit. In the first retrosynthetic step disconnection of the C=C bond is preferred, leading to **TM 4.10a**, cyclohexane 1,3-dione derivative with a quaternary C-atom in the α position to both carbonyl groups (Scheme 4.32). *retro*-Michael disconnection of the new 1,5-CO pattern at the C–C bond to the ring affords a stable carbanion of 2-methyl-1,3-cyclohexandione **TM 4.10b** and methyl vinyl ketone. Final disconnection of the methyl group leads to 1,3-cyclohexandione.

Note 1,3-Cyclohexandione (dihydroresorcinol) is available from 1,3-diphenol (resorcinol) by hydrogenation of its mono-sodium salt with one mole of hydrogen at temperatures below 50 °C catalyzed by RaNi in basic medium. Methyl vinyl ketone is a broadly used C_4 building block produced by gas-phase formylation, aldol condensation of acetone with formaldehyde catalyzed by metal oxides (Scheme 4.33) [28, 29].

Now we can propose a short synthetic route to **TM 4.10** starting from the available building blocks (Scheme 4.34).

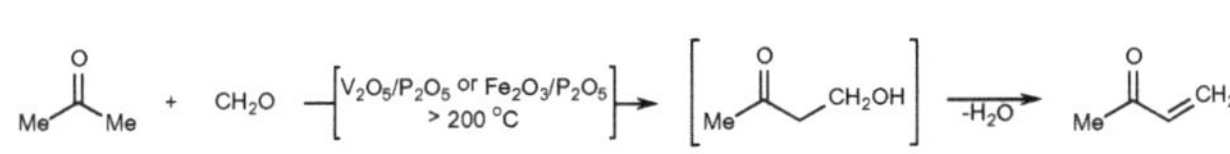

Scheme 4.33 Industrial production of methyl vinyl ketone

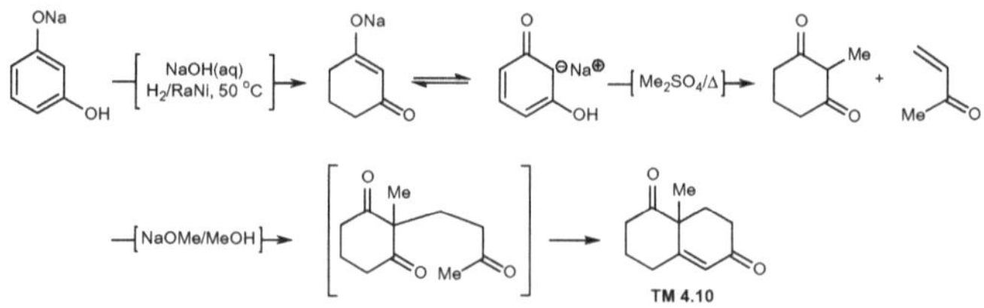

Scheme 4.34 Proposal for the synthesis of **TM 4.10**

On partial hydrogenation of resorcinol to 1,3-cyclohexanedione, C-methylation of enol is performed under standard conditions. In the next two steps, *Robinson annelation* is completed. Michael addition of methyl vinyl ketone affords an intermediate that spontaneously enters intramolecular aldol condensation to a stable six-membered ring in **TM 4.10**.

The next retrosynthetic analysis illustrates the application of the concept *of an activating group*.

Example 4.14 Propose the retrosynthesis and then synthesis of racemic natural product *piperitone* **TM 4.11**.

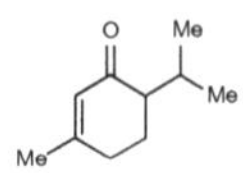

TM 4.11
Piperitone

retro-Aldol type disconnection of cyclic enone in the first step is straightforward (Scheme 4.35). In the new **TM 4.11a** either of two central C–C bonds can be disconnected according to the *retro*-Michael. However, immediate disconnection of the C–C bond to the branching point, suggested by maximal simplification of the target molecule, results in methyl vinyl ketone and methyl isobutyl ketone, an inconvenient reagent for an unstabilized carbanion. This problem solves the FGA, the addition of an *activating carbethoxy group* at the branching point affording **TM 4.11b**.

Note The concept of an *activating group* is related to the well-known concept of a *protecting group* and illustrated by two known examples in Fig. 4.5.

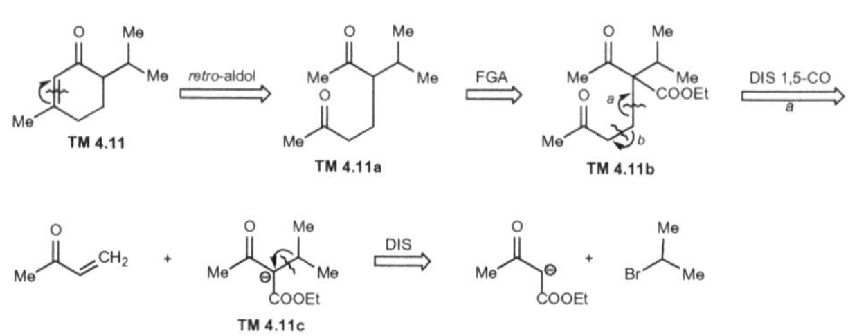

Scheme 4.35 Proposal for retrosynthetic analysis of racemic *piperitone* **TM 4.11**

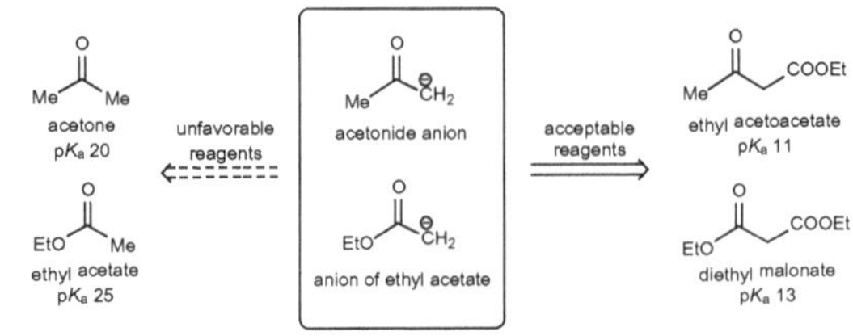

Fig. 4.5 Important C_3 and C_2 synthons and their unfavorable and acceptable reagents

Higher C–H acidity in carbethoxy derivatives enables the formation of a stabilized carbanion prone to alkylation or Michael addition.

Now 1,5-disconnection (a) in Scheme 4.35 is preferred, resulting with α-isopropyl ethyl acetoacetate, an easily available **TM 4.11c**. *retro*-Michael disconnection (b) in **TM 4.11b** is not possible since no double bond can be formed to the quaternary C-atom.

Complete retrosynthesis of racemic *piperitone* is proposed in Scheme 4.35.

Synthesis of *piperitone* can be worked out from methyl vinyl ketone, ethyl acetoacetate and 2-bromopropane, where in all steps alkylation, Michael addition and an aldol condensation medium strong base are needed, e.g., NaOEt/EtOH. For decarboxylation of intermediary **TM 4.11b**, mildly acidic conditions are convenient.

4.4.2 *Vinyl Ketones via the Mannich Reaction*

In Sect. 4.4.1, an industrial method for the production of methyl vinyl ketone is briefly described. Alkyl vinyl ketones are produced on the industrial scale by formylation of the methyl group in alkyl methyl ketones. This reaction preferably runs in gas phase at elevated temperatures and is not convenient for preparation of sensitive vinyl ketones of more complex structures and also not practical on the laboratory scale. At the laboratory level these compounds can be prepared by a three-step route that includes the *Mannich reaction* in the first step.

Practical laboratory synthesis of vinyl ketones starts with the preparation of β-amino ketones (*Mannich bases*) and then quaternization of the *tert*-amino group followed by thermal degradation of the quaternary ammonium salt (Scheme 4.36).

Before discussing examples that demonstrate the workability of the above scheme in the synthesis of compounds of academic or commercial interest, we consider the retrosynthesis of a simple Mannich base and mechanism of reaction.

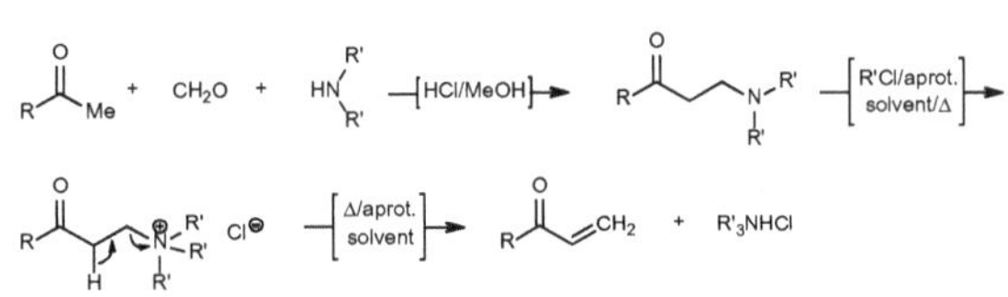

Scheme 4.36 Preparation of vinyl ketones via a Mannich base

Example 4.15 Propose the retrosynthesis and then synthesis of **TM 4.12**.

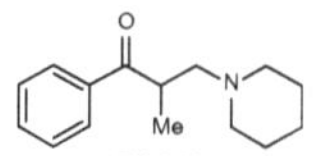

TM 4.12

The β-amino carbonyl pattern in **TM 4.12** suggests a *retro*-Mannich disconnection, contemporaneous disconnection of the internal C–C and C–N bonds. Available building blocks are obtained: propiophenone, formaldehyde and *sec* amine (Scheme 4.37).

By this disconnection, σ-electrons from the C–C bond move to the α-C atom of propiophenone and from a C–N bond to the N atom of piperidine. Two positive charges that formally appear on the central C-atom can be compensated by the double bond to the O atom; hence, the third reagent is formaldehyde.

Proper reaction conditions for the Mannich reaction are determined by its mechanism (Scheme 4.38).

The acid-catalyzed step imposes the formation of iminium salt with a strongly electrophilic C-atom of the C=N group. It reacts with propiophenone enolized under acidic conditions. This mechanism reveals the double role of acid, usually protic acid, but Lewis acids can also be used.

The synthesis of **TM 4.12** including steps for the preparation of propiophenone and piperidine is proposed in Scheme 4.39.

Scheme 4.37 *retro*-Mannich disconnection of **TM 4.12**

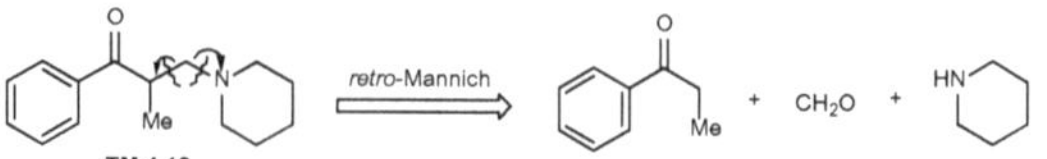

Scheme 4.38 Basic mechanism of the Mannich reaction

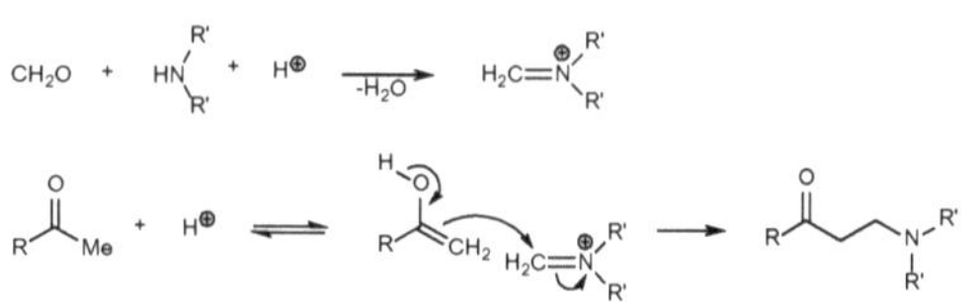

Scheme 4.39 Complete synthesis of **TM 4.12**

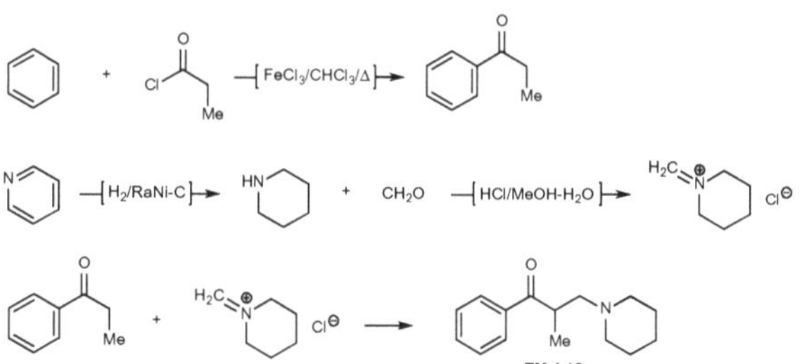

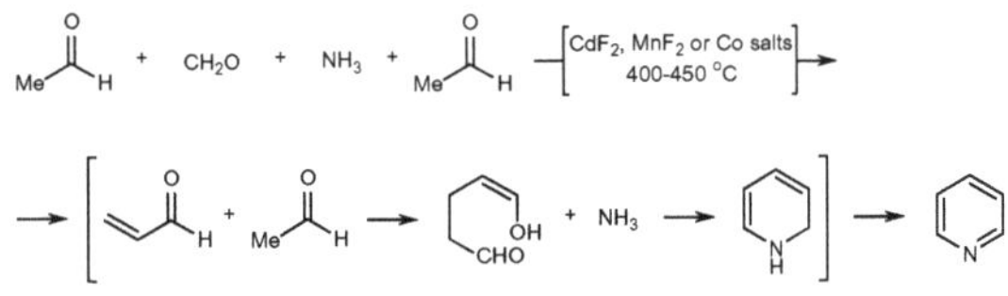

Scheme 4.40 Production of pyridine according to the Chichibabin reaction

Propiophenone is the product of F.–C. acylation; piperidine is available by reduction of pyridine as a cheap commodity.

Note For over a century, pyridine was only available from tar of pit coal, until the Chichibabin discovery of pyridine synthesis was developed into an industrial method (Scheme 4.40).

Higher products of aldol condensation are formed as the side products. Cyclization with ammonia to dihydropyridine is followed by dehydrogenation to pyridine. The outlined catalytic method is not highly selective; isomeric mono- and dimethyl pyridines are formed as side products and separated by a complex technological process [30].

Example 4.16 Suggest the retrosynthesis and then propose the synthesis of **TM 4.13**, a model compound for the synthesis of steroids.

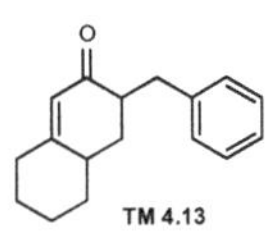

TM 4.13

On first sight, it is difficult to imagine *retro*-Mannich disconnection as the key step in the retrosynthesis of **TM 4.13**. Starting the retrosynthetic analysis with the disconnection of the C=C bond in cyclic enone we arrive at a 1,5-dicarbonyl pattern in **TM 4.13a** (Scheme 4.41). Now two possible *retro*-Michael disconnections,

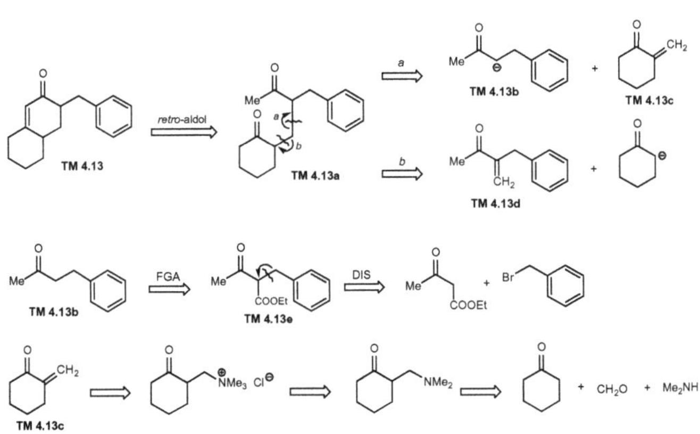

Scheme 4.41 Retrosynthetic analysis of **TM 4.13**

a and *b*, lead to enones **TM 4.13c** and **TM 4.13d**, whereby greater simplification of the target structure is achieved by disconnection *a*.

To obtain the reagent for anionic synthon **TM 4.13b**, we add an activating carbethoxy group to **TM 4.13e**, and the new target molecule can now be disconnected to benzyl bromide and ethyl acetoacetate.

Enone **TM 4.13c** can be disconnected by the sequence of steps to the Mannich base and *retro*-Mannich in the last step. Note that the Wittig reaction is less convenient for achieving **TM 4.13c** because of the presence of the second carbonyl group, and a tandem reaction offers a practical alternative.

In the next example, a similar sequence of retrosynthetic steps suggests the synthesis of biologically important triene.

Example 4.17 A vitamin D group characterizes the presence of a polycyclic 7-dehydrocholesterol frame. Biosynthesis starts with photochemically formed vitamin D3, and enzymatic hydroxylation produces a C(25)-OH derivative. The model compound in the synthesis of vitamin D is **TM 4.14**. Consider its retrosynthesis to easily available starting materials and then propose the synthesis.

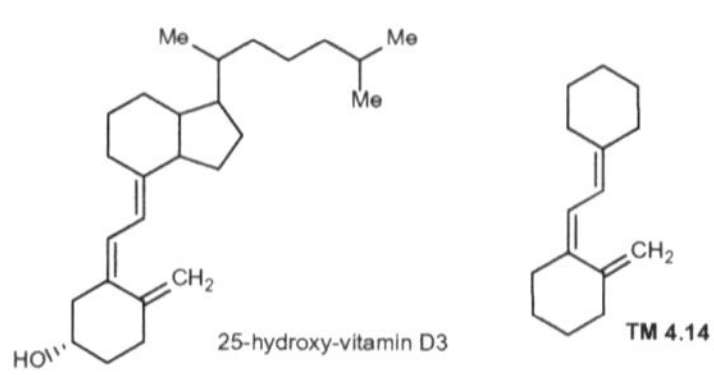

The logical first retrosynthetic step is *retro*-Wittig disconnection with maximal simplification of **TM 4.14** affording 2-methylenecyclohexanone **TM 4.14a** and Wittig reagent **TM 4.14b** (Scheme 4.42).

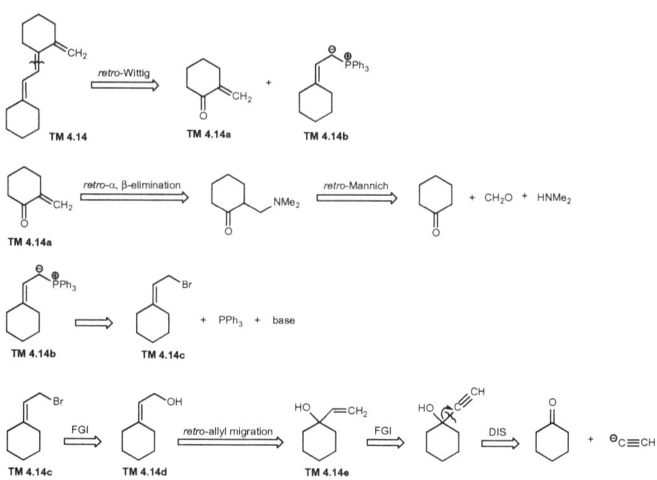

Scheme 4.42 Retrosynthetic analysis of **TM 4.42**

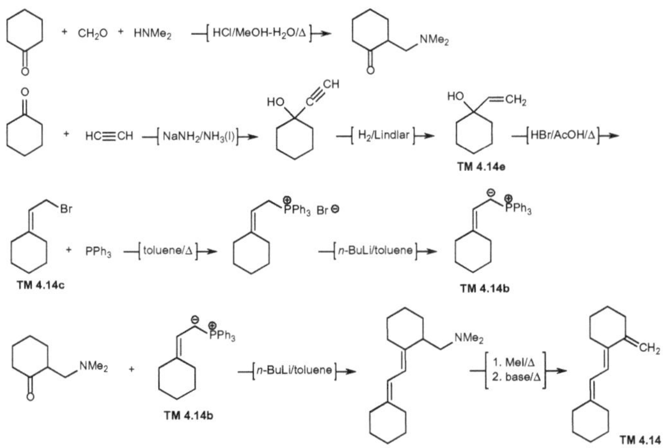

Scheme 4.43 Proposed synthesis of **TM 4.43**

TM 4.14a appears as **TM 4.13c** in Scheme 4.41. Retrosynthesis of Wittig reagent **TM 4.14b** leads over allyl bromide **TM 4.14c** to allyl alcohol **TM 4.14d** available on allyl migration from carbinol **TM 4.14e**, as we already discussed (Sect. 2.3.1, example 2.12).

The proposal of a synthetic route to **TM 4.14** requires precautions respecting the correct order of synthetic steps (Scheme 4.43).

This route avoids the isolation of unstable enone **TM 4.14a** by reacting phosphonium ylide **TM 4.14b** with the keto group of the intermediary Mannich base. Only then is quaternization to ammonium intermediate and elimination of tertiary amine performed. Reaction conditions for most synthetic steps in Scheme 4.43 are known from some previous examples.

It is proper to draw attention to the modification or reversal of the sequence of synthetic steps as compared to retrosynthetic analysis, often required to avoid the presence of incompatible functionalities under specific reaction conditions.

Example 4.18 Fluoxetine (Prozac) **TM 4.15** is a well-known antidepressive, an inhibitor of serotonin uptake in the nerve cells. This chiral drug is introduced in therapy as a racemate since numerous biological and pharmacological studies have confirmed that the *eudismic ratio* for *fluoxetine* enantiomers is near unity. *The eudismic ratio* is the ratio of the biological activity of enantiomers, formerly denoted as *eutomers* (more active) and *distomers*, less active or inactive enantiomers.

Numerous syntheses of racemic *fluoxetine* are reported. Consider the retrosynthesis and then propose the synthesis starting with the Mannich reaction and benzyl methylamine as the amino component.

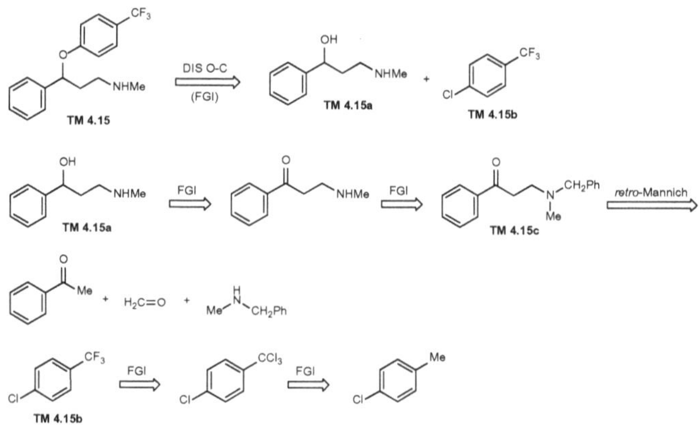

Scheme 4.44 Retrosynthetic analysis of *fluoxetine*, **TM 4.15**

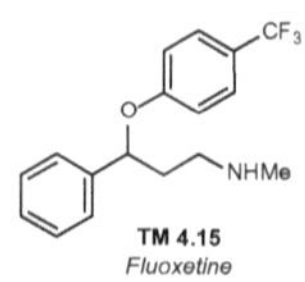

The methylamino group in the β-position to the oxygen functionality in **TM 4.15** suggests the transitory presence of a carbonyl group at the benzylic C atom, i.e., the Mannich base as an intermediate. To arrive retrosynthetically at this building block, we start with disconnection of the aryl-C–O bond to alcohol **TM 4.15a** and aryl halogenide **TM 4.15b** (Scheme 4.44). Retrosynthesis of **TM 4.15a** now conceives of two FGIs, hydroxyl to the keto group and terminal methylamino to the benzyl methylamino group in **TM 5.15c**, prone to *retro*-Mannich disconnection.

Note Retrosynthesis of *para*-disubstituted benzene **TM 4.15b** deserves comment in view of the inconvenience of its production. Direct trifluoromethylation of chlorobenzene requires the use of the toxic and not easily available trifluoromethyl group. Industrial methods are therefore developed based on *para*-chlorotoluene. A trichloro intermediate is produced by catalytic chlorination promoted by peroxides, and then substitution of Cl for F atoms is completed with HF/KF or more conveniently with CCl_2F_2 in the presence of a catalyst based on Cu/CuF_2 at elevated temperatures [31, 32].

The complete synthesis of *fluoxetine* can now be proposed as presented in Scheme 4.45.

Note that during the reduction of the keto group in the Mannich base, the *N*-benzylic group is contemporaneously hydrogenolitically removed. In the last step, nucleophilic substitution of chlorine by an alkoxide anion is promoted by an electron-withdrawing trifluoromethyl group in the *para*-position.

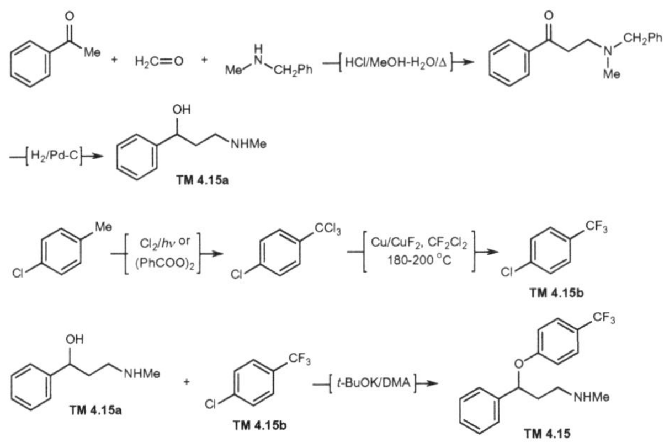

Scheme 4.45 Proposal for the synthesis of *fluoxetine*

4.4.2.1 Chiral Variants of the Mannich Reaction

When a new stereogenic center is formed in the Mannich base, the product is racemic until chiral information is included in the reacting system to orient the reaction toward the targeted enantiomer. Chiral information, as already mentioned, may reside in one of the reactants, in the catalytic system or exceptionally in the solvent. Some selected examples of non-catalytic and catalytic asymmetric Mannich reactions are discussed in this chapter.

Consideration of the basic mechanism of the Mannich reaction between the imine of prochiral aldehyde with methyl alkyl ketone serves as a useful introduction (Scheme 4.46).

With the exception of formaldehyde, aldehydes (R″#H) are prochiral, and the chiral center in the Mannich base appears on the carbonyl C atom of aldehyde. When the prochiral C–H acid component, usually a ketone, enters the Mannich reaction with formaldehyde, a chiral center is formed on the α-C atom of the C–H acid partner (Scheme 4.47).

If both aldehyde and ketone are prochiral, two new chiral centers are formed in the Mannich base, affording two racemic pairs of diastereomers.

As an illustration of achiral and chiral variants of the Mannich reaction leading to the same target molecule as a racemate or as an optically pure compound, in the next examples the retrosynthesis of the racemic form of *fenpropimorph* **TM 4.16** and asymmetric syntheses of biologically active (*S*)-enantiomer are presented.

Scheme 4.46 Mannich reaction of prochiral aldehyde with methyl ketone

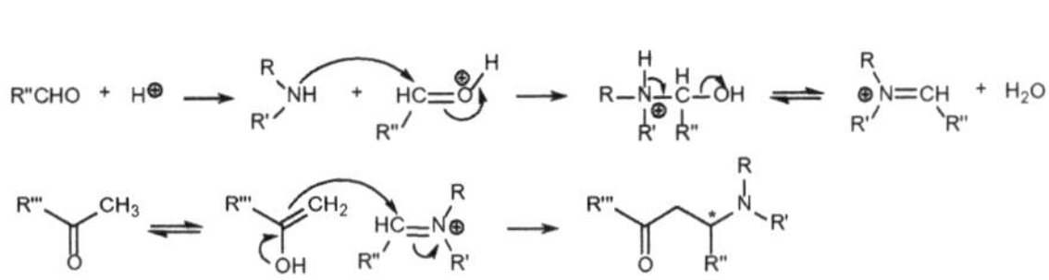

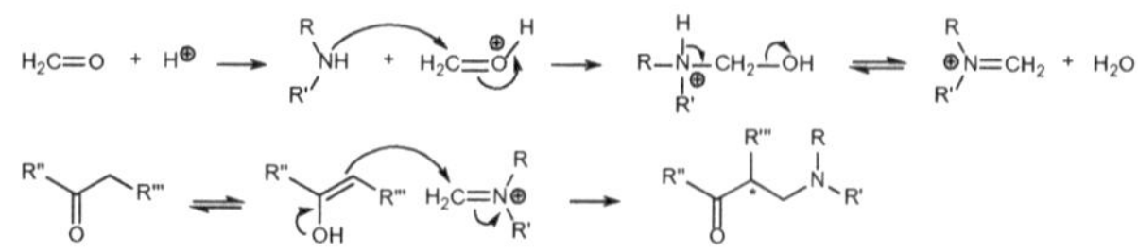

Scheme 4.47 Mannich reaction of formaldehyde with prochiral ketone

Example 4.19 Racemic *fenpropimorph* **TM 4.16** is a broadly used herbicide. Propose the retrosynthetic analysis of the racemate and synthetic route using in the key step the Mannich reaction with *cis*-3,5-dimethlymorpholine, an industrial product and easily available building block.

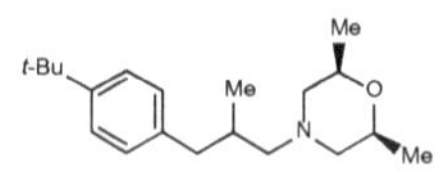

TM 4.16
Fenpropimorph

Since β-amino ketone is a product of the Mannich reaction, we conclude that this structure appears as an intermediate in the synthesis of *fenpropimorph* and start the retrosynthesis adding a carbonyl group in the β-position to the *tert*-amino group in **TM 4.16a** (Scheme 4.48).

In the *retro*-Mannich step, disconnection of two bonds affords three reagents, ketone **TM 4.16b**, formaldehyde and morpholine derivative **TM 4.16c**. Two consecutive *retro*-Friedel-Crafts disconnections of **TM 4.16b** result in the available commodities as starting materials. The disconnection of **TM 4.16c** in Scheme 4.48 is one among more possible ones and corresponds to industrial synthesis. Scheme 4.49 outlines the synthesis of *fenpropimorph* and indicates some reaction conditions on the industrial scale.

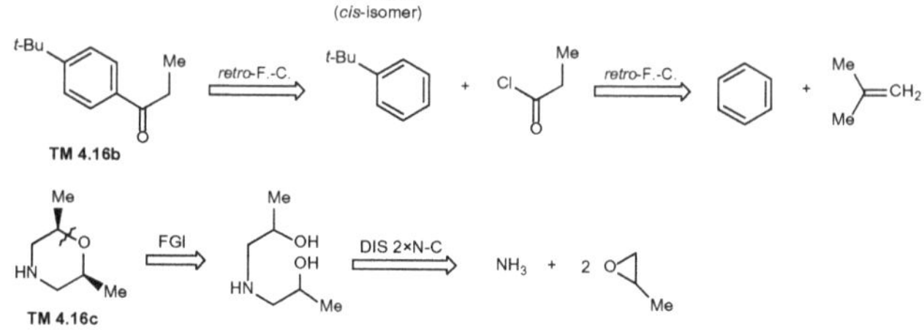

Scheme 4.48 Retrosynthetic analysis of racemic *fenpropimorph* **TM 4.16**

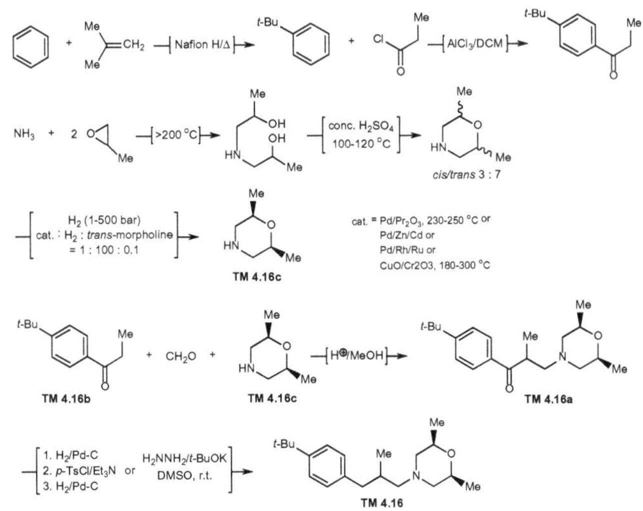

Scheme 4.49 Industrial synthesis of racemic *fenpropim*orph **TM 4.16**

In the first step, the F.–C. reaction is catalyzed by Nafion H, a broadly used industrial catalyst based on sulfonated polytetrafluoroethylene, a strong protic acid bound to polymeric support. Selective acylation in the *para*-position is controlled by the bulky *tert*-butyl group.

Note Industrial production of 3,5-*cis*-dimethylmorpholine has some instructive aspects. Double alkylation of ammonia by propylenoxide in the gas phase is controlled by the ratio of reactants and reaction temperature. Dipropanolamine is separated by distillation, and acid-catalyzed cyclization affords a *cis/trans* mixture of 3,5-dimethylmorpholines wherein *trans*-isomer prevails, a product of kinetic control. Isomerization to *cis*-isomer **TM 4.16c** is conducted at elevated temperature catalyzed by mixed metal catalysts in the presence of hydrogen (Scheme 4.49). These conditions reveal the mechanism of isomerization, which includes dehydrogenation-hydrogenation.

In the last steps, **TM 4.16b** and **TM 4.16c** undergo a Mannich reaction to **TM 4.16a** followed by the stepwise reduction of the keto to a methylene group via benzylic alcohol, activated for hydrogenolysis as a *para*-tosyl derivative.

Example 4.20 Asymmetric Mannich reaction in the synthesis of (*S*)-*fenpropimorph* is completed from the same starting materials as in Scheme 4.49. Chiral information is transitorily introduced in enamine derived from *O*-methyl D-prolinol (Scheme 4.50) [33].

To this aim the chiral secondary amine *O*-methyl D-prolinol is prepared from D-proline and condensed with propiophenone to chiral enamine using tetraethyl silane at high temperature as condensing agent. On the parallel route, the immonium

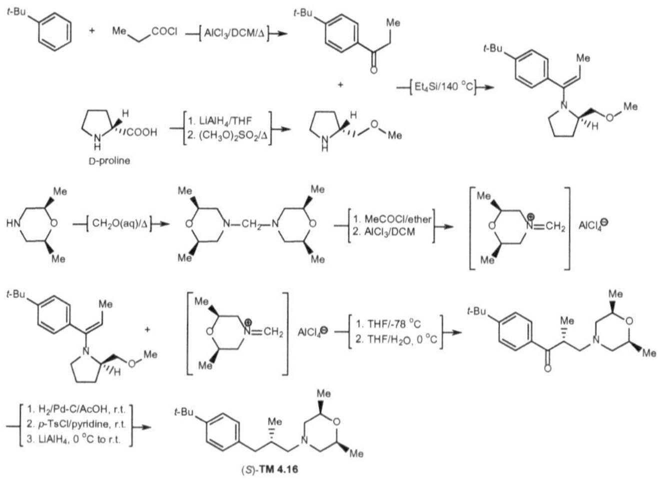

Scheme 4.50 Asymmetric Mannich reaction in the synthesis of (*S*)-*fenpropimorph*

derivative of *cis*-3,5-dimethylmorpholine with formaldehyde is isolated as a stable tetrachloroaluminum salt.

In the key step of this convergent synthesis, chiral enamine reacts with immonium salt, affording a chiral Mannich base with 100 % e.e. Due to the tendency to enolization, this intermediate has limited configurational stability. Reduction under mild conditions by a two-step protocol afforded (*S*)-*fenpropimorph* with 95.1 % e.e.

An interesting mechanism is proposed for the stereoselective step (Scheme 4.51) [33].

The transition state that determines the stereoselective direction reveals the importance of the OMe group in chiral enamine. Electronegative N and O atoms enter the dipolar Coulombic interaction with a cationic center of the immonium ion. This interaction orients the C=N bond to the *sterically more perturbed face* of the enolic C=C bond. The unexpected position of the methyl group on the backside of the plane leads to an (*S*)-configuration of the new chiral center [34].

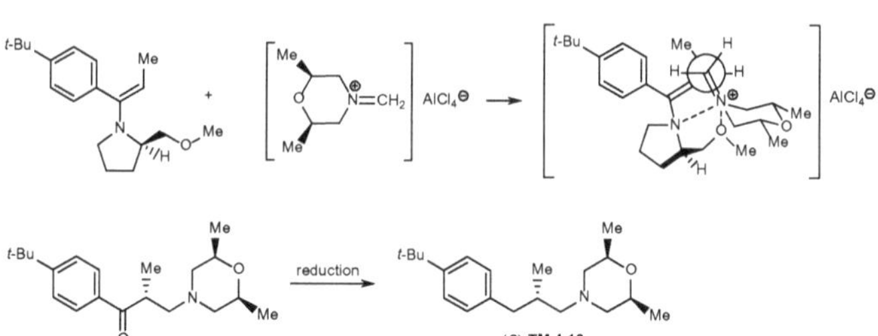

Scheme 4.51 Mechanism of the chiral variant of the Mannich reaction

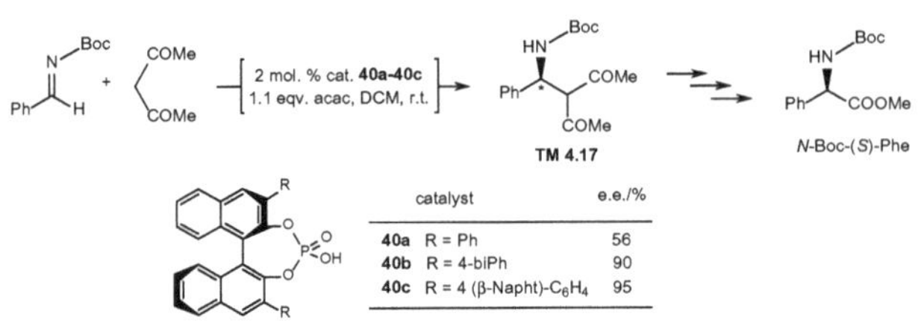

catalyst	e.e./%
40a R = Ph	56
40b R = 4-biPh	90
40c R = 4 (β-Napht)-C_6H_4	95

Scheme 4.52 Chiral organocatalytic Mannich reaction

Catalytic chiral Mannich reactions in the next examples are promoted by *organocatalysts* [35]. Characteristic of the selected *organocatalysts* is their proton or Bronsted acidity and stabilization of the catalytic complex by hydrogen bonding to enamine.

Example 4.21 Synthesis of **TM 4.17** was completed by an enantioselective reaction between aldimine and acetyl acetone (2,4-pentandione) catalyzed by *axially chiral* organophosphates **40a–40c** (Scheme 4.52) [36].

By alkylation of acetyl acetone with the Mannich-type intermediate chiral β-aminodiketone **TM 4.17** is formed. Enantioselectivity proved dependent on the steric requirements of the R group in the chiral organocatalyst as revealed by the e. e. of *N*-Boc-(*S*)-phenylglycine methyl ester. Proper acidity of the organophosphate group enables the formation of hydrogen bonding, avoiding complete proton transfer and the formation of a weakly bound ion pair. The oxygen atom of the P=O group acts as a Lewis base forming a hydrogen bond with the enolic form of β-diketone so that the phosphate group acts as a *bifunctional catalyst*. As the result, both reactants are bound into a catalytic complex as determined by the "chiral cleft" of binaphthyl ligand, similar to the transition state presented in Fig. 4.6

Further steps from **TM 4.17** to *N*-Boc-(*S*)-phenylglycine methyl ester in Scheme 4.52 are performed to determine the optical purity and absolute configuration on the chiral center. Chiral diketone derivative **TM 4.17** is transformed to *N*-Boc-phenylglycine without racemization and esterified by diazomethane.

Example 4.22 An axially chiral binaphthyl unit also characterizes chiral catalyst **41** for the asymmetric Mannich reaction presented in Scheme 4.53. The sulfonamido group acts as the proton donor to the imino N atom. Aldehydes are alkylated by an *N-para*-methoxyphenyl (PMP) derivative of oxalaldehyde as an imino component. *anti*-Isomers of α-amino-γ-keto acids **TM 4.18a–e** with (2*S*,3*R*) absolute configurations are obtained as the prevailing products with very high optical purity (Scheme 4.53) [37].

The acidic NH proton of the sulfonamide group in chiral organocatalyst **41** turned out to be of key importance for both aspects of selectivity, an *anti/syn* ratio >20:1 and e.e. >99 %. The mechanism of this reaction comprises an intermediary formation of enamine between the aldehydes and chiral catalyst and protonation of α-iminoester. In the transition state two reactants are oriented in parallel planes affording *anti*-isomers of **TM 4.18a–e** in (2*S*,3*R*)-absolute configuration (Fig. 4.6).

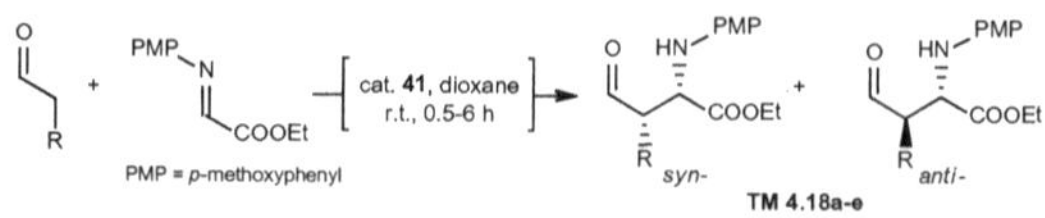

TM 4.18	R	cat./mol. %	time/h	*anti/syn*	e.e. anti/%
a	Me	1	0.5	13/1	>99
b	*n*-Bu	1	4	>20/1	99
c	Bn	1	4	11/1	>99
d	*i*-Pr	2	0.5	>20/1	>99
e	*t*-Bu	5	16	>20/1	99

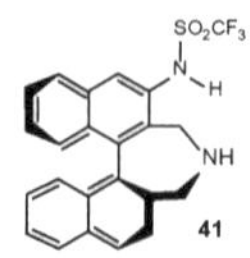

Scheme 4.53 Asymmetric and *anti*-selective organocatalytic Mannich reaction

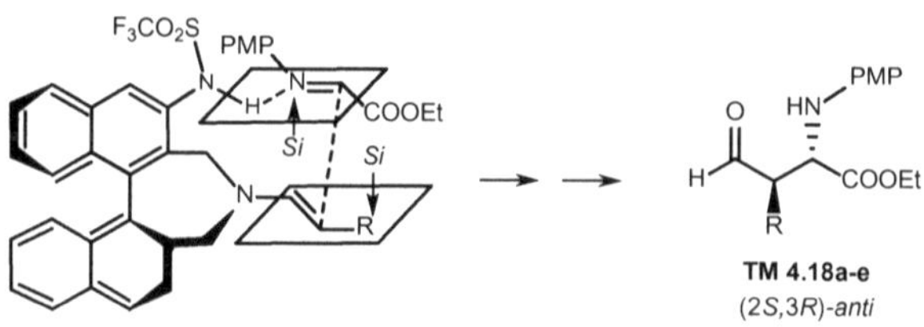

Fig. 4.6 Transition state model for organocatalytic Mannich-type reactions

The distant trifluoromethyl sulfonamide group dictates the orientation of the imino component and "chiral cleft" of the binaphthyl control orientation of the *trans* C=C bond in enamine.

In conclusion, chiral variants or asymmetric Mannich reactions are reliable synthetic methods on the laboratory scale. Chiral Mannich reactions are close to large-scale applications catalyzed by easily available derivatives of L- and D-proline, although presently limited to a small set of substrates [38, 39].

References

1. Ingold CK (1925) J Chem Soc 513–518 (ibid. pp 870–875)
2. Saltzman M (1972) J Chem Educ 49:750–752
3. Pople JA, Gordon MS (1967) J Am Chem Soc 89:4253–4258
4. Corey EJ (1967) Pure Appl Chem 14:19–37
5. Ho T-L (1991) Polarity control for synthesis. Wiley, New York, pp 5–9
6. Evans DA (1971) An organizational format for the classification of functional groups. Application to the construction of difunctional relationships. UCLA Physical Organic Chemistry Seminar
7. Seebach D (1979) Angew Chem Int Ed 18:239–258
8. Zimmerman HE, Traxler MD (1957) J Am Chem Soc 79:1920–1923
9. Xie L, Isenberger KM, Held G, Dahl LM (1997) J Org Chem 62:7516–7519
10. Evans DA, Bender SL, Morris J (1988) J Am Chem Soc 110:2506–2526
11. Cowden CJ, Paterson I (1997) Asymmetric aldol reactions using boron enolates. Org React 51:1–72

12. Walker MA, Heathcock CH (1991) J Org Chem 56:5747–5750
13. Furata K, Maruyama T, Yamamoto H (1991) J Am Chem Soc 113:1041–1042
14. Corey EJ, Cywin CL, Roper TD (1992) Tetrahedron Lett 33:6907–6910
15. Pearson RG (1968) Hard and soft acids and bases, HSAB, Part I. Fundamental principles. J Chem Educ 45:581–586
16. Pearson RG (1968) Hard and soft acids and bases, HSAB, Part I. Underlying theories. J Chem Educ 45:643–648
17. Fleming I (2009) Molecular orbitals and organic chemical reactions, Stud. edn. Wiley, New York, pp 109–111
18. Jacob A (1997) Understanding organic raction mechanism. Cambridge University Press, Cambridge, pp 44–46
19. Metzger JO (1998) Angew Chem Int Ed 37:2975–2978
20. Makosza M (2000) Pure Appl Chem 72:1399–1403
21. Starks CM (1971) J Am Chem Soc 93:195–199
22. Michael A (1894) J Prakt Chemie 49:10–20
23. Little RD, Masjedizadeh MR, Wallquist O, McLoughlin JI (1995) Org React 47:315–352
24. Carey FA, Guliano RM (2011) Organic Chemistry, 8th edn. Mc GrowHill, New York, pp 910–912
25. Musser MT (2005) Cyclohexanol and cyclohexanone, in Ullmann's encyclopedia of industrial chemistry. Wiley-VCH, Weinheim
26. Rapson WS, Robinson R (1935) J Chem Soc 1285–1291
27. Ho TL (1992) Tandem organic reactions. Wiley, New York, pp 74–82
28. Mamoru A (1987) J Catalysis 106:273–279
29. Afke G, Irion WW, Neuwirth OS (2003) Ullmann's encyclopedia of industrial chemistry, 6th edn, vol 1. Wiley-VCH, Weinheim.
30. Shimizu S, Watanabe N, Kataoka T, Shoji T, Abe N, Morishita S, Ichimura H (2005) Pyridine and pyridine derivatives, in Ullmann's encyclopedia of industrial chemistry. Wiley-VCH, Weinheim
31. May DD, Ford C (1993) US Pat 5,233,104, (for Du Pont de Namours & Co.)
32. Lavagnino ER, Molloy BB, Prane P (1979) US Pat 4,168,388 (for Eli Lilly & Co.)
33. Vinković V, Šunjić V (1997) Tetrahedron 53:689–696
34. Hajashi T, Tsuboi W, Ashimine I, Urushima T, Shoji M, Sakai K (2003) Angew Chem Int Ed 42:3677–3680
35. Maruoka K, Ooi T (2003) Chem Rev 103:3013–3028
36. Urugachi D, Terada M (2004) J Am Chem Soc 126:5356–5357
37. Ooi T, Takeuchi M, Kato D, Uematsu Y, Tayama E, Sakai D, Maruoka K (2005) J Am Chem Soc 127:16408–16409
38. Eder U, Sauer G, Wiechert R (1971) Angew Chem Int Ed 10:496–497
39. Hajos ZG, Parish DR (1974) J Org Chem 39:612–1615

Chapter 5
Illogical Disconnections with Participation of Two Groups

Abstract Compounds with a 1,2-, 1,4- and 1,6-dioxygen pattern and related bifunctional structures are presented. Disconnection of the internal bonds results in illogical synthons because of the mismatch of charges in the patterns with an even number of C atoms between the functional groups. Three-membered heterocyclic rings are presented as an important class of illogical nucleophiles in the retrosynthesis of the 1,2-difunctional pattern. Retrosynthesis of the 1,6-dicarbonyl pattern by reconnection and *retro*-Birch reduction of the aromatic building block is related to chemoselective Birch reduction and ozonolysis in the synthetic route. The retrosynthesis and synthesis of *salbutamol* and asymmetric synthesis of (−)-*frontalin* are presented.

5.1 Introduction

Disconnection of 1,3-CO and 1,5-CO patterns is supported by the matching distribution of charges in two synthons. One of them comprises a stabilized carbanion and the other the carbocation or C atom with high positive charge, both with recognizable and available reagents, as we saw in Chap. 2. Generally, patterns with an odd distance of functional groups support the preferred disconnection, different from patterns with an even distance of functional groups, 1,2-, 1,4- or 1,6. The reason rests in non-matching distribution of partial charges on two C atoms on the bond to be disconnected so that disconnection always generates one "unacceptable" or illogical synthon (Scheme 5.1).

Here the specific retroanalysis for the 1,6-CO pattern is not presented, which is discussed in Sect. 5.4.

Destabilized synthons with charges opposite to the natural polarization of the bond often have acceptable though not easily recognizable reagents or synthetic equivalents. After all, syntheses of numerous compounds with even distances of functional groups are successfully completed by formation of C–C bonds. They are important intermediates in the syntheses of complex natural compounds. In the next sections, we discuss the retrosynthetic approach to such compounds.

V. Šunjić and V. Petrović Peroković, *Organic Chemistry from Retrosynthesis to Asymmetric Synthesis*, DOI 10.1007/978-3-319-29926-6_5

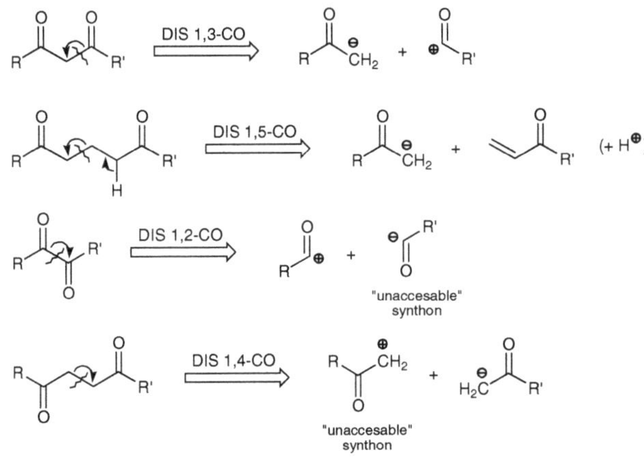

Scheme 5.1 Overview of favorable and unfavorable disconnections

5.2 1,2-Dioxygenated Pattern (1,2-CO)

5.2.1 Illogical Nucleophiles

α-Hydroxy acids belong to an important group of natural and synthetic compounds with a 1,2-dioxygenated pattern (Scheme 5.2).

Disconnection (a) corresponds to the synthesis of **I**, α-halogenation of carboxylic acid or its derivative followed by hydrolysis of halogen (Scheme 5.2). Disconnection (b) envisages building a carbon framework from $C_n + C_1$ synthons but looks unacceptable since illogical synthon $^{-}$COOH with a negative charge on the carbonyl C atom appears. Let us, however, consider the next example.

Example 5.1 Suggest the retrosynthesis and then propose the synthesis of 2-hydroxy-2-phenylpropionic acid **TM 5.1**.

Scheme 5.2 Retrosynthetic approaches to α-hydroxy carboxylic acids

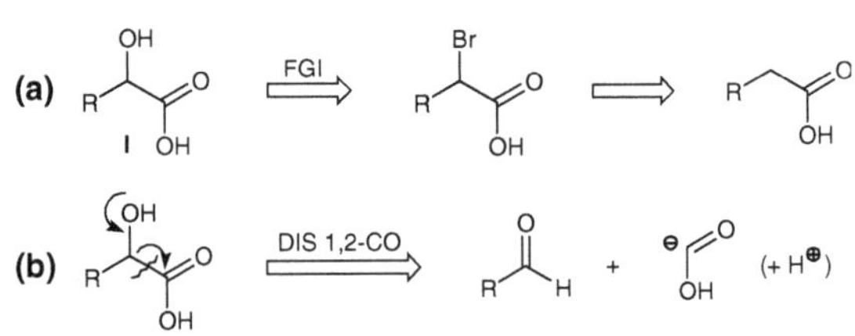

Scheme 5.3 Disconnection of **TM 5.1** as a 1,2-CO pattern

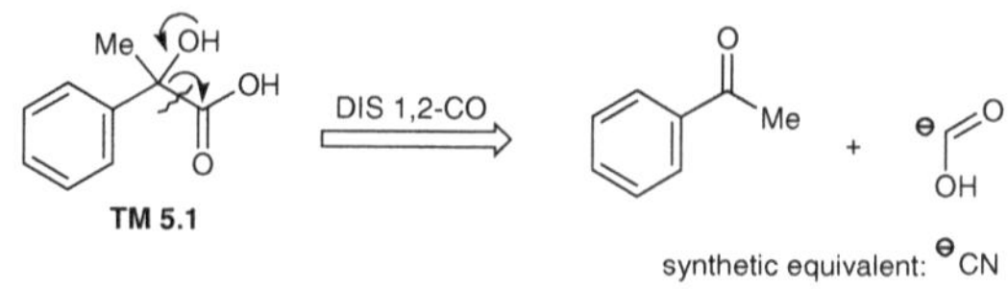

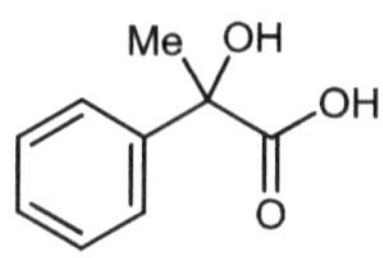

TM 5.1

Completing the disconnection of the internal C–C bond with participation of the hydroxy group, we generate the simple building block acetophenone and illogical synthon $^{-}$COOH (Scheme 5.3).

Ketone is easily available so we start the search for the synthetic equivalent of the illogical anion, and as a matter of fact this is found in the cyanide anion!

In the synthetic direction we use its salts as the source of the cyanide anion to complete the well-known *cyanohydrin reaction*. In the last step, the hydrolysis of nitrile affords the carboxylic group (Scheme 5.4).

Cyanide salts are available with either inorganic or organic cations, and a cyanide group can be conveniently prepared with radioactive ^{14}C-labeled atoms. The cyanohydrin reaction serves for the preparation of more complex ^{14}C-labeled carboxylic acids and their derivatives intended for pharmacological studies.

Example 5.2 Consider the retrosynthesis and then propose the synthesis of ^{14}C-labeled **TM 5.2.**

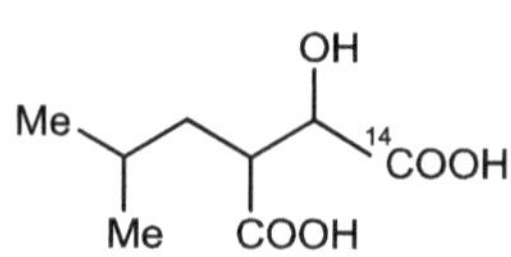

TM 5.2

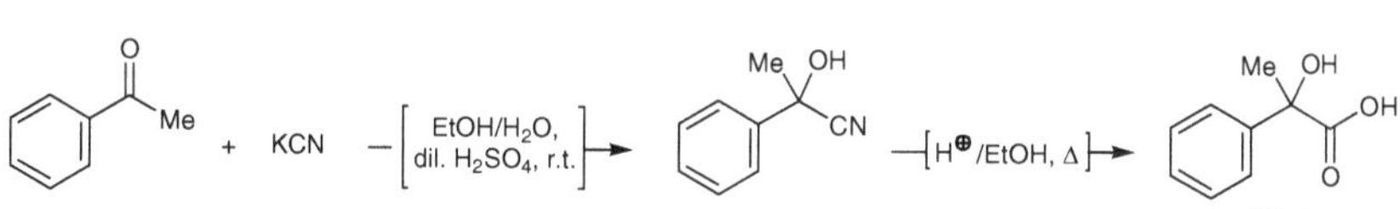

Scheme 5.4 Proposal for the synthesis of **TM 5.1**

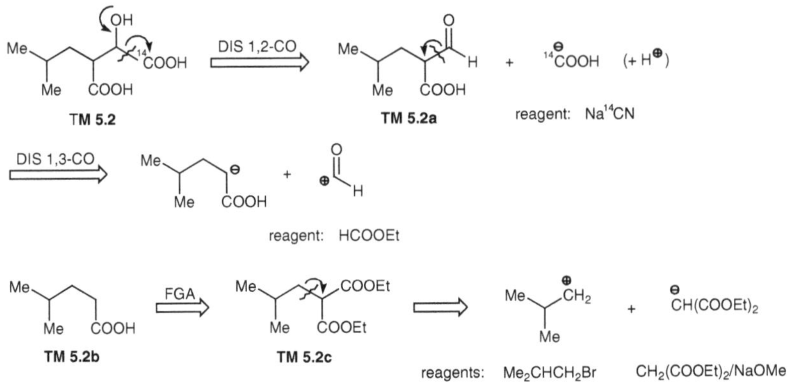

Scheme 5.5 Complete retrosynthetic analysis of **TM 5.2**

The isotopically labeled α-hydroxy carboxylic group indicates the use of the ^{14}C cyanide ion in cyanhydrin synthesis and consequently 1,2-CO disconnection in the first retrosynthetic step (Scheme 5.5).

The target molecule of the second generation **TM 5.2a** comprises a 1,3-CO pattern, and the ^{14}C-labeled cyanide anion indicates sodium cyanide as the second reagent. Disconnection of **TM 5.2a** leads to the carbanion of **TM 5.2b** and formyl carbocation, whose synthetic equivalent is ethyl formate. On first sight, the branched carboxylic acid **TM 5.2b** does not seem an easily available building block. On addition of the second activating carbethoxy group, the alkylated diethyl malonate unit appears as an available target molecule of the second-generation **TM 5.2c**. The last disconnection step leads to a stable diethyl malonate carbanion and synthon with carbocation on the primary C atom. Its synthetic equivalent is bromide, available in a few steps from isobutyraldehyde (Sect. 4.3.1, Scheme 4.3). An attempt to alkylate the diethyl malonate anion by isobutene in a Michael-type reaction fails because of the low reactivity of the terminal C atom on the double bond as an electrophile. Note that the hyperconjugative effect of methyl groups inverts the charge distribution in the $Me_2C{=}CH_2$ double bond as compared to enones.

Now we can propose the complete synthesis of ^{14}C-labeled **TM 5.2** (Scheme 5.6).

Cyanide salts are also used in the *Strecker synthesis* of α-amino acids. This well-known name reaction generates compounds with a 1,2-disubstituted pattern whose disconnection results in one illogical synthon. Example 5.3 presents some mechanistic details of this reaction, suggested by *retro*-Strecker disconnection.

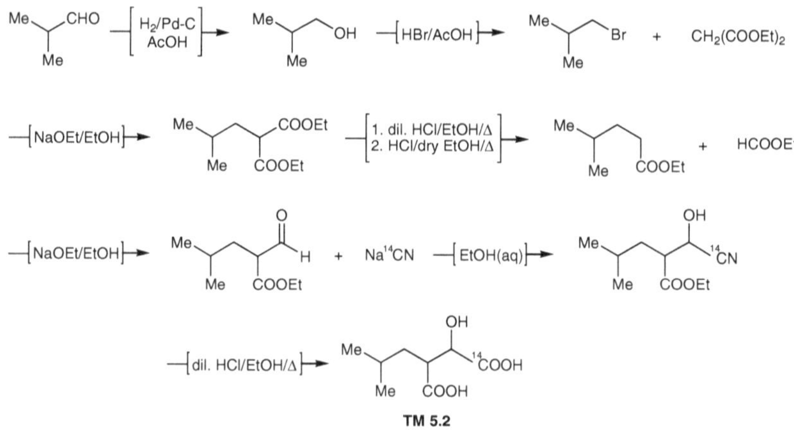

Scheme 5.6 Proposal for the synthesis of ^{14}C-labeled **TM 5.2**

Example 5.3 Propose the retrosynthesis of *para*-nitrophenyl glycine **TM 5.3**.

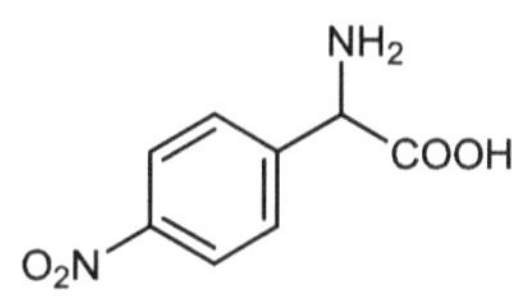

TM 5.3

In the first step, we perform 1,2-disconnection with participation of the neighboring amino group and formation of an illogical anionic synthon (Scheme 5.7).

The new target molecule is aldimine **TM 5.3a**, which is interconverted to *para*-nitrobenzaldehyde **TM 5.3b** by the logical FGI step. Aldehyde is a commodity available by the nitration of toluene and partial oxidation by chromium(IV) oxide in acetic anhydride.

To complete the Strecker synthesis of amino acid **TM 5.3**, beside cyanide salt, a proper source of ammonia is needed. To this aim, ammonium salts are more

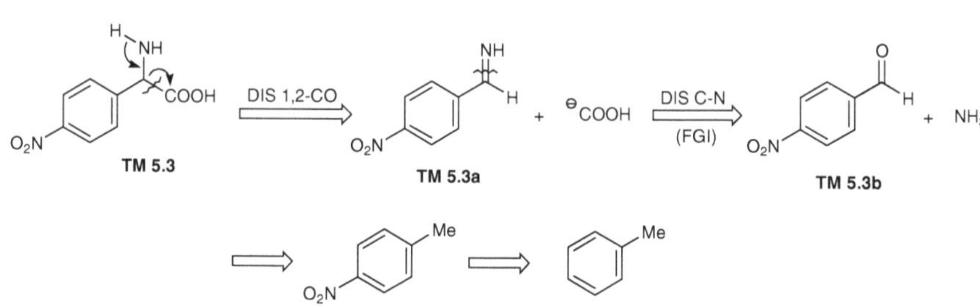

Scheme 5.7 Retrosynthetic analysis of **TM 5.3**

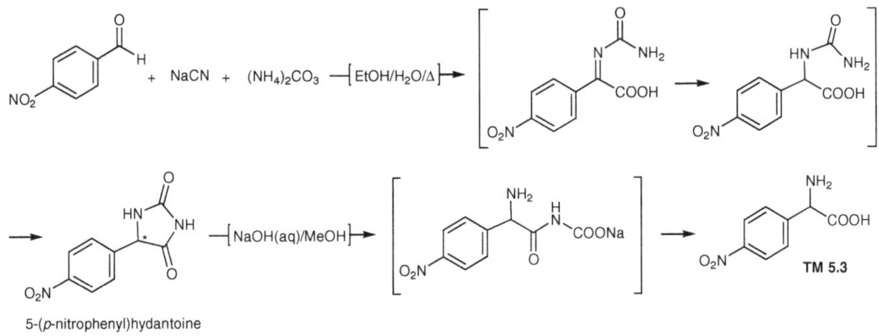

Scheme 5.8 Large-scale synthesis of *para*-nitrophenyl glycine **TM 5.3**

convenient than dissolved ammonia, and ammonium carbonate is preferred because of it is low thermal stability. The two initial steps in Scheme 5.8 are presented to illustrate the formation of the key intermediate, which is not aldimine **TM 5.3a**! Aldimines are highly unstable species and serve as a "guide" in the retrosynthetic analysis of α-amino acids.

The only stable product of the industrial method is *para*-nitrophenyl hydantoine; the other proposed but not isolated intermediates are in parentheses. The hydantoine derivative possesses a stereogenic center on the C-5 atom and is racemic. Separation of enantiomers on an industrial scale is completed by selective biocatalytic hydrolysis of *R*-enantiomer by hydantoinases, a group of hydrolytic enzymes. The "wrong" *S*-enantiomer can be easily racemized by heating in weak basic medium or by the enzyme hydantoin racemase and racemate recycled to separation.

1,2-Dioxygenated patterns do not only appear in α-hydroxy carboxylic acids, but are also important functionalities in α-hydroxy ketones and their derivatives. In the former examples, we have seen negatively polarized α-C atoms in carbonyl compounds as nucleophilic centers. As shown in Scheme 5.9, this polarization is already present in enoles. Hence, even in the absence of a strong base, α-C atoms in carbonyl compounds are nucleophilic.

To render α-C atoms in carbonyl compounds electrophilic, we need to invert the natural charge separation. This can be completed by the introduction of an electron-withdrawing group, usually a halogen atom (Scheme 5.10).

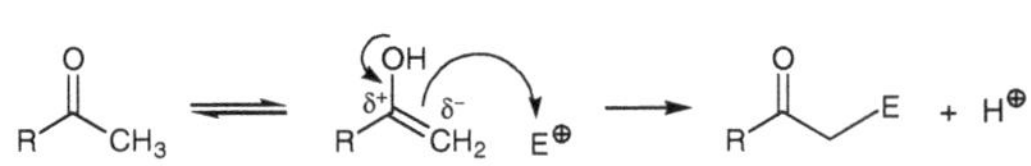

Scheme 5.9 Keto-enol tautomerism and nucleophilic character of an α-C atom

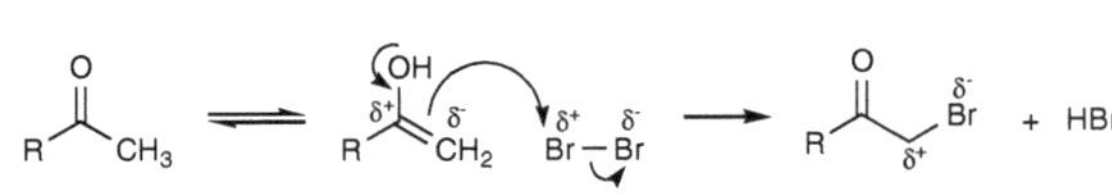

Scheme 5.10 Inversion of polarity on the α-C atom in ketones

We already saw the term *Umpolung* for the *inversion of natural polarity* to obtain electrophiles in Sect. 4.2. In Sect. 5.2.2, we consider three-membered heterocyclic rings as a synthetically particularly interesting group of illogical electrophiles.

The next example reveals how advantageous the recognition of inversion of polarity is in the course of retrosynthetic analysis, specifically in planning the synthesis of the target molecules that comprise the 1,2-CO pattern.

Example 5.4 Propose the key step in the retrosynthesis and then synthesis of **TM 5.4**.

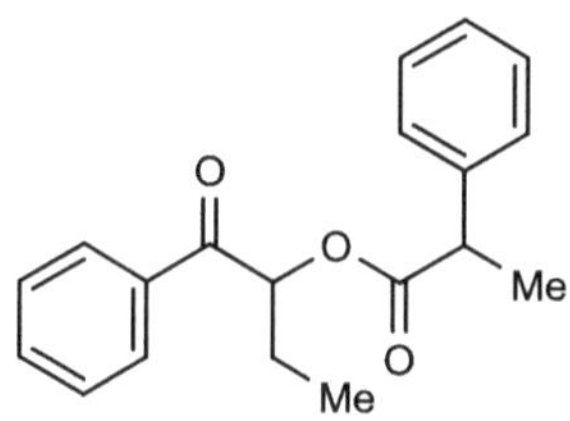

TM 5.4

Obvious disconnection *a* of the C–O bond in the ester unit represents FGI to alcohol and free carboxylic acid and in the synthetic direction corresponds to esterification of carboxylic acid (Scheme 5.11).

Interconversion *a* results in logical synthons, alkoxy anion **TM 5.4a** and acyl cation **TM 5.4b**. The reagent for anionic synthon is α-hydroxy ketone, available by α-bromination of ketone followed by hydrolysis of halogen. Now we observe that disconnection *b* offers a more simple solution. Since the reagent for **TM 5.4c** is α-haloketone, this intermediate can be directly acylated by carboxylate anions to **TM 5.4**. The complete synthetic proposal for **TM 5.4** is presented in Scheme 5.12. Synthesis of enantiomerically pure α-alkylcarboxylic acids is discussed in Sect. 3.6.3.

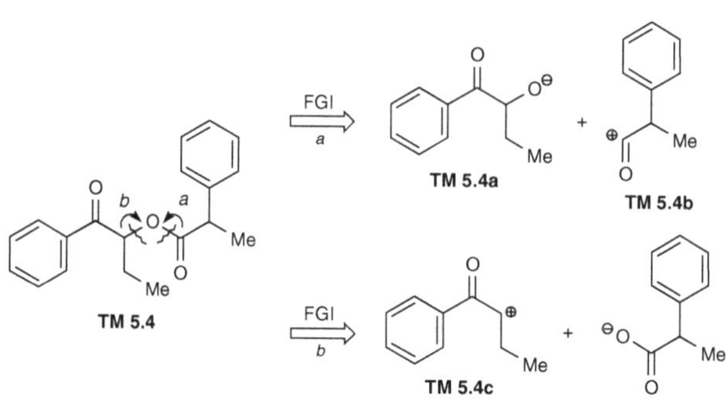

Scheme 5.11 Two possibilities for FGI of the ester group in **TM 5.4**

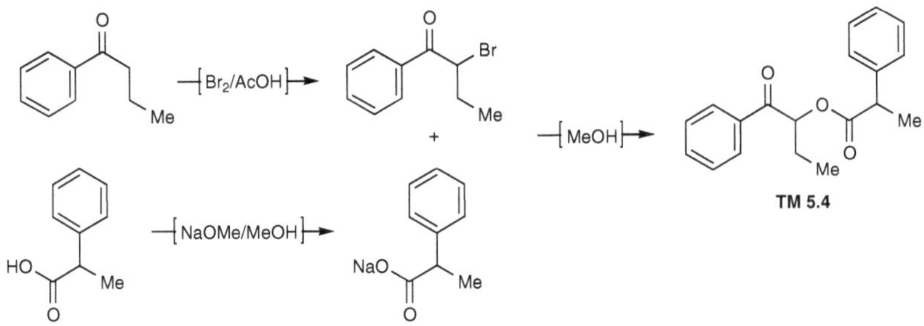

Scheme 5.12 Proposal for the synthesis of **TM 5.4**

Preparation of α-hydroxy ketones is by no means limited to α-halogenation and then hydrolysis of halogen. In the next example, we meet another possibility for the introduction of this functionality.

Example 5.5 Propose the retrosynthesis of **TM 5.5** and then its synthesis using a couple of reactions, alkynylation of ketone-hydration of the triple bond in the intermediary carbinol.

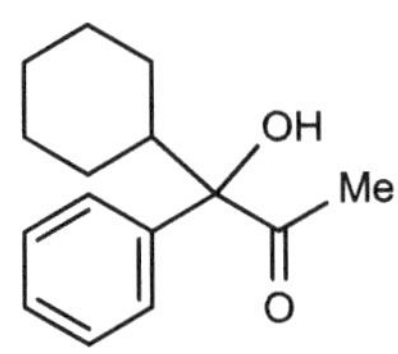

TM 5.5

We described the alkynylation of the carbonyl group with the acetylide anion in Sect. 2.2. Before the initial retrosynthetic step from **TM 5.5**, let us consider the mechanism of hydration of the triple bond activated by Hg(II) ions (Scheme 5.13).

Note the orientation of the water molecule in the complex resulting in the addition of the OH group to the higher substituted C atom in accord with the Markovnikov rule. Having this reaction in mind, now we can approach the disconnection of **TM 5.5** as in Scheme 5.14.

Such 1,2-CO disconnection results in ketone and an illogical synthon, the acetyl anion. Since we discovered sodium acetylide as the proper reagent for anionic C_2 synthons, a short and elegant synthetic scheme for **TM 5.5** can be proposed (Scheme 5.15).

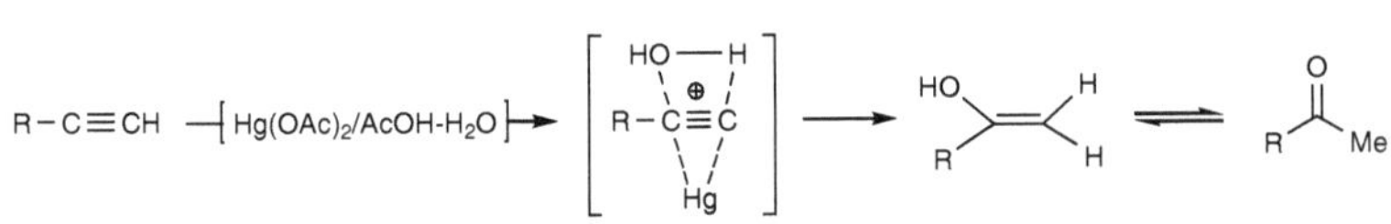

Scheme 5.13 Hydration of alkynes catalyzed by Hg(II) ions

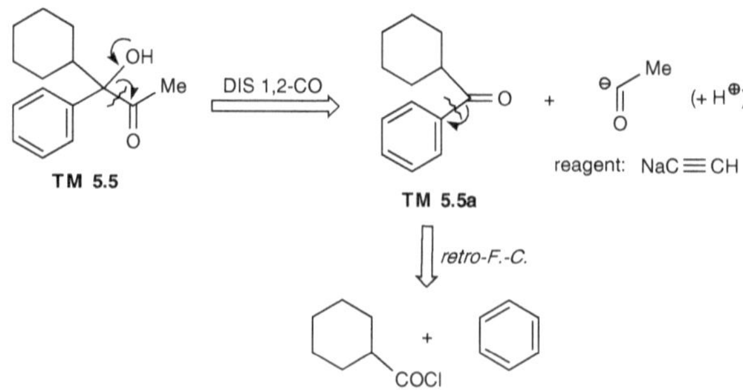

Scheme 5.14 Retrosynthetic analysis involving the illogical disconnection of the 1,2-CO pattern in **TM 5.5**

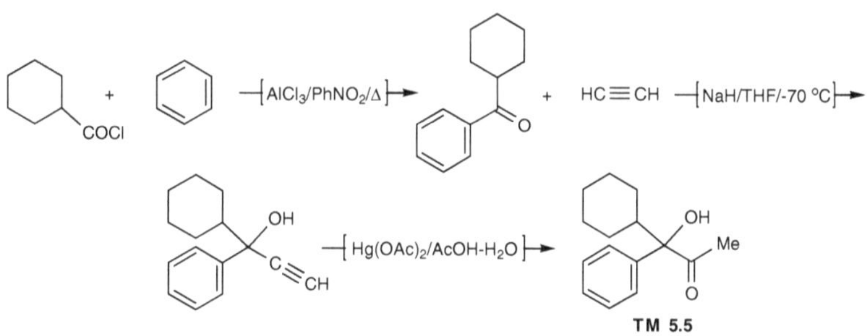

Scheme 5.15 Proposed synthesis of **TM 5.5**

The suggested retrosynthetic analysis of targeted α-hydroxy ketone might seem artificial, especially since it invokes a few less-known reactions, alkynylation-hydration of the triple bond. It is difficult to conceive of another practicable synthetic route to **TM 5.5**.

The next example reveals the power of retrosynthetic analysis where an α-hydroxycarbonyl pattern in the target molecule requires a sophisticated approach.

Example 5.6 Propose the retrosynthetic analysis and then synthesis of cyclic ether **TM 5.6**.

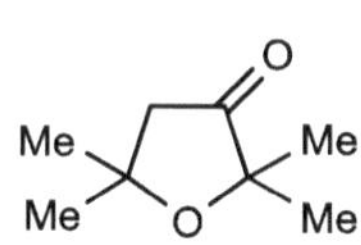

TM 5.6

The target furanone molecule comprises C–O bonds to *tert* C-atoms, which suggests disconnection forming two *tert* OH groups (Scheme 5.16).

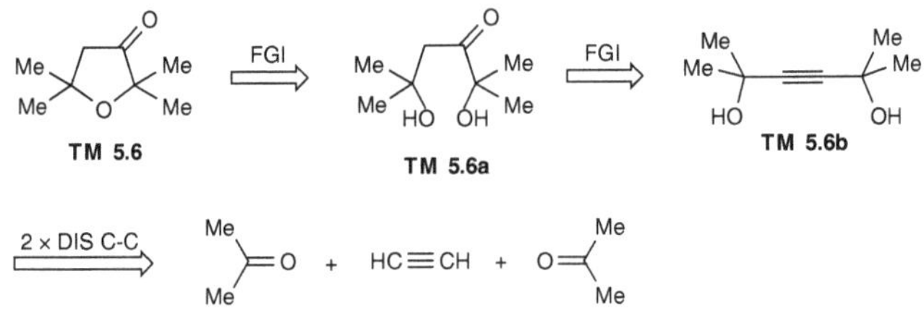

Scheme 5.16 Retrosynthetic analysis of **TM 5.6**

The new **TM 5.6a** is apparently a more complex acyclic diol than the target **TM 5.6**! An emerging issue is elegantly solved by the creative retrosynthetic approach to this intermediate. Interconversion of the keto group to the central triple bond affords symmetric **TM 5.6b**, and therefore in the synthetic direction the hydration step need not be regioselective! In the last retrosynthetic steps, two consecutive disconnections of the single C–C bonds by participation of OH groups afford acetylide anions and two moles of acetone, a surprisingly simple set of starting materials.

Synthesis of **TM 6.6** is straightforward and needs two moles of a strong base to create the two alkyne anions needed for the alkynylation of acetone (Scheme 5.17).

The next example confirms the importance of retrosynthetic analysis of the α-hydroxycarbonyl pattern conceiving sodium acetylide as a reagent for the acetyl anion as an illogical synthon.

Example 5.7 Propose the retrosynthetic analysis for **TM 5.7**.

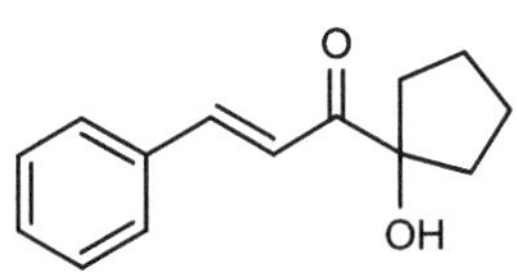

TM 5.7

The target molecule is characterized by the presence of enone and a 1,2-CO pattern that includes the *tert* OH group on the cyclopentane ring. In the first step,

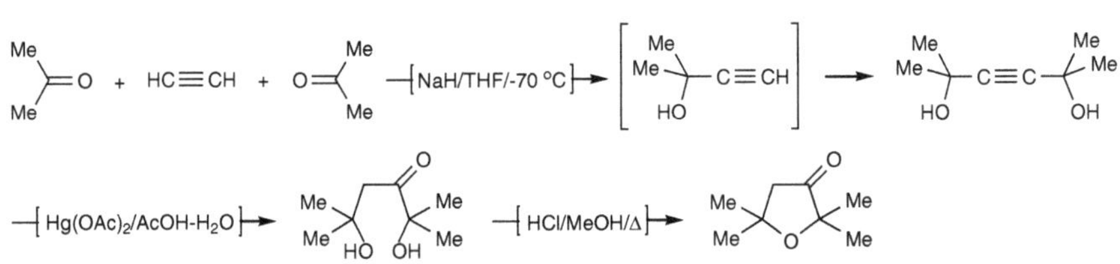

Scheme 5.17 Proposed synthesis of **TM 5.6**

Scheme 5.18 Retrosynthetic analysis of **TM 5.7**

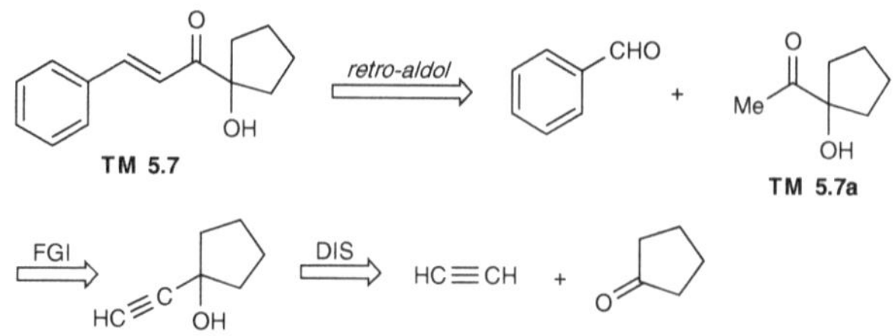

the maximal simplification of this structure is available by *retro*-aldol disconnection of the central C=C bond (Scheme 5.18).

In the target molecule of the second-generation **TM 5.7a**, we follow a two-step retrosynthetic pattern from the former example. First, the FGI and then DIS of the central C–C bond end up with acetylene and cyclopentanone as convenient starting materials. Synthesis of **TM 5.7** can be completed under the reaction conditions indicated in Scheme 5.17.

Example 5.8 In an early synthesis of camphor, the important intermediate **TM 5.8** was prepared from easily available 2,2-dimethlycyclopentanone. Perform the retrosynthetic analysis of **TM 5.8** and propose the three-step synthesis.

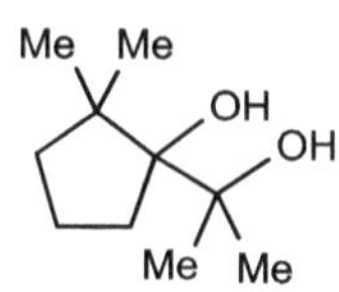

TM 5.8

The 1,2-CO pattern in the target molecule appears as a vicinal diol. The starting 2,2-dimethylcyclopentanone indicates the formation of an OH group on the ring by alkylation of the keto group. It remains open, however, how the second carbinol OH group is introduced in the side chain. The retrosynthetic solution to this puzzle is hidden in *retro*-Grignard disconnection to **TM 5.8a** and interconversion of methyl ketone to an acetylenic unit (Scheme 5.19).

The last target molecule is logically disconnected to 2,2-dimethylcyclopentanone and acetylide anions.

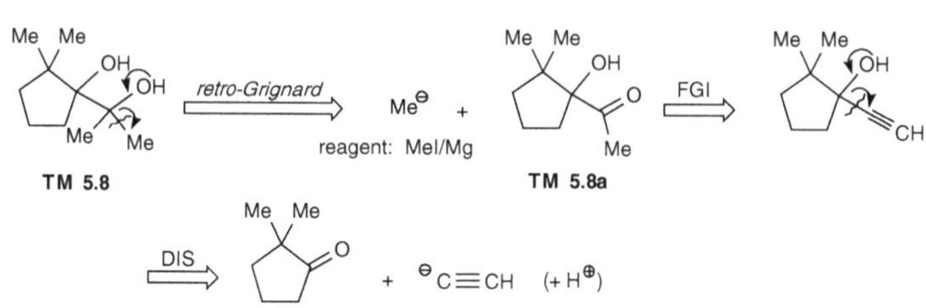

Scheme 5.19 Retrosynthetic analysis of **TM 5.8**

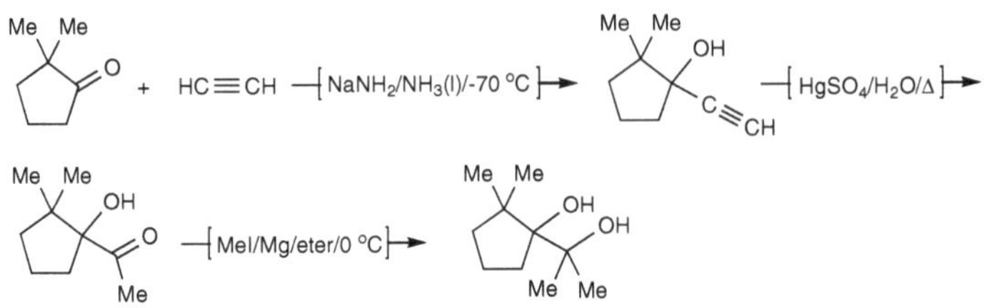

TM 5.8

Scheme 5.20 Proposed synthesis of **TM 5.8**

Alkynylation of the cyclopentanone derivative by the acetylide anion leads to carbinol. Hg(II)-catalyzed hydration in the next step affords α-hydroxy ketone, which in the last step is methylated by the Grignard reagent, completing the short synthesis of **TM 5.8** (Scheme 5.20).

It is important to note that dihydroxylation of the tetrasubstituted C=C bond in the conceivable intermediate formed by the Wittig reaction with 2,2-dimethylcyclopentanone is not a workable route because of the steric perturbation of the *gem*-dimethyl group in the ring. Write and analyze this possible reaction route!

5.2.2 *Three-Membered Heterocyclic Rings, Illogical Electrophiles*

Basic organic chemistry approaches three-membered heterocycles as strained and somehow exotic structures. Beside some methods of preparation, most attention is devoted to the ring strain and electronic character of the σ bonds. These very properties, however, render small heterocycles with general formulae C_2X (X=O, N, S) valuable reagents and building blocks. The basic reaction of the three-membered heterocycles can be outlined as in Scheme 5.21.

Since anionic and even neutral nucleophiles react with three-membered heterocycles with a ring opening, they are regarded as "illogical electrophiles." This property enables the original retrosynthetic analysis and proposals for the elegant synthesis of target molecules with a 1,2-disubstitution pattern. Before discussion of the retrosynthetic utilization of three-membered heterocycles, a brief review of the synthetic methods for their construction is appropriate. Three-membered rings can in principle be disconnected in three different modes (Scheme 5.22).

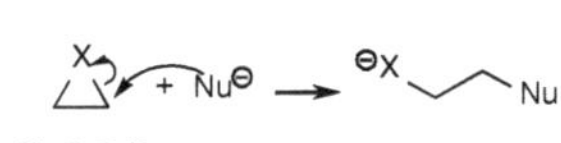

Scheme 5.21 Basic reaction of three-membered heterocycles

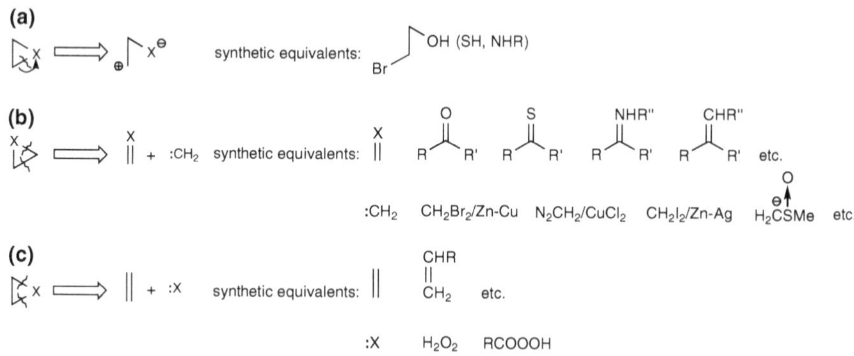

Scheme 5.22 Basic modes of disconnection of three-membered rings

The intramolecular nucleophilic substitution corresponds to disconnection (a) in the synthetic direction. In addition, the heteroatom as a nucleophile can act as a carbanion, resulting in cyclopropanation.

Disconnections (b) and (c) are specific for three-membered rings since they generate synthons with an electron sextet, electrophiles toward unsaturated C=X or C=C bonds. Contemporaneous disconnection of two bonds can be completed in two ways, by disconnection of one C–X and one C–C bond or of both C-heteroatom bonds (Scheme 5.23).

The above scheme presents disconnections of oxiranes (epoxides), aziridines and thiiranes (thioepoxides) leading to six-electron synthons and their synthetic equivalents.

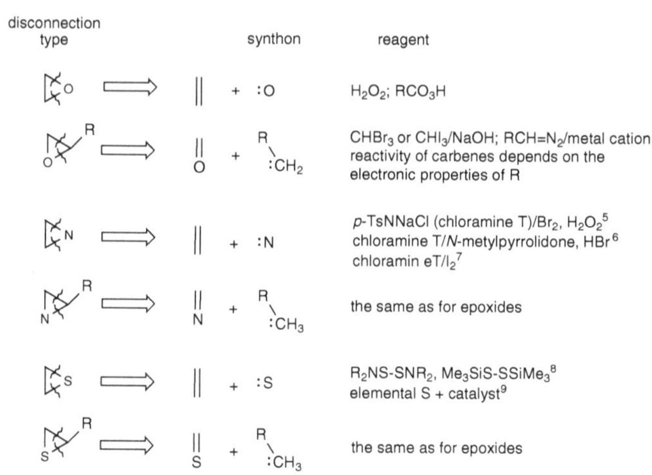

Scheme 5.23 Schematic presentation of two-bond disconnections in three-membered heterocycles

Disconnection of both C-heteroatom bonds leads to a neutral alkene and six-electron heteroatom as a strong π-electrophilic synthon. Disconnection of one C–C and one C-heteroatom bond leads to C=O, C=N or C=S functional groups and six-electron carbene. Their stability and use as electrophiles depend on the presence of an EWG (Sect. 2.2.2). Many syntheses are reported for the reagents in Scheme 5.23 containing N or S atoms with an electronic sextet prone to addition to alkene forming aziridines or thiiranes [1–5].

The following examples illustrate the practicability of syntheses of tricycles with reagents based on illogical synthons. First we discuss the epoxidation of the C=C bond in its achiral and chiral variant.

Example 5.9 Consider the retrosynthetic analysis of **TM 4.9**, a key intermediate in the synthesis of the anti-asthmatic drug *salbutamol,* and then propose the synthesis. Use the *dimsyl anion* ($MeSOCH_2^-Na^+$) as the source of carbene for cyclization of epoxide.

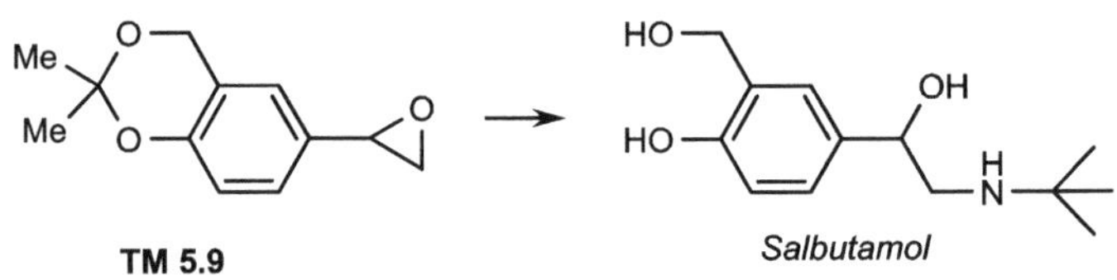

First, we complete contemporaneous disconnections of C–C and one C–O bond in oxirane to end up with an aldehyde and carbene (Scheme 5.24).

Continuing the retrosynthetic analysis, we perform FGI, interconversion of the protecting acetonide unit to the hydroxymethyl and phenol group. The hydroxymethyl group is introduced by acid-catalyzed hydroxymethylation of phenol and is therefore disconnected in the last step, affording the easily available *para*-hydroxybenzaldehyde and formaldehyde.

Synthesis of *salbutamol*, including preparation of the dimsyl anion, is presented in Scheme 5.25.

Electrophilic hydroxymethylation of *ortho* to the phenolic group is catalyzed by protic acids. In the next step, both hydroxy groups are protected as acetonide, cyclic ketal, and then the dimsyl anion reacts with the C=O group as a six-electron electrophile forming epoxide. In the final steps, regioselective opening of epoxide

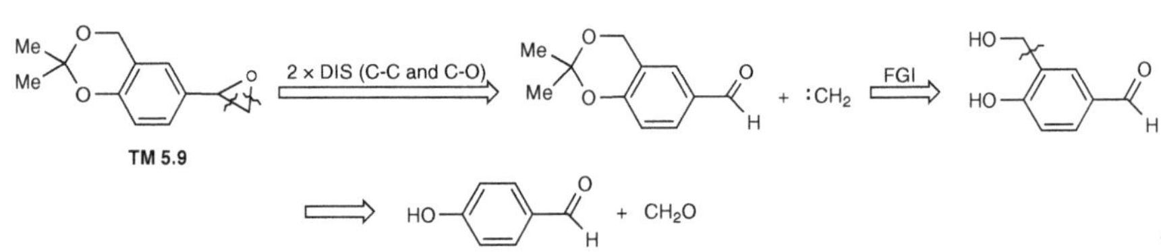

Scheme 5.24 Proposed retrosynthetic analysis of **TM 5.9**

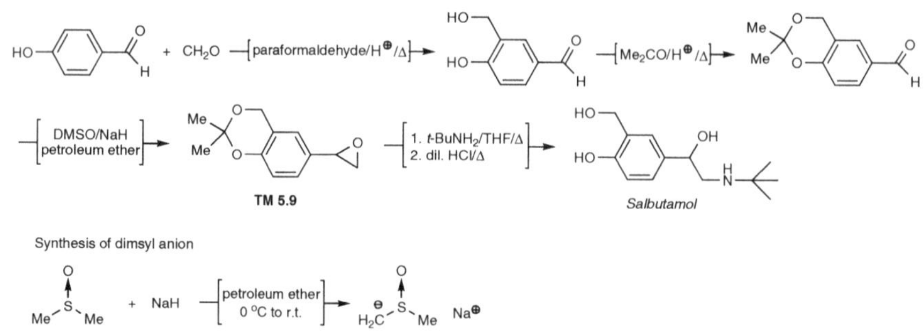

Scheme 5.25 Proposed synthesis of *salbutamol*

by nucleophilic attack of *tert* butylamine on the less substituted C-atom and then removal of the protecting group afford *salbutamol*.

Example 5.10 Propose the disconnection of the three-membered ring in **TM 5.10–5.13** and suggest some reagents for synthetic reaction.

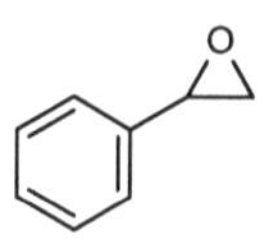

TM 5.10

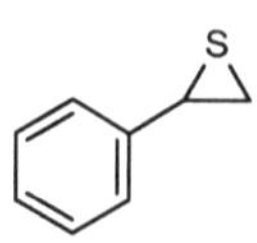

TM 5.11

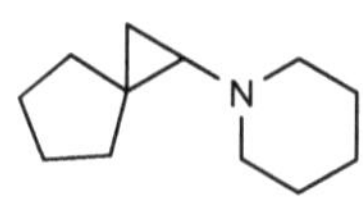

TM 5.12

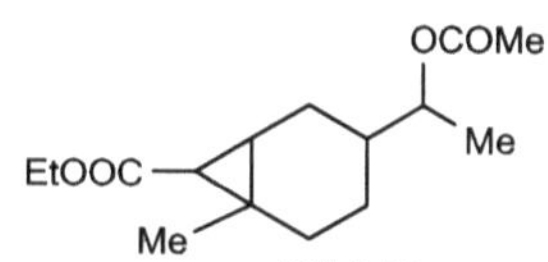

TM 5.13

For the oxirane and thiirane derivatives, different types of disconnections are preferred (Scheme 5.26). The epoxide in **TM 5.10** can be disconnected to alkene and a source of oxygen. In the synthetic direction, convenient epoxidation agents are peracids, in particular *meta*-chloroperbenzoic acid (MCPBA). This reagent is produced on the industrial scale by chlorination and then oxidation of benzoic acid and, due to its selectivity and solubility in most organic solvents, is used for epoxidation of complex molecules even on the large scale [6].

Reagents with the six electrones are good electrophiles but rather complex and require controlled reaction conditions (Scheme 5.23). Thiirane in **TM 5.11** is therefore conveniently interconverted by one-bond disconnection and regioselective ring opening, and then proper FGI affords styrene as the starting material. Formation of the thiirane ring in **TM 5.11** needs more synthetic steps from styrene, bromination, substitution of more reactive benzylic bromine by sulfide anions and cyclization in the last step.

Great simplification of **TM 5.12** is achieved by disconnection of cyclopropane to diazomethane as the source of carbene and the C=C bond in enamine. The target molecule of the second generation is an obvious condensation product of cyclopentanecarbaldehyde and piperidine.

Retrosynthetic analysis of **TM 5.13** starts with the logical disconnection of the cyclopropane ring to ethyl diazoacetate as the source of carbene and cyclohexene

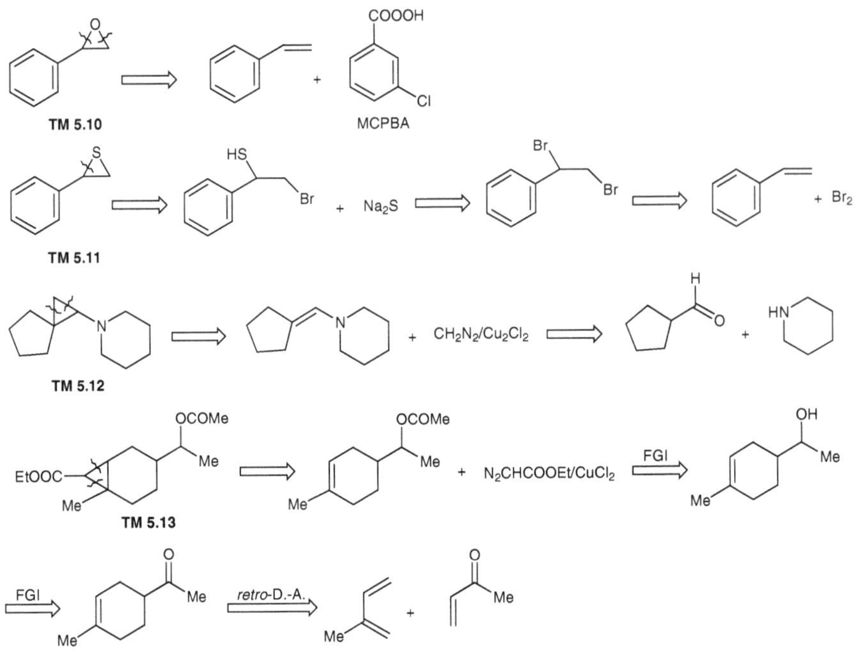

Scheme 5.26 Retrosynthetic analyses of **TM 5.10–5.13**

derivative. This new TM is transformed by two obvious FGIs to cyclohexenyl methyl ketone, which is convenient for *retro*-Diels-Alder disconnection.

Example 5.11 Perform retrosynthetic analysis and then propose the synthesis of **TM 5.14.**

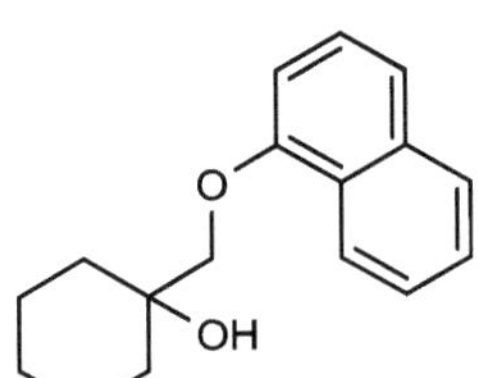

TM 5.14

The 1,2-dioxigenated pattern in **TM 5.14** comprises two single C–O bonds, one of them as an ether functionality. Having in mind the known property of three-member heterocycles as kinetically preferred and relatively thermodynamically unstable structures, we invoke the disconnection of the central C–O bond and epoxide **TM 5.14a** as the target molecule of the next generation (Scheme 5.27).

Disconnection of the central C–O bond generates an unstable cation on the primary C atom beside the α-naphthyloxy anion as the second synthon. Here we

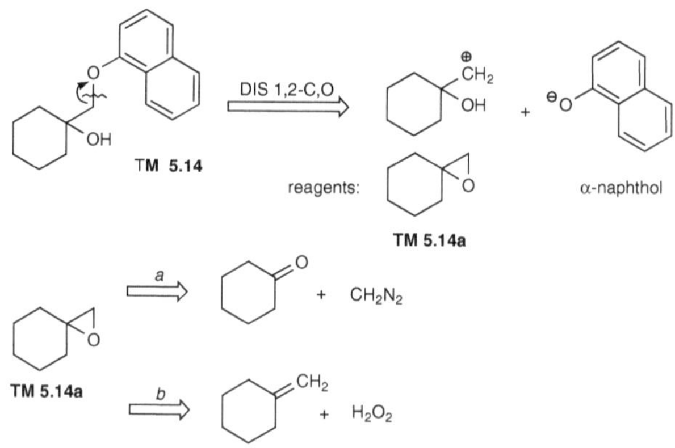

Scheme 5.27 Retrosynthetic analysis of **TM 5.14**

have the case in which the cationic synthon has an acceptable reagent in the corresponding epoxide **TM 5.14a**. Vicinally substituted 1-hydroxy-1-bromomethyl cyclohexane is the wrong choice since it spontaneously cyclizes to epoxide **TM 5.14a**! Disconnection of the three-membered ring can be completed according to either *a* or *b*. In the first case, the resulting reagents are cyclohexanone and diazomethane, in the second methylenecyclohexane and peroxide. Both disconnections suggest synthetically feasible reactions.

Note α-Naphthol is available by oxygenation of tetraline (tetrahydronaphthalene) to 1-tetralone followed by dehydrogenation (aromatization). The technological process is catalyzed by zeolites and oxygenation completed by oxygen. Intermediary 1-tetralone can be isolated or dehydrogenated to 1-naphthol by the same catalytic system.

The proposed synthesis of **TM 5.14** in Scheme 5.28 characterizes the regioselective ring opening of epoxide with an anion of α-naphthol under basic conditions. This bulky nucleophile approaches the less-substituted C atom of epoxide.

Related to the regioselective ring opening in **TM 5.14a**, the next simple example is instructive.

Example 5.12 Which product can be expected in the reaction of epoxy propane with methanol in the basic medium and in the presence of acid?

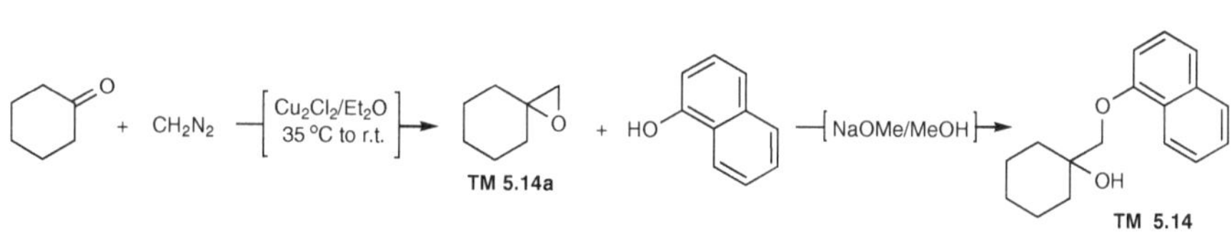

Scheme 5.28 Proposal for the synthesis of **TM 5.14**

Scheme 5.29 teaches that the regioselectivity of the ring opening depends on the reaction conditions, respectively, the reaction mechanism.

In the basic medium, the methoxy anion approaches the less-substituted C atom. In the acidic medium, methanol as a neutral nucleophile approaches protonated epoxide at the *sec* C-atom with a partially localized positive charge.

The next example demonstrates how complex epoxidation is in its chiral variant with a chiral reagent used in stoichiometric quantities.

Example 5.13 Preparation of enantiomerically pure *trans*-epoxides **4a–4e** is completed by non-catalytic asymmetric epoxidation and the addition of chiral sulfonium ylide **2** to the C=O bond of aromatic aldehydes (Scheme 5.30) [7].

Optically pure sulfonium ylide **2** is available from sulfonium salt **1**. The transfer of chirality to epoxides **4a–4d** is completely *trans*-selective and highly enantioselective, e.e. 98–99.9 %, affording (1*R*, 2*R*)-enantiomers. Recycling of the chiral reagent into the process is indicated in the scheme. A peculiar feature of this process is the availability of chiral bicyclic sulfonium ylide, on the first glance not related to any compound from the "chiral pool of nature." However, its three-step synthesis starts from (+)-(*R*)-pulegone and sesquiterpene available from an extract of *Mentha pulegium* and requires common reagents under standard conditions (Scheme 5.31).

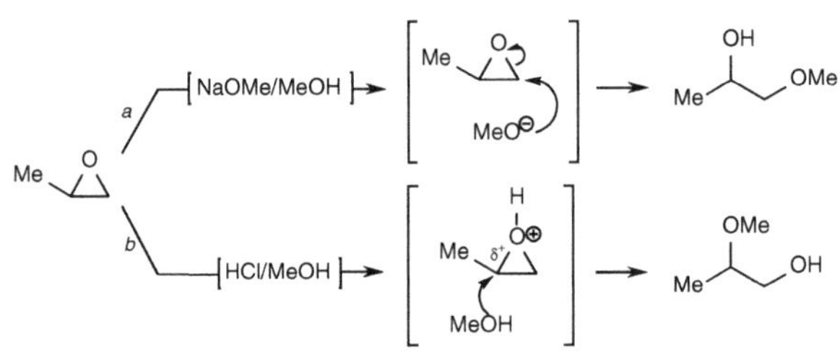

Scheme 5.29 Regioselective opening of the epoxy propane ring

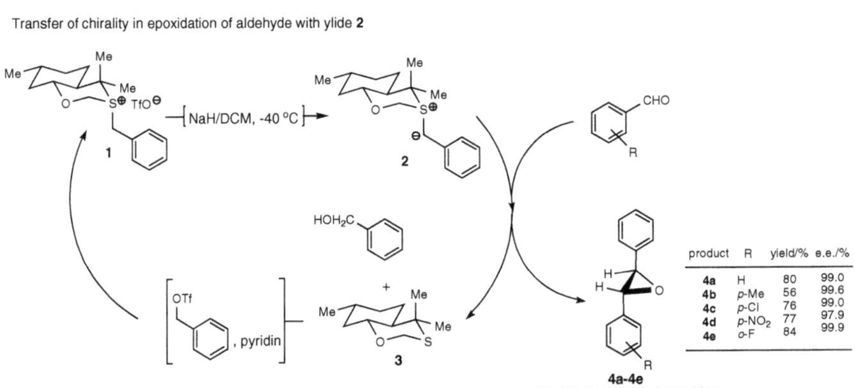

product	R	yield/%	e.e./%
4a	H	80	99.0
4b	*p*-Me	56	99.6
4c	*p*-Cl	76	99.0
4d	*p*-NO_2	77	97.9
4e	*o*-F	84	99.9

Scheme 5.30 Non-catalytic epoxidation of benzaldehydes

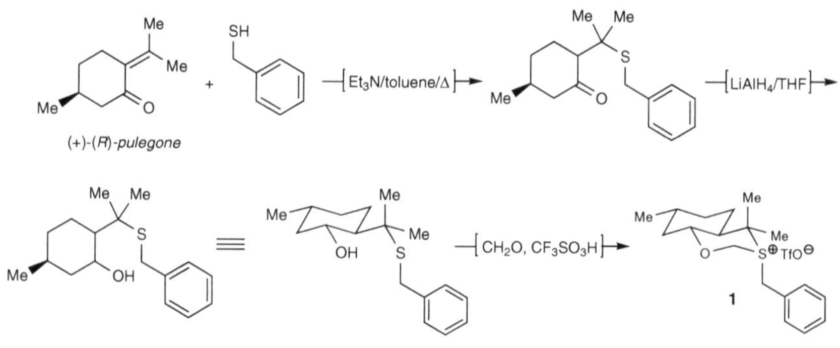

Scheme 5.31 Synthesis of chiral sulfonium salt **1**

5.2.3 1,2-Dihidroxy Pattern, Vicinal Diols

In this chapter we analyze 1,2-diols or *vicinal* diols, with a 1,2-CO pattern with two hydroxy groups (Scheme 5.32).

In all three cases, FGIs are proposed since no disconnection of the central C–C bond gives a couple of synthons for which available reagents exist. In other words, 1,2-diols are not available by direct formation of the central, single C–C bond.

FGIs (a) and (b) involve retrosynthetic transformation of the hydroxy to carbonyl group; the third FGI (c) implies elimination of both OH groups to obtain alkene, the target molecules of the next generation. In the synthetic direction, this means that reduction of the C=O group(s) or dihydroxylation of the C=C bond is required for rich 1,2-diols.

Note There is an exceptional formation of C–C bonds in derivatives of vicinal diols. An interesting stereoselective method for the synthesis of *anti*-vicinal diols uses organometallic alkoxyallyl tins (γ-metallated enol-ethers) and aldehydes in the presence of $BF_3 \cdot Et_2O$ at −78 °C [8]. The reagents and reaction conditions limit this method to the laboratory scale.

Widely used methods for the dihydroxylation of alkenes are discussed in the examples that follow.

Scheme 5.32 Possible FGIs in the retrosynthetic analysis of 1,2-diols

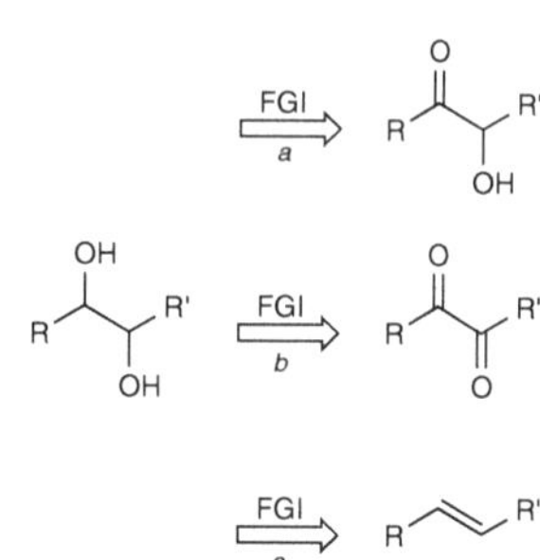

Example 5.14 Consider the retrosynthesis of **TM 5.15**, a model structure for the pheromone of the bark beetle, and then propose its multistep synthesis.

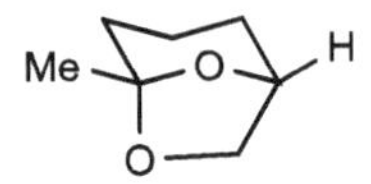

TM 5.15

First, we observe the presence of intramolecular ketal, which is formed in the last step of synthesis. Therefore, the first FGI leads to **TM 5.15a**, a straight chain molecule with ketone and vicinal diol functionalities (Scheme 5.33).

By the next FGI, we introduce the terminal C=C bond interconverting diol in alkene **TM 5.15b**. On the route to the next target molecule convenient for disconnection with great simplification, the FGI of the C=C bond appears in the *retro*-Wittig manner. It does not greatly simplify the general structure, but provides **TM 5.15c** with a comfortable 1,5-CO pattern! *retro*-Michael disconnection results in the well-known starting materials acrolein and acetonide anion, for which ethyl acetoacetate is repeatedly accentuated as a proper reagent.

Although this retrosynthesis seems elegant and straightforward, it is not workable in a synthetic direction without certain modifications (Scheme 5.34).

A chemoselective Wittig reaction with one of the two carbonyl groups in **TM 5.15c** is not possible! It is also not possible to selectively protect only the keto group. The preferred solution is to introduce the aldehyde group only after protecting the keto group. The proposed synthesis solves this problem; still, seven steps are needed to **TM 5.15**. This route requires Michael addition on ethyl acrylate, decarboxylation and then esterification of the remaining carboxylic group, protection of the keto group and two further steps for transformation of the carboxylic ester to aldehyde. Only now is there convenient introduction of the seventh C-atom by the Wittig reaction. After dihydroxylation of the C=C bond, there is no need to remove the protecting group in a separate step since intramolecular transketalization occurs spontaneously.

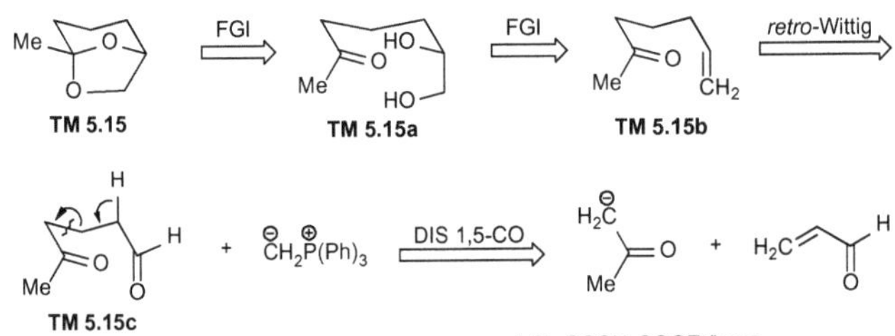

Scheme 5.33 Retrosynthetic analysis of **TM 5.15**

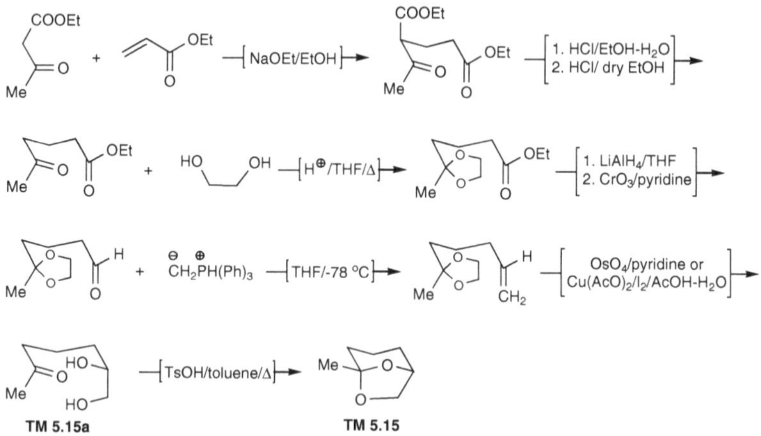

Scheme 5.34 Proposed synthesis of **TM 5.15**

5.2.3.1 Asymmetric Synthesis of (−)-Frontalin

The former retrosynthetic analysis reveals a relatively simple though lengthy multistep synthesis of the racemic frontalin analog **TM 5.15**. Since this product contains two stereogenic centers, four stereoisomers are formed. By asymmetric synthesis of (−)-*frontalin* **TM 5.16**, only one of four stereoisomers is targeted. Two stereogenic centers possess a (1*S*, 5*R*) configuration, and asymmetric synthesis represents a much more demanding enterprise.

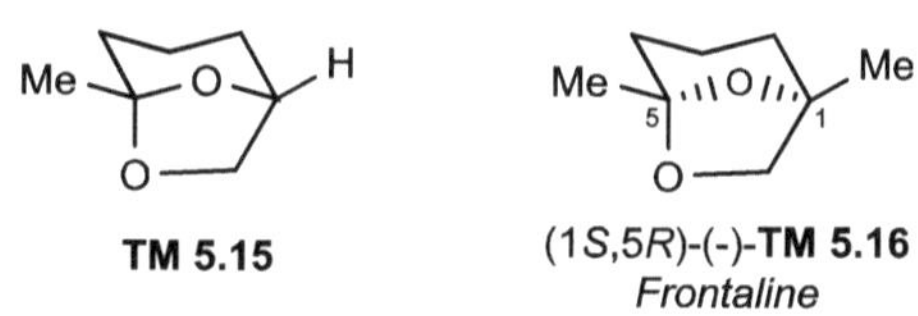

Both stereogenic centers in (−)-*frontalin* sit on *tert* C atoms. The one on C-5 is a constituent of the ketal group and hence formed in the moment of ketalization.

Note (−)-*Frontalin* is an active component of the aggregating pheromone of pine beetle, which devastates whole pine-tree forests, and is therefore used to kill this insect. Over 30 asymmetric syntheses of (−)-*frontalin* have been published, which due to their complexity, many steps and low yields remain of academic interest. The synthesis we discuss enabled preparation of 10 g of (−)-*frontalin* in ten steps with a total yield of 7.8 % [9]. This was the quantity needed for the field studies aimed at controlling various insect populations.

Scheme 5.35 outlines the first four steps and Scheme 5.37 the final steps of this synthesis. Cyclic keto ester **5** is easily available by Dieckmann condensation

(Sect. 4.3.3, Scheme 4.23). Compound **5** is racemic, and an interesting reduction step affords enantiomerically pure ethyl (1*S*, 2*S*)-2-hydroxycyclopentanecarboxylate **6**. Complete conversion of racemic **5** into optically pure **6** under the indicated reaction conditions is possible because of the *continuous racemization* of the non-reduced *R*-enantiomer of **5** and contemporaneous reduction of the *S*-enantiomer from the racemic mixture (Scheme 5.35).

The reduction of racemic **5** is a nice example of *product-selective biocatalytic kinetic resolution*. This means that the enantiomer that is reduced affords only one of two possible diastereomers. Enzyme reductase from the cells of *Saccharomyces cerevisiae* (baker's yeast) selectively reduces the carbonyl group in the *S*-enantiomer of **5** into alcohol with an *S* configuration on the new stereogenic center, hence producing (1*S*, 2*S*)-**6**. Due to the C–H acidity of the α-C atom in **5**, it easily tautomerizes to enols **5a** and **5b** in equilibrium (Scheme 5.36).

The stereogenic center in (*R*)-**5** inverts over planar enols affording racemate and enables recycling of this "wrong" enantiomer into reduction. The combination of selective reduction of one enantiomer accompanied by contemporaneous racemization of the second enantiomer assures complete "chiral economy" of the process and the formation of only one enantiomeric product from the racemic substrate; see also Sect. 9.4.2, Example 9.6.

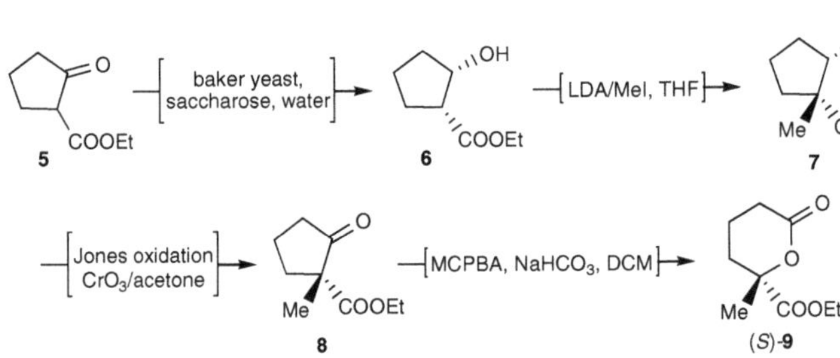

Scheme 5.35 Synthesis of key intermediate (*S*)-**9**

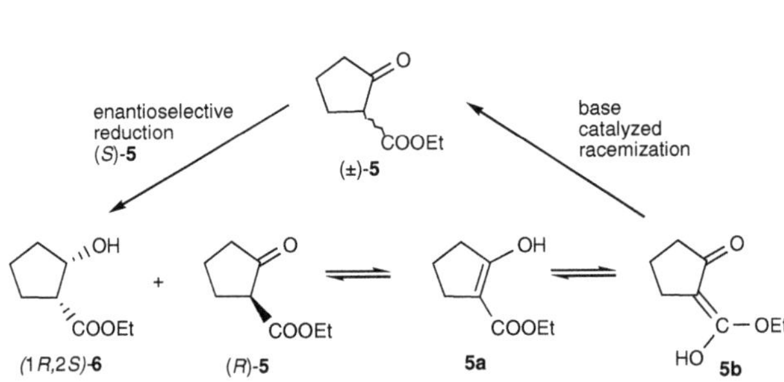

Scheme 5.36 Stereoselective reduction of (*S*)-**5** and tautomeric equilibria

The next step in Scheme 5.35 represents another stereochemical peculiarity of the process, the methylation of the carbanion of hydroxyester **6** with *retention of the configuration* at stereogenic center C-2. Detailed discussion of this phenomenon known as the *memory of chirality* is reported in the literature [10, 11]. The hydroxyl group in ester **7** is re-oxidized by Jones reagent to ketone **8**, revealing the importance of its transitory presence for methylation with retention of the configuration. In the last step, Bayer-Villiger rearrangement of keto ester **8** to lactone **9** is completed. The detailed mechanism of this reaction is discussed in Sect. 8.7 and according to the proposed mechanism the oxygen atom of the lactone group in **9** inserts to a higher substitute C atom.

All reaction steps in Scheme 5.35 are stereoselective. Biocatalytic reduction affords (1*S*, 2*S*)-**6** with 97.7 % e.e. and 99 % d.e., whereas methylation affords (1*S*, 2*S*)-**7** with 90.4 % e.e. The optical purity of (*S*)-**9** on oxidation and Bayer-Villiger rearrangement remains unchanged.

The last step of the synthesis of (−)-*frontalin* is presented in Scheme 5.37. Lactone and the ester group in **9** are reduced to C_6 triol, where two vicinal OH groups are protected as ketal **10**. This permits selective oxidation of the terminal OH group to aldehyde **11**, which in the Grignard reaction afforded *sec* alcohol **12**. Oxidation of this alcohol under conditions used for the oxidation of **10** to **11** afforded ketone **13**. In the last step, acid-catalyzed transketalization afforded (−)-*frontalin* **TM 5.16** where a (5*R*) stereogenic center is generated under asymmetric induction.

This example is the best illustration of the substantial difference in complexity of the synthetic approach to racemic compounds and asymmetric synthesis of the preferred enantiomer.

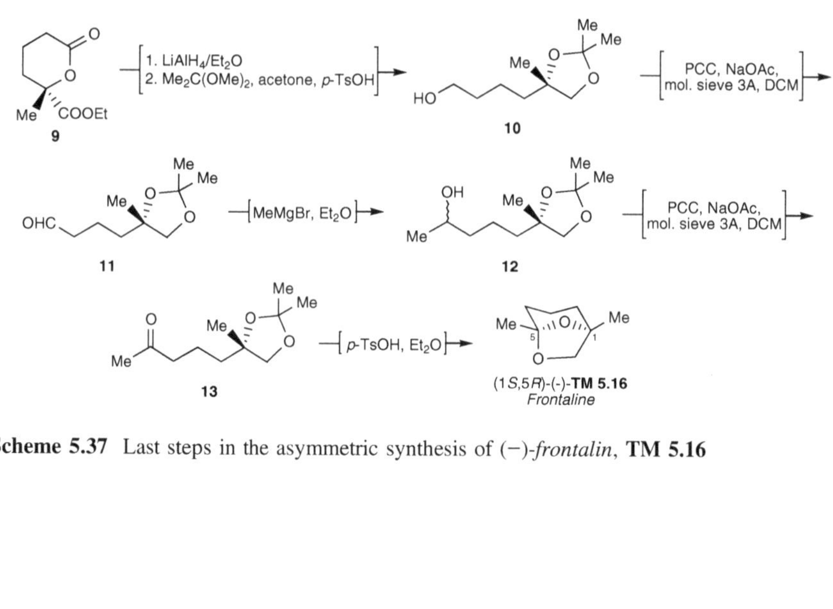

Scheme 5.37 Last steps in the asymmetric synthesis of (−)-*frontalin*, **TM 5.16**

5.3 1,4-Dioxygenated Pattern (1,4-CO)

5.3.1 1,4-Dicarbonyl Compounds

As presented in Sect. 5.1, an even number of C atoms between oxygen functionalities results in a mismatch of partial charges on the atoms of the central C–C bond. Accordingly, disconnection of the central C–C bond in 1,4-dicarbonyl compounds results in an acceptable anionic synthon and illogical cationic synthon with a positive charge on the α-C atom (Scheme 5.38).

We have repeatedly introduced acceptable reagents for anionic synthons, while those for cationic synthons are discussed in Sect. 5.2.2. Inversion of polarity on α-C atoms in cationic synthons is achieved by introduction of an σ-electron acceptor group, as analyzed in Example 5.4.

Example 5.15 Tricarbonyl, branch-chained **TM 4.17** is an important intermediate in the synthesis of certain psychopharmaca. Perform the retrosynthetic analysis and then propose the synthesis of this compound.

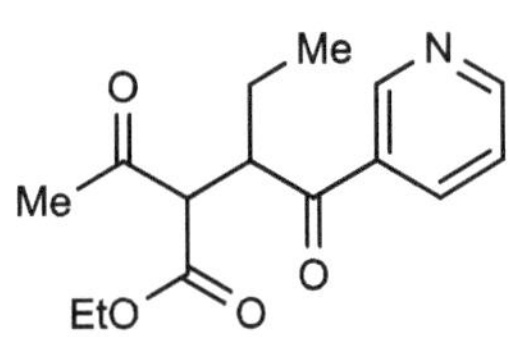

TM 5.17

The decision to first disconnect the 1,4-dicarbonyl pattern is challenging. It results with in a carbanion of ethyl acetoacetate and cationic synthon charged on the α-C atom to the carbonyl group (Scheme 5.39). We therefore propose α-bromo ketone **TM 5.17a** as a new target molecule.

The next obvious target is **TM 5.17b**, the result of FGE, the elimination of bromine to pyridyl propyl ketone. *retro*-Friedel-Crafts disconnection of **TM 5.17b** is not acceptable for the good reasons we discussed in Sect. 1.3.1. We also learn that the *retro*-Grignard disconnection of **TM 5.17b** presented in the above scheme results in *tert* alcohol if it starts from ester and not from the corresponding nitrile. Another convenient carboxylic acid derivative for acyl cations in the Grignard reaction is *Weinrib's amide*, a less known reagent presented in Scheme 5.39. Weinrib's amide reacts with Grignard reagent, affording an intermediary chelate that decomposes to ketone **TM 5.17b** (Scheme 5.40).

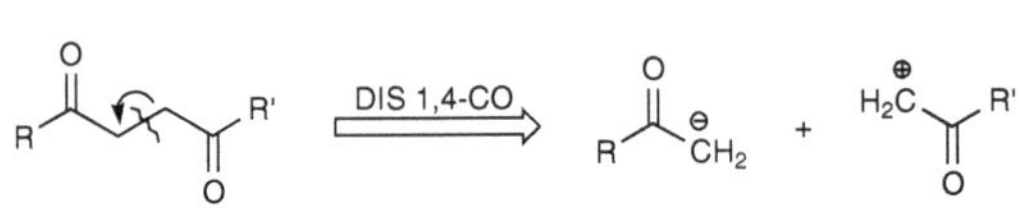

Scheme 5.38 General scheme of disconnection of 1,4-dicarbonyl compounds

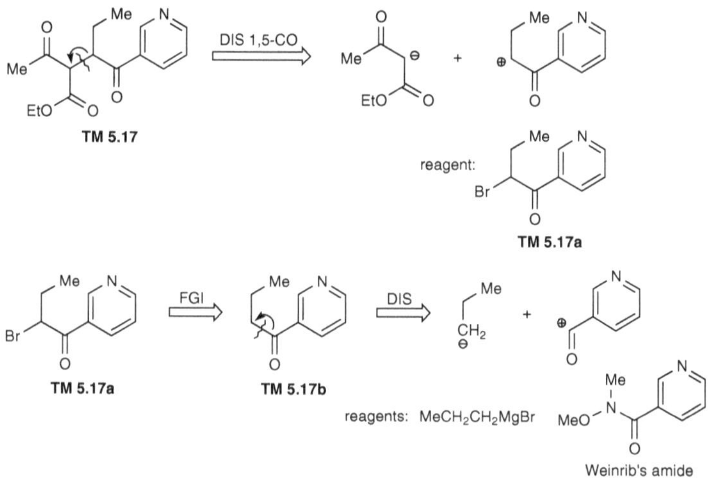

Scheme 5.39 Complete retrosynthetic analysis of **TM 5.17**

The intermediary chelate is stable at low temperatures and decomposes on workup in aqueous medium. Ketone **TM 5.17b** is brominated under standard conditions, with bromine in weak acidic medium, and used for alkylation of ethyl acetoacetate anions to **TM 5.17**.

As an introduction to the next subject, an interesting phenomenon deserves comment. The acidity of the α-C–H bond in ketones (pK_a about 20) is usually lower than that of the α-C–H bond in α-halogenated esters (pK_a about 17) since halogen atoms enhance C–H acidity. We shall see in the next two examples what happens when both species are expected to react and a strong base is added.

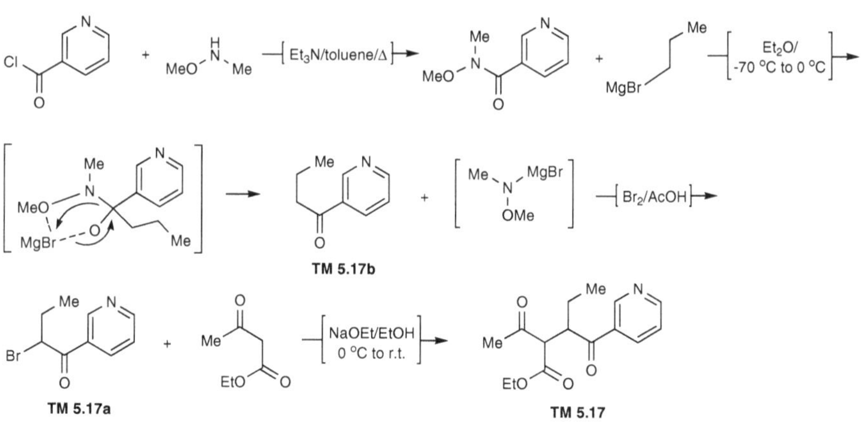

Scheme 5.40 Proposal for the synthesis of **TM 5.17**

Example 5.16 Suggest an obvious disconnection of **TM 5.18** and evaluate the feasibility of the corresponding one-step synthesis.

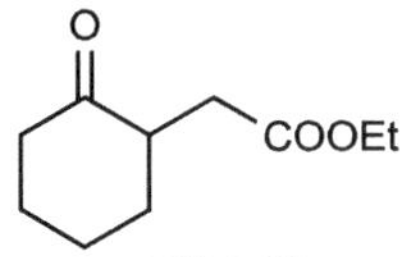

TM 5.18

An obvious 1,4-CO disconnection is presented in Scheme 5.41 using reagents for the conceived synthons.

Alkylation of cyclohexane is expected to start with deprotonation of the α-C atom by a strong base. However, the more acidic α-C atom of ethyl α-bromoacetate will preferably be deprotonated. This deprotonation triggers the sequence of reactions presented in Scheme 5.42.

The formation of α,β-epoxy carboxylic esters from carbonyl compounds and α-halo esters is not just an undesired side reaction, but is also an important synthetic method known as *Darzens reaction*.

To solve the issue with the workable synthesis of **TM 5.18** according to the retrosynthetic step in Scheme 5.4, we need to convert the α-C atom of the ketone to a neutral nucleophile. This is achieved by fixing the enolic C=C bond in enamine, available by condensation of ketone and *sec* amine. The mechanism of this acid-catalyzed reaction is presented in Scheme 5.43.

The proposal for the workable synthesis of **TM 5.18** comprises the preparation of cyclohexanon-enamine as an uncharged nucleophile (Scheme 5.44).

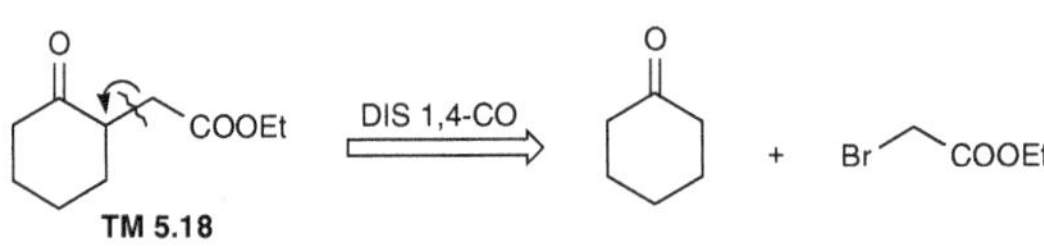

Scheme 5.41 1,4-CO Disconnection of **TM 5.18**

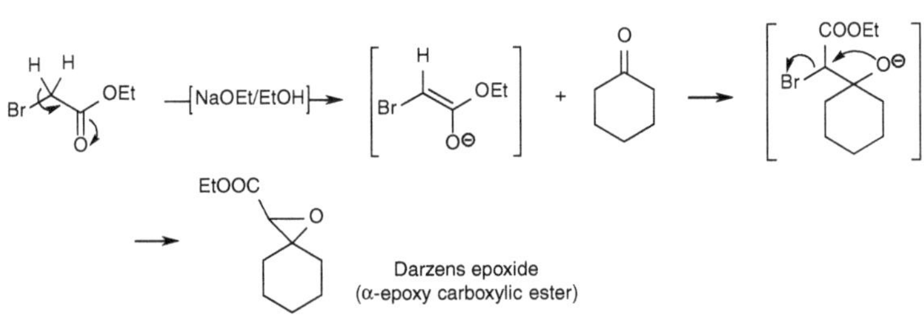

Scheme 5.42 Base-promoted reaction between cyclohexanone and enolate of ethyl α-bromoacetate

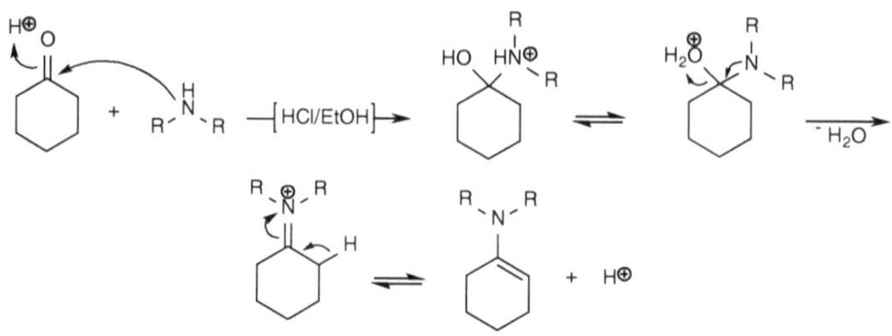

Scheme 5.43 Mechanism of the formation of cyclohexanon-enamine

Scheme 5.44 Proposal for the synthesis of **TM 5.18**

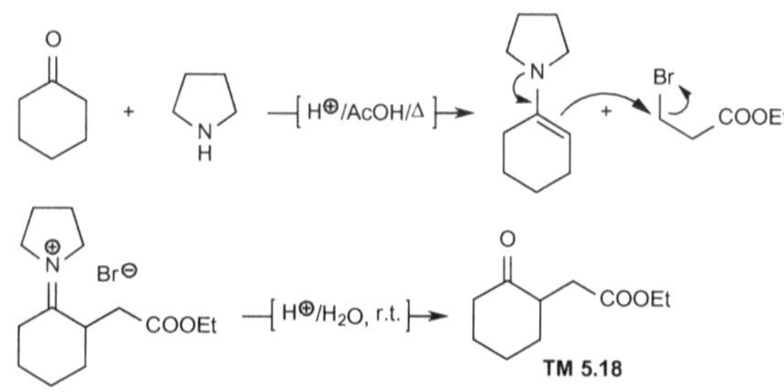

Example 5.17 Using this concept, complete the retrosynthetic analysis and suggest a workable synthesis of **TM 5.19**.

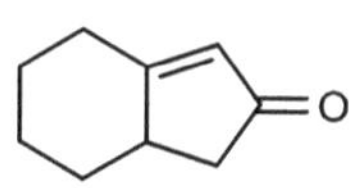

TM 5.19

Retro-aldol disconnection of α,β-unsaturated ketone leads to 1,4-diketone **TM 5.19a**. By disconnection of the central C–C bond, we generate two synthons, α-carbanion of cyclohexanone and α-carbocation in acetone. Synthetic equivalents for two synthons are the enamine of cyclohexanone **TM 5.19b** and α-chloroacetone **TM 5.19c** (Scheme 5.45).

Now the question of why the alkylation electronegative of the *tert* N-atom does not compete arises, as indicated in Scheme 5.46, instead of exclusive alkylation of the terminal C-atom of enamine **TM 5.19b**.

To answer this question, we consider the distribution of electronic density inside an aza-allylic system of enamine (Scheme 5.46b). The electron pair on the N-atom is delocalized, and the highest electron density is found on the terminal C-atom, which is therefore a strong nucleophile. Besides, it should be taken into account that quaternization of the *tert* N-atom is reversible under the thermal conditions used for C-alkylation. Both characteristics of the enamine structure favor C-alkylation.

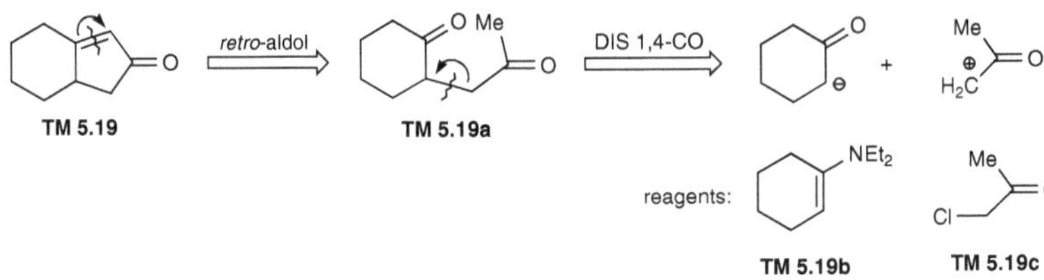

Scheme 5.45 Retrosynthetic analysis of **TM 4.19**

Example 5.18 Propose the disconnection and then synthesis of diketone **TM 5.20**.

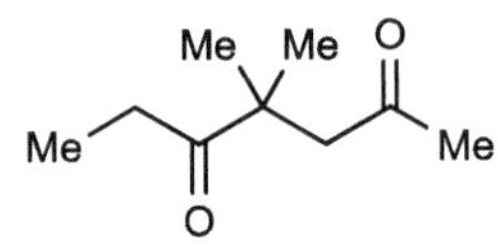

TM 5.20

In this 1,4-dioxygenated pattern, we can expect that all disconnections of central bonds will result in one preferred and one illogical synthon. Let as first consider two variants of 1,4-CO disconnection *a* (Scheme 5.47).

An anionic synthon from disconnection *a1* requires deprotonation at the *tert* α-C atom in the presence of a more acidic α-bromomethyl ketone, hence having unfavorable chemoselectivity. Disconnection of the same C–C bond with inverted flow of σ-electrons *a2* leads to the acetonide anion as the preferred synthon and unfavorable carbocation on the α-C atom to the carbonyl group. A reagent with bromine on the *tert* C atom is not available since bromination preferably occurs on the methylenic *sec* α-C atom.

A surprising solution is offered by *retro*-Mannich type disconnection *b* where the acidic C–H group participates and the carbanion appears as an illogical synthon on the acyl C atom. We showed the transformation of the nitro to a carbonyl group already in Sect. 2.4. For the illogical C_3 synthon, an unexpected reagent exists, 1-nitropropane, whose α-C–H atom has a pK_a of ca. 10. Taking into account this

Scheme 5.46 Presumed alkylation of the N-atom in enamine **TM 5.19b**

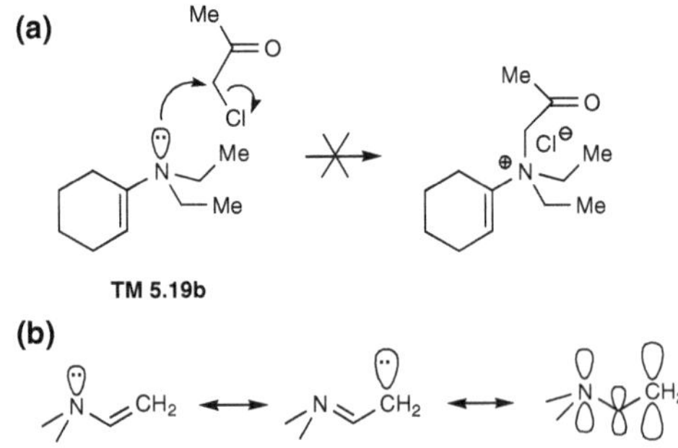

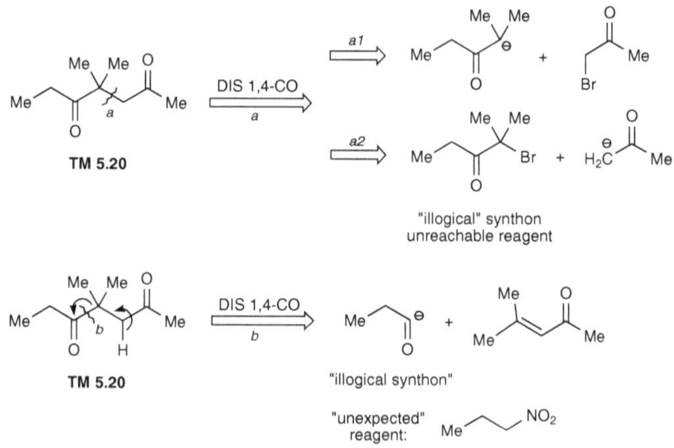

Scheme 5.47 Retrosynthetic analysis of **TM 5.20**

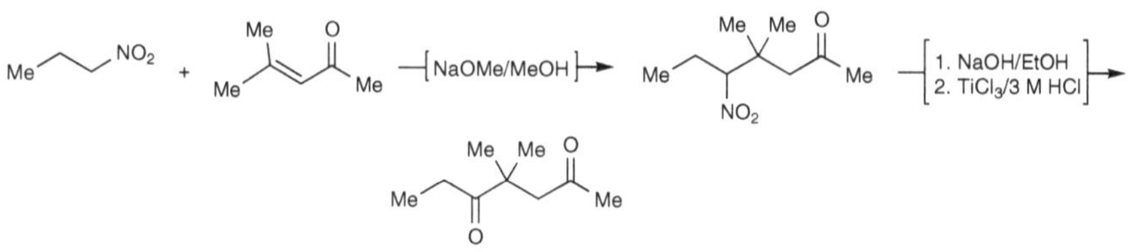

Scheme 5.48 Proposal for the synthesis of **TM 5.20**

transformation and the availability of mesityl oxide, a dimer of acetone, as the second product of disconnection *b*, we can now propose a short synthesis of **TM 5. 20** (Scheme 5.48).

By Michael-type addition of stable carbanion α– to the nitro group intermediary, 1,4-nitroketone is obtained, which is oxygenated to 1,4-diketone **TM 5.20** by $TiCl_3$ in acidic medium. An overview of the methods for the oxygenation of the nitro to keto group, known as the Nef reaction, is available [12, 13], and some are cited in Sect. 2.5.

5.3.2 1,4-Hydroxy Carbonyl Compounds

1,4-Hydroxy carbonyl compounds seem available by selective reduction of one carbonyl group in 1,4-dicarbonyl compounds. Such chemoselectivity usually is not workable, and the retrosynthetic step where the hydroxy group in 1,4-hydroxycarbonyl compounds is interconverted to a keto group does not suggest

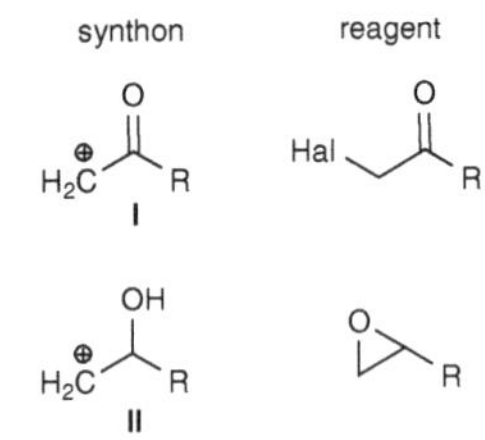

Scheme 5.49 Illogical carbocationic synthons and acceptable reagents

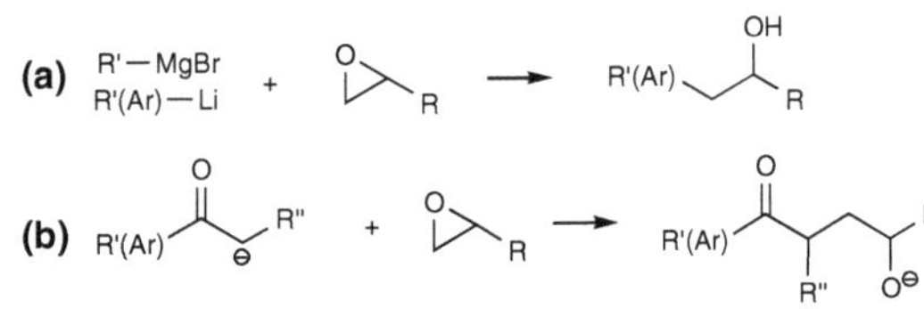

Scheme 5.50 Alkylation of epoxides by (a) unstable and (b) stabilized carbanions

any appealing solution to the synthetic problem. Therefore, we retrosynthetically consider the disconnection of 1,4-hydroxycarbonyl compounds to the available building blocks.

In the previous chapter, we saw α-halo carbonyl compounds as acceptable reagents for illogical synthons **I** with carbocation α- to the carbonyl group (Scheme 5.49).

Synthon **II**, an analog of **I** with an oxygen atom at a lower oxidation state, is also conceivable. Here, the positive charge can be compensated intramolecularly by an electron pair from an oxygen atom, i.e., by the formation of epoxide. In other words, epoxides are ideal reagents for illogical synthons **II**!

In Sect. 5.2.2, Example 5.12, we saw the electrophilic character of epoxide toward the alkoxide ion. The carbanion is another charged nucleophile that opens a three-membered ring in reaction with epoxides. When such an anion is unstabilized, its synthetic equivalent is an organometallic compound affording α-alkylated or arylated alcohols as products (Scheme 5.50a).

When as the nucleophilic component α-carbanion stabilized by a carbonyl group or "masked" as enamine reacts, the reaction products are 1,4- or γ-hydroxycarbonyl compounds. The next example illustrates the utility of such retrosynthetic considerations.

Example 5.19 Complete the retrosynthetic analysis and then propose the synthesis of **TM 5.21**.

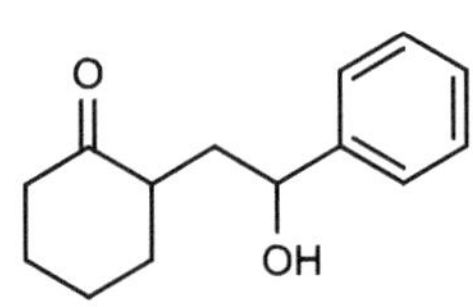

TM 5.21

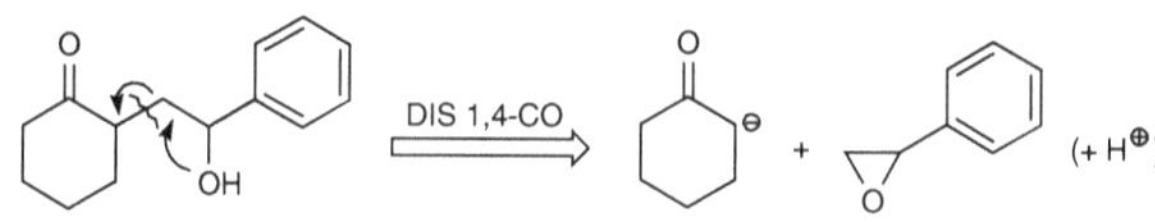

TM 5.21

Scheme 5.51 Retrosynthetic analysis of **TM 5.21**

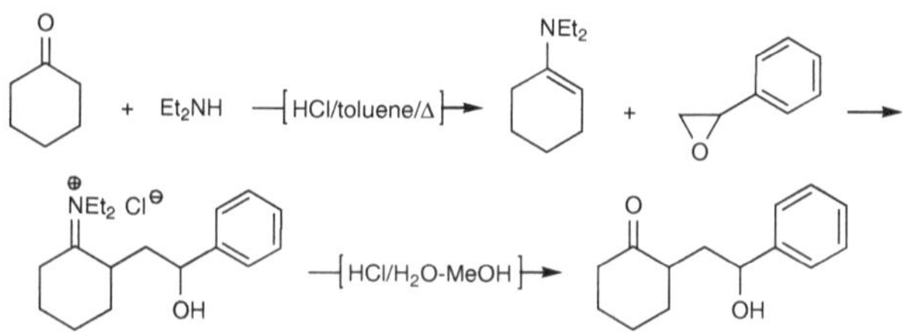

TM 5.21

Scheme 5.52 Proposal for the synthesis of **TM 5.51**

Proper disconnection of the central C–C bond of the 1,4-hydroxycarbonyl pattern generates a carbanion on the α-C atom to carbonyl, not an α-C atom to the hydroxy group, and styrene epoxide (Scheme 5.51).

An anionic synthon for cyclohexanone is inconvenient since a strong anionic base is needed for the generation of a carbanion, which acts like a competitive nucleophile and attacks epoxide. Neutral enamine is therefore the preferred nucleophile (Scheme 5.52).

On the completed ring opening, intermediary immonium salt hydrolyzes easily to the keto group in **TM 5.21**.

5.4 1,6-Dicarbonyl Pattern (1,6-CO)

The large through-bond distance between two oxygenated functional groups in a target molecule with a 1,6-CO pattern leaves the impression that no workable C–C bonding reaction can bind two building blocks since disconnection of any central C–C bond unavoidably results in one acceptable and one illogical synthon (Scheme 5.53).

Disconnection *a* is feasible only if protection of carbonyl group in carbanionic synthon is completed before the preparation of the Grignard reagent as a synthetic equivalent. Disconnection *b* indicates the synthetic route that starts with alkylation of the stabilized α-carbanion. Disconnection *c* does not offer any good solution for the synthon where the carbanion appears on the β-C atom. For the cationic synthon, instead, enone $RCOCH{=}CH_2$ is a convenient reagent.

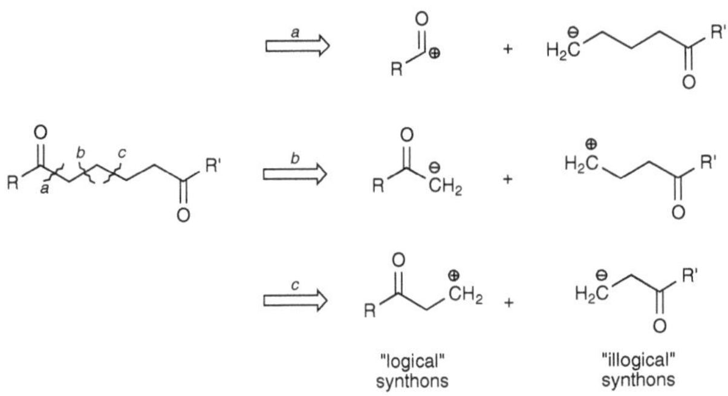

Scheme 5.53 Overview of disconnections of the central C–C bonds in the 1,6-CO pattern

The preferred retrosynthetic solution for 1,6-dicarbonyl compounds invokes a completely different concept of *reconnection* (RCN). This retrosynthetic step connects two functional groups into a cyclic, easily available structure, a target molecule of the next generation. Reconnection of the 1,6-dicarbonyl pattern results in a cyclohexene ring. Scheme 5.54 presents the synthetic step, the oxidative split of the C=C bond on the route to 1,6-dicarbonyl compounds and the corresponding retrosynthetic step, reconnection of this acyclic structure.

This scheme indicates that reconnection of the 1,6-dicarbonyl compound to a cyclic alkene corresponds to the ozonolysis of the cyclic alkene in the synthetic direction. It is important to observe that no real synthetic reaction corresponds to the reconnection. Recognition of the derivatives of cyclohexene as target molecules of the next generation for the 1,6-dicarbonyl pattern is of evident retrosynthetic utility. This concept is discussed in the next three examples.

Example 5.20 Propose the retrosynthetic analysis and then synthesis of **TM 5.22**.

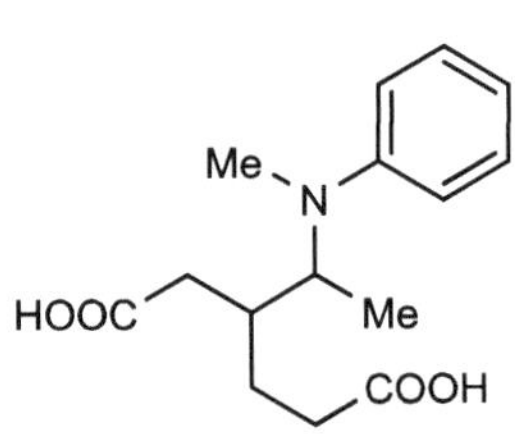

TM 5.22

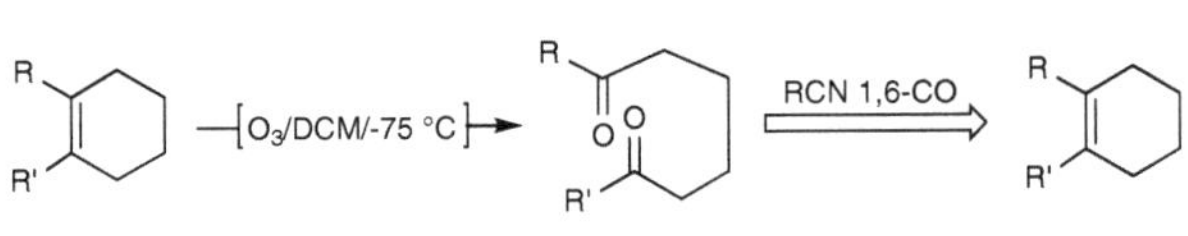

Scheme 5.54 Correlation between the synthetic step and reconnection of the 1,6-dicarbonyl compound

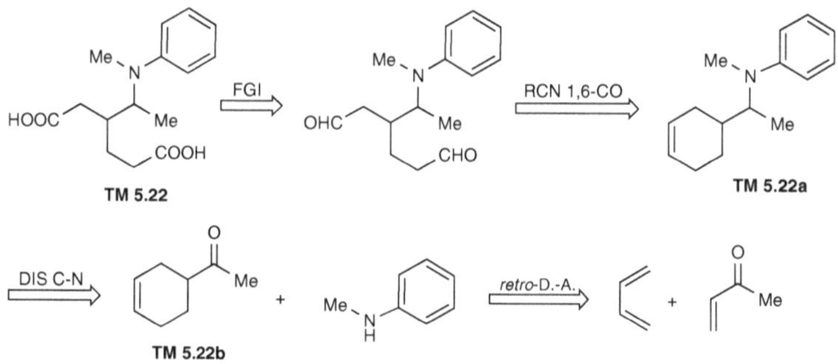

Scheme 5.55 Retrosynthetic analysis of **TM 5.22**

On recognition of the 1,6-dicarbonyl pattern in dicarboxylic acid **TM 5.22**, two retrosynthetic steps lead to **TM 5.22a** (Scheme 5.55). The first step is interconversion of 1,6-carboxylic to aldehyde groups that on reconnection afford cyclohexene derivative **TM 5.22a**.

Further retrosynthetic steps lead to *N*-methylaniline, butadiene and methyl vinyl ketone, three available building blocks.

In Scheme 5.56, oxidation of dialdehyde, a product of the ozonolysis of **TM 5.22a**, is indicated as an additional step on the route to **TM 5.22** [14, 15]. There is an important aspect of ozonolysis: under oxidative conditions, aldehydes are spontaneously oxidized to carboxylic acids.

Note Ozonolysis is a specific reaction that requires an ozonizator to produce ozone in the electric arc. Equipment is available in laboratory and industrial construction

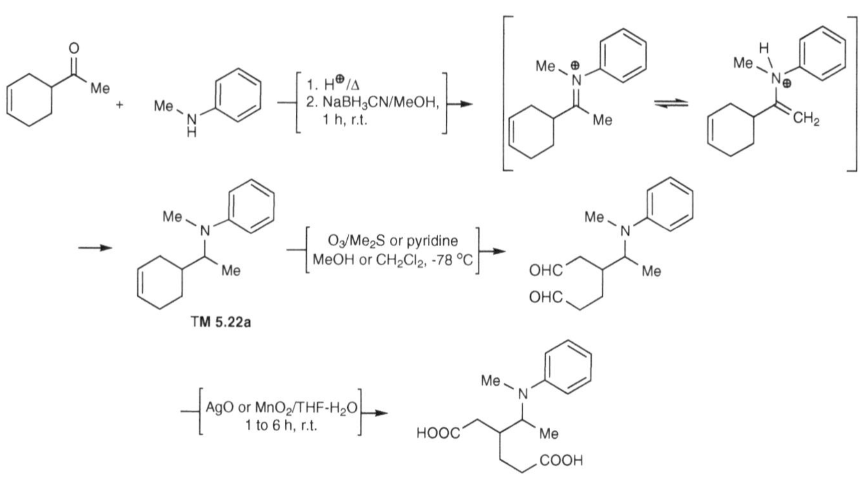

Scheme 5.56 Proposal for the synthesis of **TM 5.22**

Scheme 5.57 Electronic structure of ozone and mechanism of ozonolysis

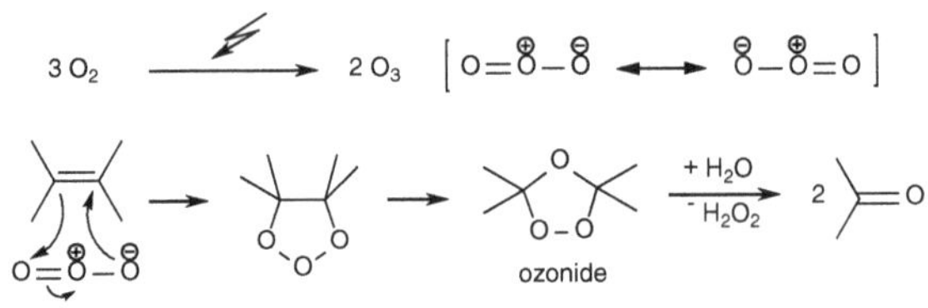

revealing the usability of this reaction on the technological scale. The mechanism of ozonolysis of the C=C bond is presented (Scheme 5.57).

The Criegee mechanism of ozonolysis characterizes the interaction of dipolar, electronically deficient ozone with the π-electron-rich C=C bond forming a cyclic intermediate [16, 17]. It rearranges with insertion of an O atom into the C–C bond and finally hydrolyzes in two moles of carbonyl compound with the formation of one mole of hydrogen peroxide. This mole of peroxide acts in situ as an oxidant of aldehyde to the carboxylic group. Oxidation of aldehydes can be completed with a number of reagents; two are cited in Scheme 5.56.

The reaction that corresponds to the disconnection of the C–N bond in **TM 5.22a** is also of considerable synthetic importance (Scheme 5.55). This is *reductive amination*, a reaction between aldehydes or ketones and primary or secondary amines in the presence of reducing agents leading to targeted amines without isolation of intermediary imines or enamines.

For preparation of *N*-methylaniline, an important commodity, industrial methods of reductive methylation by the mixture formaldehyde/formic acid, catalytic alkylation with methanol, reductive methylation of nitrobenezene or catalytic arylation of methylamine by phenol have been developed. Producers use various technologies depending on the range of the related products they offer to the market.

In the next example, an issue with the synthesis of 1,3-disubstituted benzene is hidden. Introduction of the second substituent into an aromatic ring is often a non-trivial problem in the synthesis of complex target molecules. There are probably other acceptable synthetic approaches to **TM 5.23**, and the reader is encouraged to consider the alternatives.

Example 5.21 Propose the retrosynthetic analysis and synthesis of **TM 5.23**.

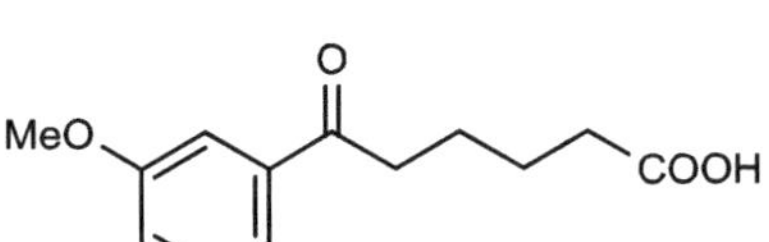

TM 5.23

The 1,6-CO pattern suggests the reconnection of the side chain to the cyclohexene ring as the first retrosynthetic step (Scheme 5.58).

The presence of 1-phenylcyclohexene in **TM 5.23a** suggests the FGI of the C=C bond to OH group **TM 5.23b** followed by *retro*-Grignard disconnection. Grignard

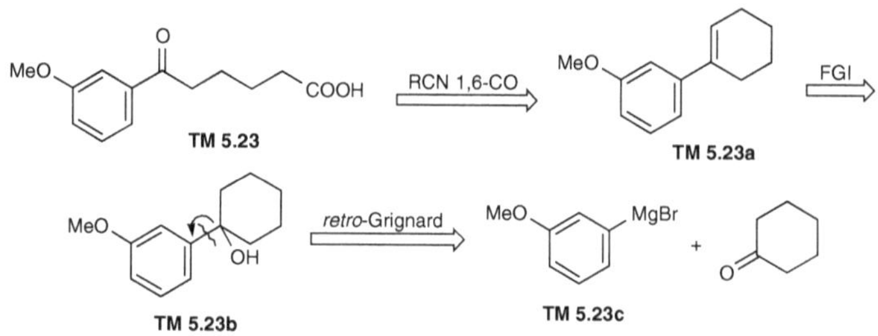

Scheme 5.58 Proposal for the retrosynthetic analysis of **TM 5.23**

reagent **5.23c** is available from 3-bromoanisole. An approach to this *meta* substituted benzene derivative requires a few steps as outlined in Scheme 5.29. Both substituents are *of the first order* characterized by positive Hammett σ-values (Table 5.1), and both direct the next substituent in the *ortho/para* position. In other words, none of them can orient a second substituent into the *meta*-position. Preparation of 3-bromoanisole follows the route where the first substituent, the *meta*-directing nitro group, after introduction of the second substituent, is transformed into a substituent of the first order present in the target molecule (Scheme 5.59).

Note Knowledge of the electronic effects of substituents in the aromatic ring, expressed by σ-values, and the Hammett equation helps in planning the syntheses of polysubstituted aromatic compounds [18–20]. Orientation of electrophilic aromatic substitution is determined by the electronic properties of the present substituents. Electron-acceptor substituents (with positive σ-values) deactivate *ortho/para* positions, resulting in the orientation of electrophilic substitution to the *meta*-position. This effect is quantitatively expressed by Hammet σ-values and for most usual substituents is presented in Table 5.1 [21].

Table 5.1 Hammett values for the most frequent substituents in aromatic compounds

Substituent	σ_p	σ_m
NH_2	−0.62	0
OCH_3	−0.29	0.11
H	0	0
F	0.05	0.34
I	0.23	0.35
Cl	0.22	0.37
Br	0.23	0.40
$COOCH_3$	0.45	0.33
$COCH_3$	0.49	0.37
CN	0.67	0.62
NO_2	0.77	0.73

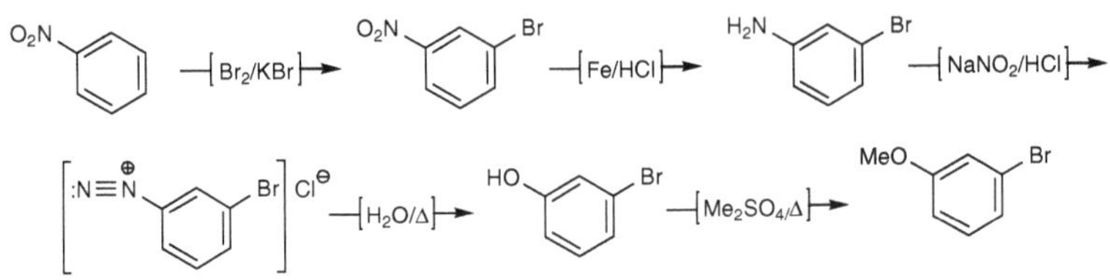

Scheme 5.59 Synthesis of 3-bromoanisole from nitrobenzene

The above data can be summarized as follows:

- electron donor groups have negative σ-values
- electron-attracting groups have positive σ-values
- no σ-values are available for the *ortho*-position because of the strong steric effects
- σ-*para* < σ-*meta* for substituents with an inductive electron-attracting effect
- σ-*para* > σ-*meta* for conjugating substituents

The complete synthesis of **TM 5.23** from *meta*-bromophenol and cyclohexanone as available raw materials is presented in Scheme 5.60.

Example 5.22 How would you approach the retrosynthesis of unsaturated ω-hydroxy ester **TM 5.24**?

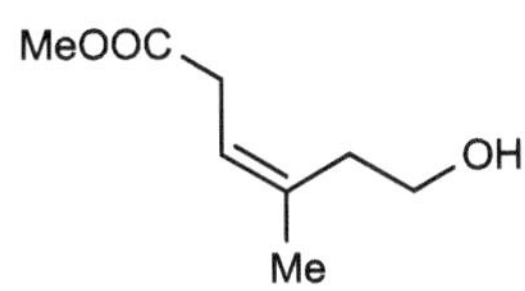

TM 5.24

The terminal hydroxyl and carboxyl groups are the highest and the lowest oxidation states of functional groups with oxygen atoms. Having in the mind easy

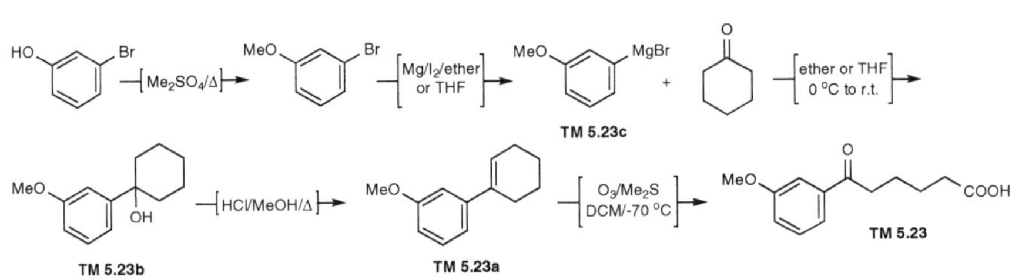

Scheme 5.60 Proposal for the synthesis of **TM 5.23**

interconversions of oxygen functional groups presented in Sect. 1.2.1, Scheme 1.14 and the 1,6-distance of these functionalities, we can enter retrosynthetic analysis (Scheme 5.61).

Interconversion of the hydroxyl group in **TM 5.24** to aldehyde in **TM 5.24a,** both presented "folded" to visualize cyclization, enables 1,6-CO reconnection. The carbomethoxy group ends up as enolic OMe in 1,4-cyclohexadiene **TM 5.24b**. Aromatization of **TM 5.24b** results in *para*-methylanisole (*para*-methoxy toluene). Now comes the dilemma: is it possible to reduce only one double bond in the aromatic ring selectively to obtain a non-conjugated isomer of cyclohexadiene? In other words, does such a retrosynthetic step have a corresponding synthetic reaction? A positive answer suggests the long-known *Birch reduction* [22, 23].

Note In this reaction aromatic compounds are reduced to non-conjugated cyclohexadienes by sodium or lithium in ammonia or in *sec* amines, preferably in morpholine. The mechanism of this important reaction is illustrated with benzene (Scheme 5.62).

Birch reduction is an important example of the process that includes *single-electron transfer* (SET). On the transfer of the first proton, the most stable intermediary radical is the one most distant from the σ-bond of the methylene group, controlling selective 1,4-addition of the second proton and formation of non-conjugated diene [24, 25].

Based on this new knowledge, we can now propose the synthesis of **TM 5.24** (Scheme 5.63).

Birch reduction and ozonolysis of the C=C bond are completely regio- and chemoselective according to the previously discussed mechanisms of both reactions. Birch reduction affords non-conjugated cyclohexadiene with both C=C bonds connected to electron-donating groups. This is a consequence of the greater stabilization of the radical anion on the non-substituted C atoms, which are then protonated under kinetic control, and also due to the higher stability of the more substituted double bonds in **TM 5.24b** (thermodynamic control). One C=C bond is stabilized by a resonant n–π interaction with the methoxy group and the other by hyperconjugative σ–π interaction with the methyl group. Ozonolysis occurs preferably on the more π-electron-rich C=C bond, and this one is connected to the methoxy group as a part of an enol ether unit.

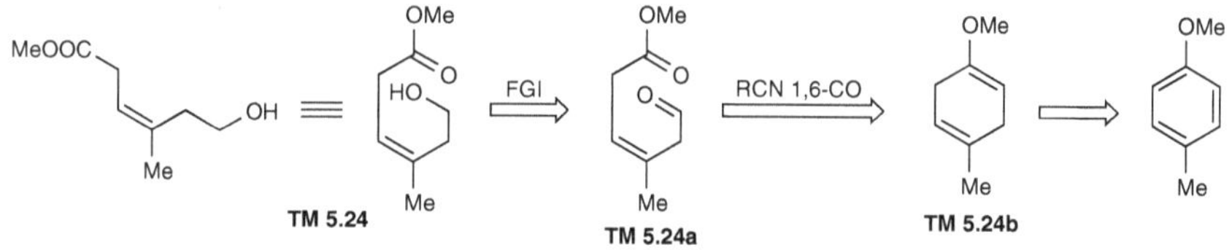

Scheme 5.61 Retrosynthetic analysis of **TM 5.24**

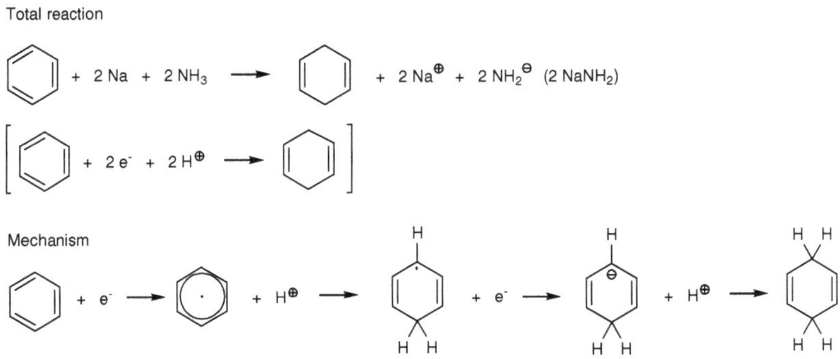

Scheme 5.62 Mechanism of the Birch reduction of benzene to cyclohexa-1,4-diene

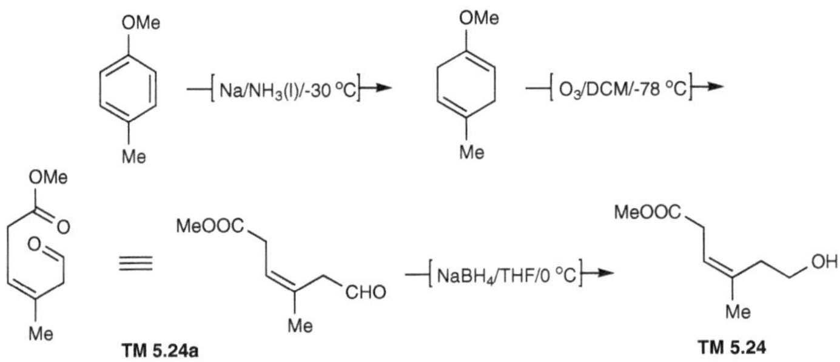

Scheme 5.63 Proposal for the synthesis of **TM 5.24**

In conclusion, retrosynthetic analysis of the 1,6-CO pattern reveals the value of reconnection when no productive route to the C_6 carbon chain is available from the $C_m + C_n$ building blocks.

References

1. Jain SL, Sain B, Sharma VB (2004) Tetrahedron Lett 45:8731–8732
2. Jain SL, Sharma VB (2006) J Mol Catal A 256:16–20
3. Ando T (1998) Tetrahedron 54:13485–13494
4. Capozzi F, Capozzi G, Menichetti S (1988) Tetrahedron Lett 29:4177–4180
5. Adam W, Bargon RM, Schenk WA (2003) J Am Chem Soc 125:3871–3876
6. McDonald RN, Steppel RN, Dorsey JE (1988) Org Synth Coll 6:276–279
7. Solladie-Cavallo A, Roje M, Giraud-Roux M, Chen Y, Berova N, Šunjić V (2004) Chirality 16:196–203
8. Koreda M, Tanaka Y (1987) Tetrahedron Lett 28:143–146
9. Nishimura Y, Mori K (1998) Eur J Org Chem 233–236
10. Fuji K, Kawabata T (1998) Chem Europ J 4:373–376

11. Brewster AG, Jayatissa J, Mitschel MB, Schofield A, Stoodlay RJ (2002) Tetrahedron Lett 43:3919–3922
12. Grierson DS, Husson HP (1991) Comp Org Synth 6:937–944
13. Lahner K, Ayed T, Bellassoued M, Amri H (2005) Beilst J Org Chem 1:11–13
14. Van Ornum SG, Champeau RM, Pariza R (2006) Chem Rev 106:2990–3001
15. Ishmuratov G, Legostaeva Y, Botsman L, Tolstikov G (2010) Russ J Org Chem 46:1593–1621
16. Geletneky C, Berger S (1998) Eur J Org Chem 8:1625–1627
17. Criegee R (1975) Angew Chem Int Ed Engl 14:745–752
18. Hansch C, Leo A (1979) Substituent constants for correlation analysis in chemistry and biology. Wiley-Interscience, NY
19. Hansch C, Leo A, Taft RW (1991) Chem Rev 91:165–195
20. Takahata Y, Chong DP (2005) Int J Quant Chem 103:509–515
21. Clayden J, Greeves N, Warren S, Wothers P (2001) Organic chemistry. Oxford University Press, pp 1090–1097
22. Birch JA, Subba Rao G (1972) Adv Org Chem 8:1–65
23. Zimmerman HE, Wang PA (1993) J Am Chem Soc 115:2205–2216
24. Birch JA (1996) Pure Appl Chem 68:553–562
25. Smith MB, March J (2007) March's advanced organic chemistry, 6th edn. Wiley, New York, pp 1066–1070

Chapter 6
Specific Synthetic Methods

Abstract Some innovative synthetic methods in organic chemistry are concisely presented, *multicomponent reactions, specifically the Ugi multicomponent reaction, parallel syntheses* and *combinatorial chemistry, mechanochemically promoted organic reactions, organic reactions promoted by microwave irradiation* and *syntheses in ionic liquids*. Examples of chemoselective or asymmetric syntheses completed by one of the presented specific methods are presented for the antihypertensive drug *nifedipine,* the alkaloid *tropinone,* the local anesthetic *xylocaine* and the HIV inhibitor *tipranavir*.

6.1 Introduction

This chapter is more informative then educative and to a certain extent departs from the main concept of this book. Familiarity with principles of less-known synthetic methods, though not correlated to retrosynthetic analysis, is expected to expand the imaginative capacity of synthetic organic chemists.

Organic synthetic chemistry is progressing thanks to the development of new and specific experimental methods. Among them, the largest interest has been attracted by *multicomponent reactions, parallel syntheses* and *combinatorial chemistry, mechanochemically promoted organic reactions, organic reactions promoted by microwave irradiation* and *syntheses in ionic liquids*. In the next sections we briefly describe the basic characteristics of these methods and offer examples of their successful application.

6.2 Multicomponent Reactions

Multicomponent reactions (MCRs) have long been known in laboratory praxis, but not recognized as a general concept. In the last decades their industrial utility has been recognized and multicomponent syntheses developed for large-scale

V. Šunjić and V. Petrović Peroković, *Organic Chemistry from Retrosynthesis to Asymmetric Synthesis*, DOI 10.1007/978-3-319-29926-6_6

production. Before consideration of specific examples, it is important to emphasize that in MCR you should *not react all reagents at the same time* since this is a highly unfavorable process entropically. What actually happens in MCR is a well-defined sequence of bimolecular reactions where an intermediate formed in the first step reacts with the third component in the second step and in rare cases with the fourth component in the third step. Intermediates of the whole process usually are not known and regularly are not isolated.

According to SciFinder, over 2700 multicomponent reactions were published just in the 2005–2010 period. They are particularly useful for parallel synthesis of libraries of compounds, the concept briefly discussed in the next chapter. Multicomponent reactions might be amenable to a retrosynthetic approach if sequential disconnection of more bonds can be completed in the synthetic direction from recognizable building blocks reacting in the expected order.

6.2.1 General Concept of Multicomponent Reactions

Multicomponent reactions represent a flexible tool for the synthesis of a large number of target molecules from three or more starting molecules. They are one-step, one-pot reactions, economic regarding resources and today considered to be close to what is defined as "ideal synthesis" [1]. Their general synthetic value was recognized when I. Ugi and collaborators reported on some important variants of four-component reactions [2] and their application in the production of known drugs; an illustrative example is presented in the next section.

It is interesting that three multicomponent reactions were broadly used over 150 years, known as *name reactions* according to their inventors: Strecker synthesis of amino acids [3], Hantsch synthesis of 1,4-dihydropyrimidines [4] and particularly the important Mannich reaction [5].

Hantsch four-component synthesis of 1,4-dihydropyrimidines entered laboratory praxis long ago, but only recently became an industrial method. As an example, synthesis of *nifedipine* **3**, a broadly used drug in the therapy of hypertension, is presented in Scheme 6.1 [6].

ortho-Nitrobenzaldehyde **1**, ammonia and two moles of ethyl acetoacetate **2** combine on heating in an aprotic solvent to fully substituted 1,4-dihydropyridine **3**.

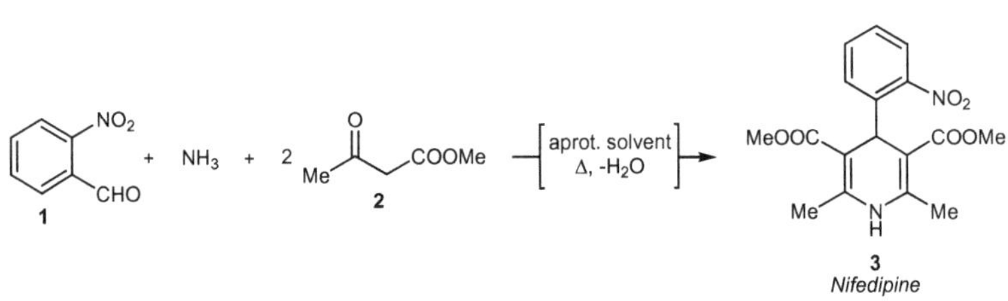

Scheme 6.1 Four-component Hantsch reaction in the synthesis of *nifedipine* **3**

Scheme 6.2 Synthesis of *tropinone* **7** by the three-component Mannich reaction

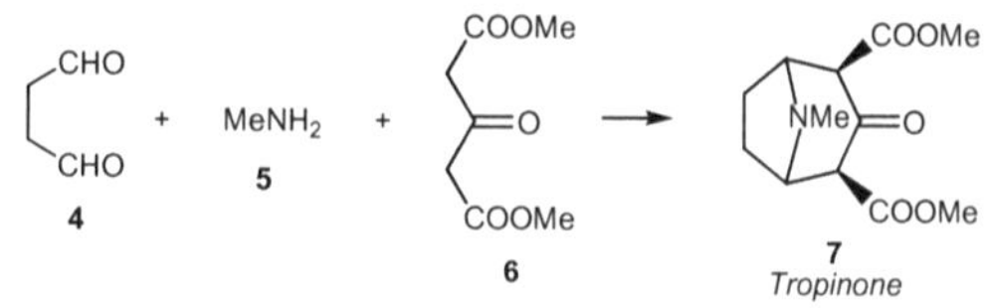

We already described the application and mechanism of the Mannich reaction in Sect. 4.4.2. An elegant application of this reaction is Robinson synthesis of the bicyclic alkaloid *tropinone* **7** starting from simple building blocks **4–6** and completed with great atom economy (Scheme 6.2) [7].

6.2.2 Ugi Multicomponent Reactions

Ugi reactions are characterized by the use of isocyanides as a strong electrophilic component. In the next example, formaldehyde, dimethylamine and 2,5-dimethylphenylisocyanide are combined in one step to *xylocaine* **11**, an important local anesthetic (Scheme 6.3) [8].

The next example explains why most multicomponent reactions are not easily amenable to rational retrosynthetic analysis.

Example 6.1 Recognition of **TM 6.1** as an anilide of *N*-alkylated and *N*-acylated α-amino acid valine suggests "obvious" retrosynthetic analysis. Retrosynthesis leads over three consecutive disconnections of C–N bonds to valine and the three additional building blocks (Scheme 6.4).

A corresponding synthesis can be proposed stating from valine, which is in the first step acylated by activated benzoic acid, and then alkylated and in the last step coupled with aniline to amide **TM 6.1**.

A completely different approach offers an Ugi four-component reaction, whose mechanism is presented in Scheme 6.5.

On mixing isobutyraldehyde, ethylamine, phenyl isocyanide and benzoic acid, all steps in the above reaction occur in "one-pot." Intermediates **I–III** are not isolated, and the presence of some of them was confirmed by spectroscopy.

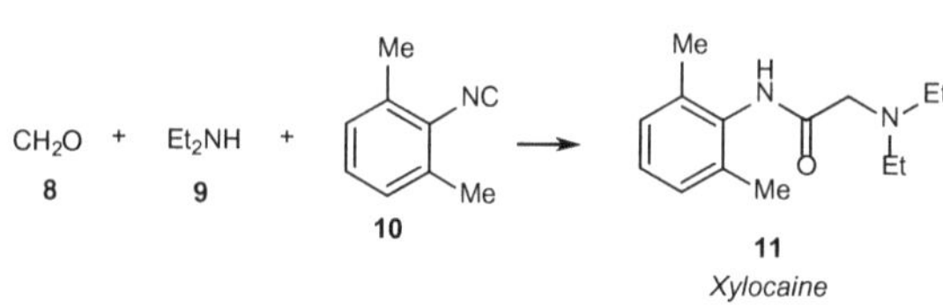

Scheme 6.3 Three-component Ugi reaction in the synthesis of *xylocaine* **11**

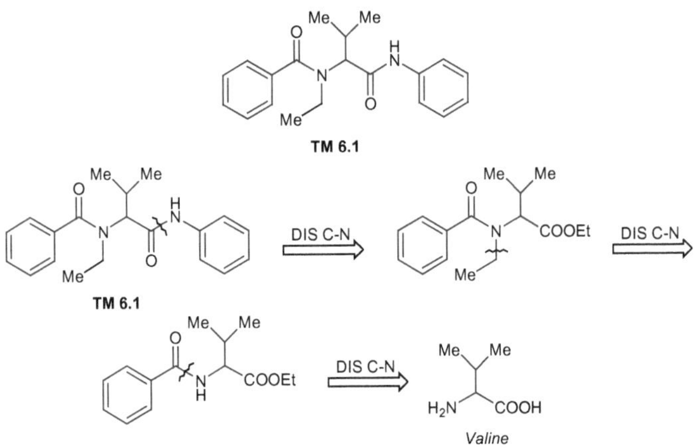

Scheme 6.4 "Obvious" retrosynthesis of **TM 6.1**

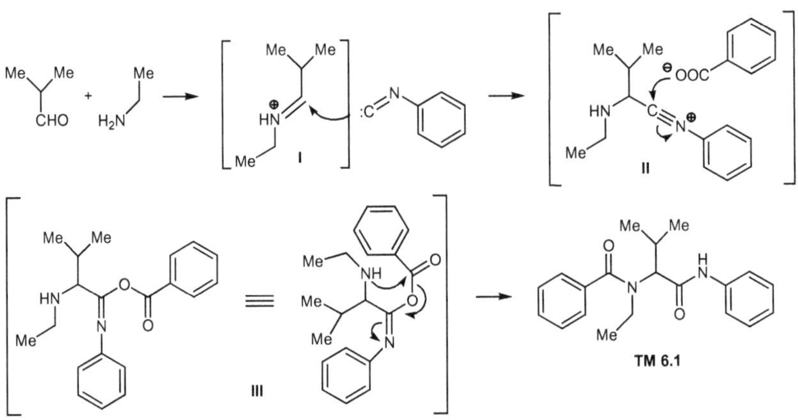

Scheme 6.5 Ugi four-component synthesis of **TM 6.1**

Immonium ion **I** is formed in the benzoic acid-catalyzed condensation of aldehyde and amine. This intermediate reacts with the strongly electrophilic C-atom of phenyl isocyanide forming a C–C bond in nitrilium intermediate **II**. The formal triple bond in **II** is highly susceptible to nucleophilic attack of weak nucleophiles such as the carboxylate anion forming intermediary acyl-imine **III**, an analog of an anhydride. This unstable species spontaneously rearranges into stable diamide **TM 6.1**. Such a complex sequence of bond-forming and bond-breaking events is nearly impossible to conceive retrosynthetically.

6.3 Parallel Synthesis and Combinatorial Chemistry

Parallel synthesis and *combinatorial chemistry* are two closely related concepts, result of intentions to automatically perform more synthetic reactions. *Parallel synthesis* enables preparation of a set of defined compounds in a number of physically separated reaction vessels or micro-departments. *Combinatorial chemistry* instead uses a combinatorial process for preparation of a large number of compounds from a defined set of building blocks. Combinatorial chemical synthesis generates a large number of compounds, so-called libraries, at the same time and in predictable mode.

Productive methods in organic synthesis, also called technologies, are born in the pharmaceutical industry, then developed in academic institutions and applied in broad synthetic fields. The pharmaceutical industry was the first to suffer from slow organic syntheses as the "narrow throat" in the process of the discovery of a new drug. It was estimated that synthetic chemists are able to prepare one preparation–one compound by the "classic method," approximately 50–100 compounds/year, with a cost $5000–7000 per compound. An automated combinatorial synthesis can generate up to 100,000 compounds/year per chemist, with an average price of $5–7 per compound. The many new compounds in the pharmaceutical industry substantially enhance the chance to identify *lead compounds* (LCs) on the route to a *new drug entity* (NDE). The availability of a large number of compounds prompted *high-throughput screening* (HTS) of biological activity. There are numerous examples of successful parallel synthesis of heterocyclic compounds, e.g., for indoles [9] and benzofuranes [10]. Here we refrain from a detailed discussion of combinatorial syntheses combined with high throughput screening of biological profiles. From the large number of monographs and review articles on this topic here, we select citing only a few [11–15].

Parallel synthesis does not require sophisticated equipment and software support. It is therefore the method of choice for the preparation of a dozen compounds with well-defined purity for biological testing in academia and industry. Some examples of parallel synthesis are presented to support this. Typically they are performed in 1–30 ml reactors with shaking or magnetic stirring and cooling/heating, which allows working temperatures between −78 and 250 °C.

Example 6.2 Using parallel synthesis, a series of 4(3H)-pyrimidinone derivatives **IV** was prepared. According to previous knowledge from these compounds, anti-HIV activity was expected [15].

From **IV** as the starting compound, where R′ = H, Y = OH in R″ and R‴ = Me, a virtual library for parallel synthesis was generated (Fig. 6.1), and then, by combinatorial synthesis, a library of 522 compounds **V** was prepared (Scheme 6.6).

The synthetic protocol characterizes the use of solid support with a carbonate unit and flexible linker (Merrifield resin) to construct library **V**. All compounds are screened on binding to the *non-nucleoside binding pocket* (NNBP). In vitro binding constants >5 kcal/mol are regarded as indicators of potential in vivo anti-HIV activity.

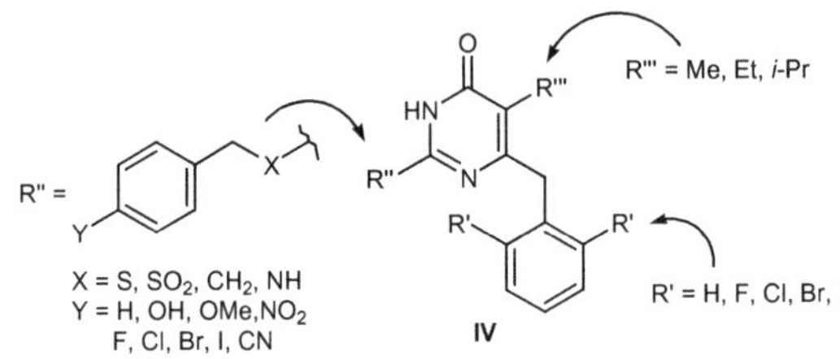

Fig. 6.1 Fragments used for generation of a virtual library of 4(3H)-pyrimidinones **IV**

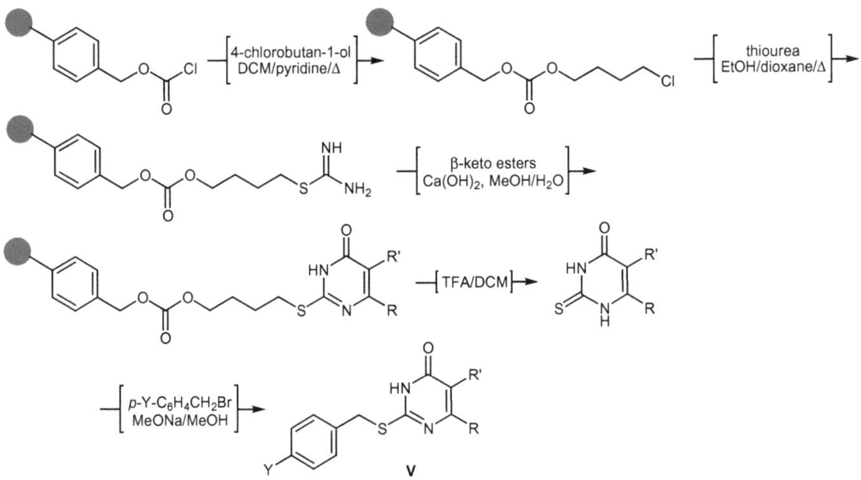

Scheme 6.6 Preparation of the library of 2-thio-3(4H)-pyrimidinones **V** by parallel synthesis

Example 6.3 On cyclization of 1,2- and 1,3-hydroxyalkylazides in the presence of Lewis acid, the library of dihydrooxazolines **VI** is prepared by one-step parallel syntheses (Scheme 6.7) [16].

This method characterizes the use of polymer-bound phopshine as a scavenger of the excess of hydroxyl azide in the reaction solution. The library of ca. 60 compounds is prepared with ca. 40 % average yield and ca. 90 % purity of isolated compounds meeting the criteria for *high-throughput screening* of their biological activity.

Scheme 6.7 Parallel synthesis of dihydrooxazoline derivatives **VI**

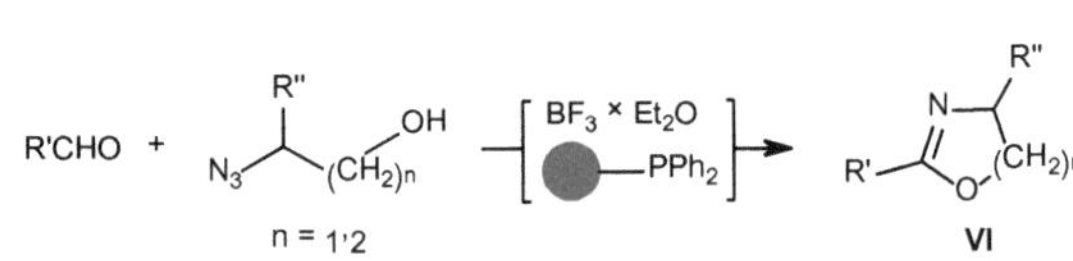

6.4 Mechanochemistry in Organic Synthesis

According to the accepted definition, mechanochemical reactions are those performed by direct absorption of mechanic energy [17]. Chemical reactions promoted by mechanical force are ever more frequently explored in organic synthesis [18–21]. It was demonstrated that mechanochemical reactions can be performed in the absence of any solvent and therefore meet the challenges of modern chemistry that require economic use of energy and auxiliary materials and fulfill the requirements for environmentally acceptable processes (Sect. 1.3.2).

Mechanochemical reactions are performed in planetary mills, moving reactors where the collision of metal balls of a defined radius transfers mechanical energy to the reacting species, prompting their chemical reactivity. Since no solvent is used, isolation of the products is regularly simplified. Mechanochemical reactions have been traditionally explored in the synthesis of organometallic compounds, porous materials, salts and cocrystals. More recently, this technology has found application in the production of active pharmaceutical compounds, known as *active pharmaceutical ingredients*, APIs [22].

The next two examples illustrate the successful application of mechanochemical conditions in organic synthesis.

Example 6.4 The Suzuki–Miyaura reaction is today the most valuable method for coupling aryl components into biaryl structures. While traditional protocols for this reaction require large quantities of expensive nonpolar aprotic solvents and often proceed with low yields, the mechanochemical variant of this reaction affords biaryls **14** in high yields (Scheme 6.8) [22].

Interestingly, when in substrates **13** Hal=Cl, they proved completely nonreactive under standard conditions in apolar organic solvents, revealing insufficient Pd-catalyzed activation of the aryl–Cl bond. This energetically demanding step is surmounted in the mechanochemical variant, optimizing some parameters such as rotation speed and ball diameter [23].

Example 6.5 This example illustrates non-catalytic Knoevenagel condensation completed under mechanochemical conditions (Scheme 6.9) [24].

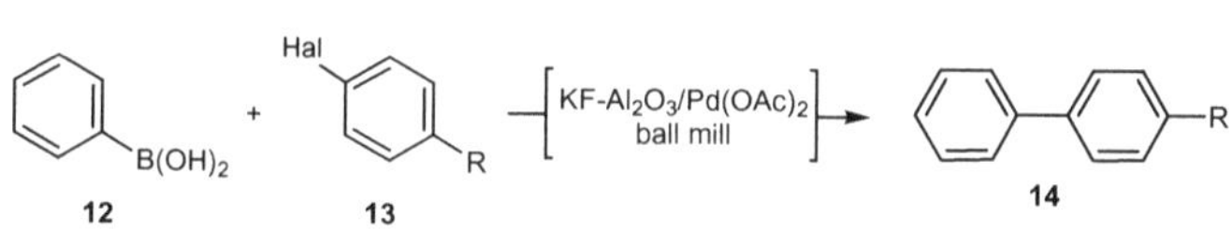

R	Hal	yield/%
$COCH_3$	Br	94
$COCH_3$	I	97
OCH_3	Br	93
OCH_3	I	30

Scheme 6.8 Example of the Suzuki–Miyaura reaction in the mechanochemical variant

Scheme 6.9 Knoevenagel reaction in the mechanochemical variant

*Conditions:

i) 400 rpm, 60 min
ii) 800 rpm, 60 min

R	*i)* yield/%	*ii)* yield/%
H	31	97
4-Me_2N	2	68
2,3-OMe	6	94
4-Cl	10	99
4-OH	77	96

The authors revealed that aromatic aldehydes **15** react with malononitrile, affording geminal dicyano derivatives **17** in high yield. Interestingly, using 15-mm-diameter steel balls, the yields increased substantially when the rotation was doubled from 400 to 800 rpm.

6.5 Organic Synthesis Promoted by Microwave Radiation

Microwaves act as a high-frequency electric field; hence, they warm up all materials possessing a mobile electric charge, such as polar molecules in an apolar solvent. Polar solvents are also warmed up by microwaves since their molecules are forced to rotate in the electric field and lose energy by collisions.

In the last decades, microwaves were occasionally used in chemical laboratories and soon found application in organic synthesis [25–28]. Microwave radiation enables heating of reactants without heating of the reaction vessel by heating a sample through its volume and not over the walls of the reactor. It allows uniform heating and great energy savings. Moreover, different compounds transform microwave radiation into energy to different extents, enabling faster *and selective heating of the components* of the reaction medium [29]. The advantages of microwave heating over a traditional oil bath or vapor jacket are speeding up the reaction, milder reaction conditions, higher chemical yields, lower energy consumption and a change in chemical selectivity as compared to conventional heating.

The following examples illustrate the application of microwave technology in organic synthesis.

Example 6.6 The Heck reaction, coupling aryl halides with a vinyl group, is promoted by microwaves and in situ prepared Pd(II) complex with P(*ortho*-tolyl)$_3$ ligand in ionic liquid 1-butyl-3-methylimidazolium hexafluorophosphate (bmim) [30].

Use of ionic liquids in microwave-promoted reactions is particularly preferred because of the fast heating of the medium, 10 °C/s, and very small enhancement of pressure, a problem often encountered in microwave reactors. The reaction outlined in the Scheme 6.10 is very fast, for X=I completed in 5 min and for X=Br in 20 min, and for less effective phosphine ligands it needs up to 45 min. Besides, it

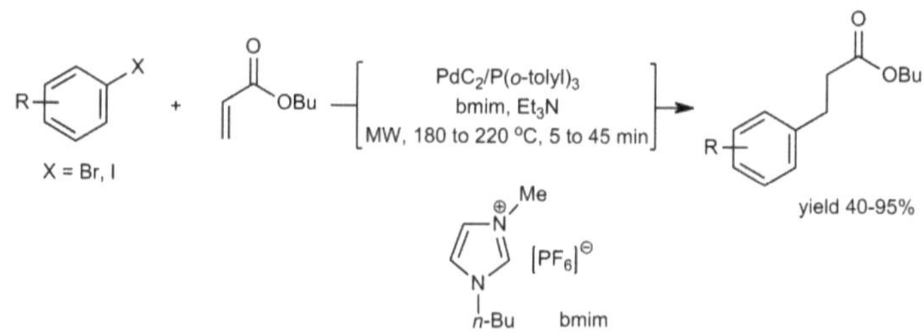

Scheme 6.10 Heck reaction promoted by microwave radiation

was possible to recycle the catalytic system ionic liquid-($PdCl_2$/bmim) five times, maintaining a high reaction rate and yield.

High pressure in microreactors is often solved using dry media or media without solvents. This process starts by adsorption of reagents or catalyst on one or more inorganic supports transparent to microwaves, such as silicates, alumosilicates or clays, or acts as a strong absorber of microwaves such as graphite.

Example 6.7 This example demonstrates the promoting effect of microwaves on an asymmetric catalytic reaction. Alkylation of dimethyl malonate carbanion is completed by a racemic allylic alcohol derivative in the presence of the complex of the Mo(III) ion with chiral bidentate nitrogen ligand **VII**. By this method, the key intermediate in the synthesis of the oral HIV inhibitor *tipranavir* was prepared in 95 % yield and 94 % e.e. (Scheme 6.11) [31]. Note that the racemic substrate is completely transformed into one enantiomer of the alkylated product revealing catalytic racemization of the non-reactive enantiomer!

As yet, nearly all important synthetic reactions are performed under microwave conditions on a laboratory scale with the aim to become the method of choice in kilo laboratories of the pharmaceutical industry using commercially available microwave reactors [32].

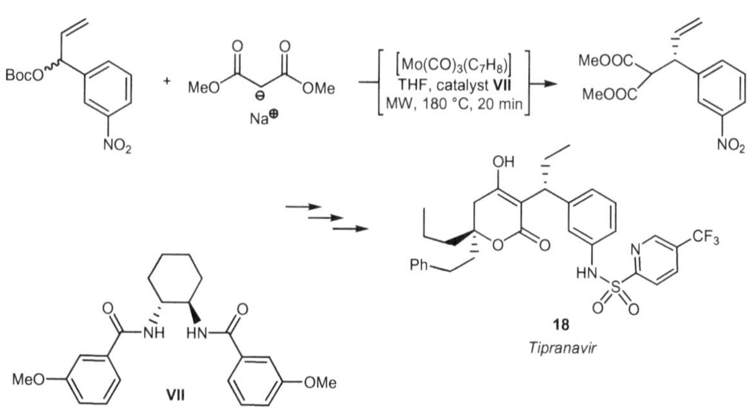

Scheme 6.11 Enantioselective alkylation in microwave reactor

6.6 Syntheses in Ionic Liquids

In Sect. 6.5 we described an example of the use of ionic liquid in combination with microwave technology. Now we briefly discuss reactivity in ionic liquids and examples of their application as a reaction medium in organic synthesis.

Ionic liquids are salts in liquid state with a melting point around or just above room temperature. Usually these are the salts between organic cations and complex inorganic anions. As a rule they are colorless, non-volatile liquids of low viscosity and therefore easy to handle. Moreover, ionic liquids demonstrate high solvation of dissolved molecules, are easy to recycle and can be stored over longer periods without decomposition, rendering them attractive solvents for laboratory and industrial application. Ionic liquids are inconvenient for distillation still their recycling can be elegantly completed. It includes separation of solids by filtration, washing of the collected filter cake with toluene and evaporation of toluene from the ionic liquid, which can be reused.

The chemical stability, negligibly low volatility and easy recycling have placed ionic liquids at the top of "green chemistry" solvents [33–35]. Figure 6.2 presents the structures of the ionic liquids most frequently used as solvents in organic synthetic chemistry.

Combining the cationic component, usually quaternary ammonium cations of nitrogen heterocycles and inorganic or organic anions, numerous ionic liquids are available (Fig. 6.3).

The application of pyridinium- and imidazolium-based ionic liquids in organic synthesis is presented in the next examples.

Example 6.8 Using pyridinium-based ionic liquid (bmpy)Cl and $AlCl_3$ as the catalyst, an interesting insertion of acetylene into the C–Cl bond of acyl chlorides was achieved (Scheme 6.12) [36].

An alkaline medium is controlled by the formation of [bmpy]$AlCl_4$, which selectively promotes the reaction without the formation of side products. Acetylene is introduced as a continuous gas stream, and products **VIII** are isolated by distillation in over 90 % yield.

Example 6.9 Catalytic asymmetric Michael-type addition of aldehydes to nitroolefins was successfully completed in imidazolium-based ionic liquid **XI** (Scheme 6.13) [37].

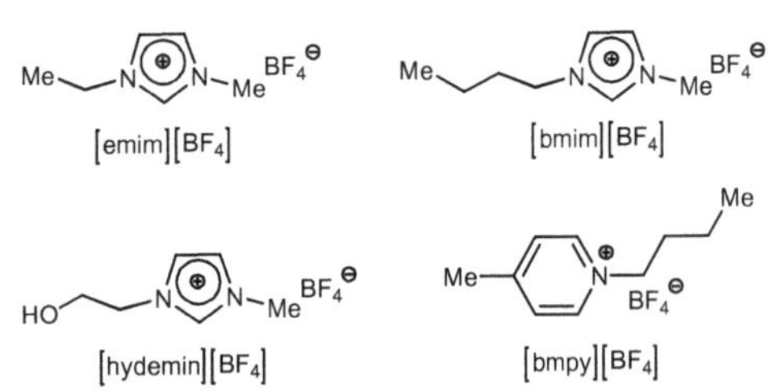

Fig. 6.2 Most frequently used ionic liquids for solvents in organic synthesis

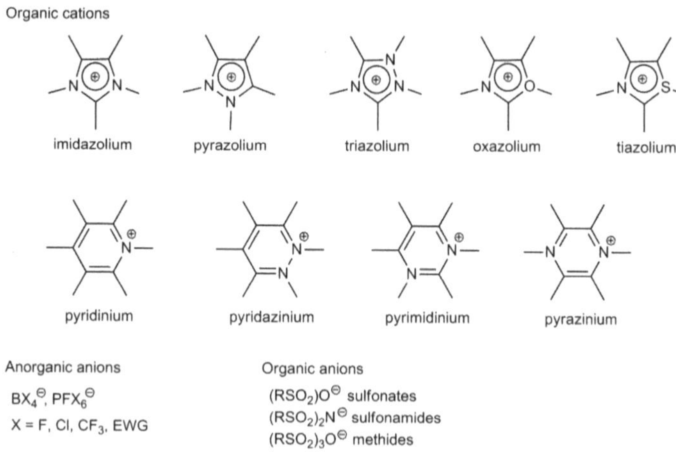

Fig. 6.3 Organic cations and inorganic or organic anions in ionic liquids

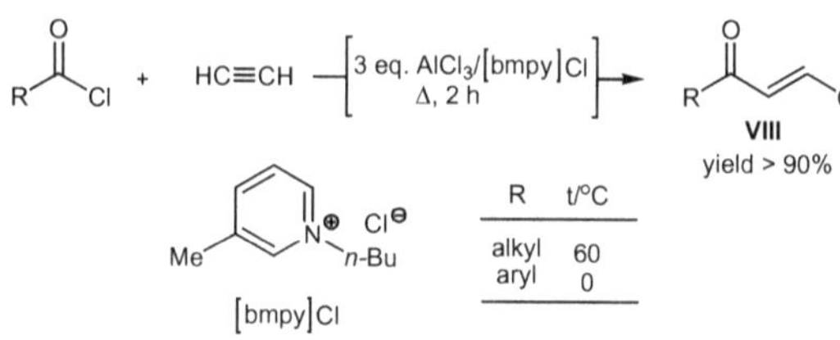

Scheme 6.12 Preparation of β-chlorovinyl ketones **VIII** in ionic liquid

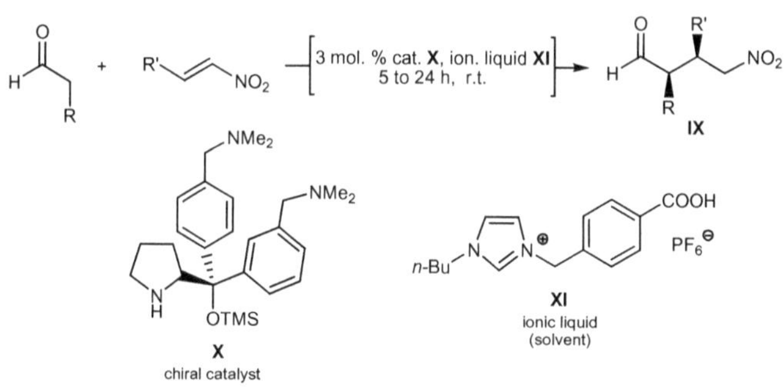

Scheme 6.13 Asymmetric Michael addition in ionic liquid

Yields of **IX** in this reaction vary between 85–95 %, *syn/anti* ratio 96:4, and the e.e. is regularly over 95 %. It is peculiar that chiral catalysts act after 12 recyclings without significant loss of enantioselectivity.

In conclusion, a broad application of ionic liquids at the laboratory scale still awaits extension to the industrial scale. Because of their properties described this chapter, ionic liquids are presently being tested as the solvents of choice for many industrial processes compatible with the standards of environmental protection.

References

1. Wender PA, Handy ST, Wright DL (1997) Chem Ind 765–772
2. Ugi I, Steinbruckner C (1961) Chem Ber 94:734–742
3. Stecker A (1850) Liebigs Ann Chem 75:27–45
4. Hantsch A (1882) Liebigs Ann Chem 215:1–82
5. Tramontoni M, Angiolini L (1994) Manich bases-chemistry and use. UCR Press, Boca Raton
6. Bossert F, Meyer H, Wehinger R (1981) Angew Chem Int Ed 20:762–769
7. Robinson R (1917) J Chem Soc 111:876–899
8. Ugi I, Steinbruckner C (1959) DE-B,1,103,337 (for A. B. Astra)
9. Heravi MM, Baghernejad B, Oskooie HA, Hekmatshoar R (2008) Tetrahedron Lett 49:6101–6103
10. Chen RE, Wang YL, Chen ZW, Su WK (2008) Can J Chem 86:875–880
11. Ley SV, Baxendale IR (2002) Nat Rev Drug Discov 1:573–586
12. Nicolau KC, Hanko R, Hartwig W (2002) Handbook of combinatorial chemistry, vols 1 and 2. Wiley-VCH, London
13. Bannworth W, Hinzen B (2006) Combinatorial chemistry-from theory to application, 2 edn. Wiley-VCH, London
14. Taber DF (2008) Organic synthesis: state of the art 2005–2007. Wiley Interscience, London
15. Petracci E (2006) Archivoc 452–478
16. Cibašek P, Bevk D, Pirc S, Stanovnik B, Svete J (2006) J Comb Chem 8:95–102
17. McNaught AD, Wilkinson A (1997) IUPAC compendium of chemical terminology, 2nd edn. Blackwell Scientific Publications, Oxford
18. Stolle A, Ondruschka B, Krebs A, Bolm C (2012) Catalyzed organic reactions in ball mills. In Andersson PG (ed) Innovative catalysis in organic synthesis. Wiley-VCH, London
19. Ma X, Yuan W, Bell SEJ, James SL (2012) Chem Soc Rev 41:413–447
20. Margitić D (2005) Kem Ind 54:351–358
21. Chauhan P, Chimni SS (2012) Beilst J Org Chem 8:2132–2141
22. Friščić T, Halasz I, Strukil V, Eckert-Maksić M, Dinnebier RE (2012) Croat Chem Acta 88:367–378
23. Schneider F, Stolle A, Ondruschka B, Hopf H (2009) Org Proc Res Dev 13:44–48
24. Trotzki R, Hoffmann MM, Ondruschka B (2008) Green Chem 10:767–772
25. Loupy A (2006) (ed) Microwaves in organic synthesis. Wiley-VCH, Weinheim
26. de la Hoz A, Diaz-Ortiz A, Moreno A (2005) Chem Soc Rev 164–178
27. Strauss C, Trainor R (1995) Aust J Chem 48:1665–1674
28. Zrinski I, Eckert-Maksić M (2005) Kem Ind 54:469–476
29. Kappe CO (2004) Angew Chem Int Ed 43:6250–6284
30. Vallin KSA, Emilsson P, Larhed M, Hallberg A (2002) J Org Chem 67:6243–6246
31. Trost BM, Andersen NG (2002) J Am Chem Soc 124:14320–14321
32. Lehmann H, LaVecchia L (2010) Org Proc Res Dev 14:650–656
33. Siodmiak T, Marszal MP, Proszowska A (2012) Ionic liquids: a new strategy in pharmaceutical synthesis. Mini-Reviews Org Chem 9:203–208
34. Cserjesi P, Belafi-Bako K, Nemestothy N, Gubicza L (2008) Hung J Ind Chem 36:27–34
35. Sowmiah S, Subbiah CI, Chu Y-H (2012) Ionic liquids for green organic synthesis. Current Org Synth 9:74–95 Bentham Science Publishers
36. Snelders DJM, Dyson PJ (2011) Org Lett 13:4048–4051
37. Sarkar D, Bhattarai R, Headley AD, Ni B (2011) Syn Lett 12:1993–1997

Chapter 7
Retrosynthetic Consideration of Heterocyclic Structures

Abstract An overview of two-heteroatom reagents and the retrosynthesis of five- to seven-membered heterocycles to building blocks with heteroatoms O, N and S in the 1,2-, 1,3- and 1,4-position is presented. Their use in the cyclization of medium-large rings is exemplified. Retrosyntheses and syntheses of biologically active heterocyclic compounds, *lanycil, sulfisoxazole, sulfamethoxazole, primicarb, hydralazine* and *glutethimide*, are presented. Cyclizations to three- to seven-membered heterocycles with one or more heteroatoms in the ring obey the Baldwin rules. The mechanistic origin of the *favored route* and heuristic value of the Baldwin rules are discussed.

7.1 Introduction

In this chapter, the retrosynthetic analysis and proposals for the synthesis of heterocyclic compounds with one to three heteroatoms in five- or six-membered heterocyclic rings are presented. Such heterocyclic rings are constituents of thousands of biologically active compounds. Compendia on the synthesis and structural properties of all classes of heterocyclic compounds are available [1–4]. Numerous syntheses of heterocycles and descriptions of their physico-chemical properties are presented in separate chapters of organic chemistry handbooks cited in the introduction chapter of this book.

We consider the key building blocks with two heteroatoms as the results of contemporaneous disconnections of two C-heteroatom bonds in the heterocyclic ring. This retrosynthetic step corresponds to cyclization by contemporaneous formation of two C-heteroatom bonds, a usual step in the synthesis of heterocyclic compounds. Depending on 1,2-, 1,3- or 1,4 relation of hetereoatoms in the heterocyclic ring, the corresponding building blocks possess heteroatoms bound directly in the germinal or vicinal position (Fig. 7.1).

V. Šunjić and V. Petrović Peroković, *Organic Chemistry from Retrosynthesis to Asymmetric Synthesis*, DOI 10.1007/978-3-319-29926-6_7

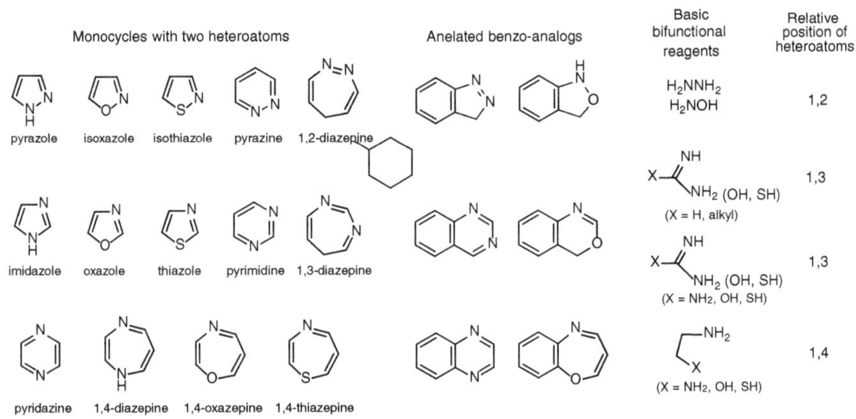

Fig. 7.1 Examples of heterocycles with two heteroatoms and corresponding bifunctional reagents

7.2 Retrosynthetic Considerations, Examples

It is interesting that the retrosynthetic approach to heterocycles is not considered in monographs or review articles on the synthesis of individual classes of heterocyclic compounds. The examples that follow illustrate the retrosynthetic approach to the five- and six-membered heterocyclic structures often encountered in natural or synthetic biologically active compounds. Examples 7.2, 7.3, 7.6, 7.11 and 7.12 demonstrate syntheses of biologically active compounds developed to large-scale production by chemists at PLIVA Co. (Croatia).

Example 7.1 Consider the retrosynthesis and then propose the synthesis of *lanicyl* **TM 7.1**, an inhibitor of the synthesis of chlorophyll.

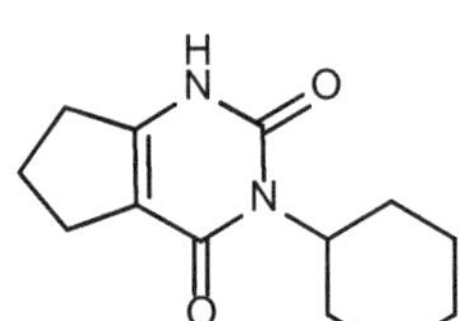

TM 7.1
Lanicyl

We start the retrosynthetic analysis with the following observations:

- in the six-membered 1,3-diazin-dione, an element of urea is incorporated
- a second carbonyl group is present in the amide of cyclopentene carboxylic acid
- the enone C=C bond is part of the aza-allylic system and can migrate to form an exocyclic C=N bond
- cyclohexylamine, easily available from aniline, is incorporated into the imide unit

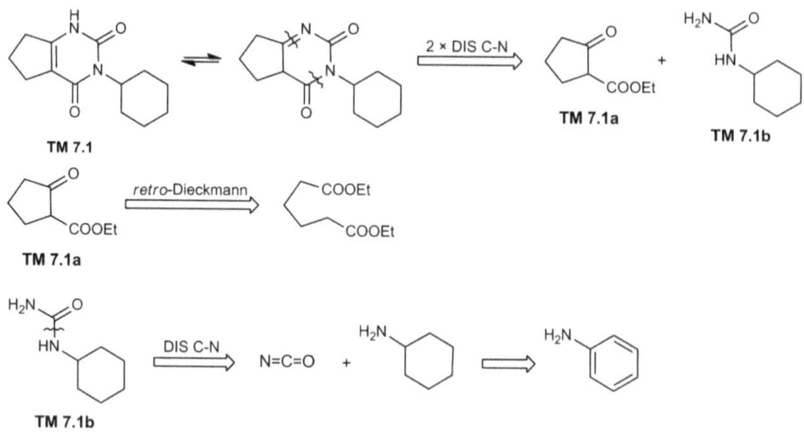

Scheme 7.1 Proposal for the retrosynthetic analysis of *lanicyl* **TM 7.1**

By the first retrosynthetic step, we disconnect two "strategic" C–N bonds in the heterocyclic ring of imino-tautomer of **TM 7.1** to recognizable building blocks (Scheme 7.1).

The result of the double C–N disconnection in the heterocyclic ring is the two target molecules of the next generation, α-keto-cyclopentane carboxylic acid **TM 7.1a** and *N*-cyclohexylurea **TM 7.1b**. On *retro*-Dieckmann disconnection, **TM 7.1a** leads to diethyl adipate (Sect. 4.3.3, Scheme 4.23) and **TM 7.1b** affords the isocyanate anion and cyclohexylamine.

In the proposed synthesis, silicon tetraisocyanate is used as a convenient source of the isocyanate anion (Scheme 7.2) [5].

Example 7.2 Complete the retrosynthetic analysis and then propose the synthesis of the antibacterial sulfonamide *sulfisoxazole* **TM 7.2.**

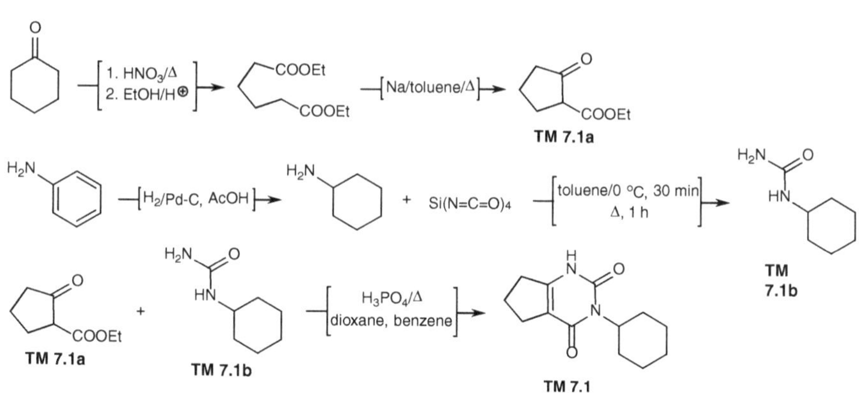

Scheme 7.2 Proposal for the synthesis of *lanicyl* **TM 7.1**

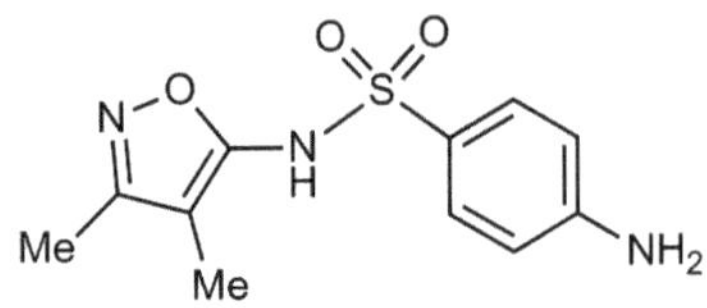

TM 7.2
Sulfisoxazole

The first FGI is needed to introduce acetyl as the amino protecting group. Disconnection of the N–S bond in the sulfonamido group of **TM 7.2a** leads to maximal simplification and results in trisubstituted isoxazole **TM 7.2b** and *para*-acetylaminobenzensulfonyl chloride **TM 7.2c**, a commodity product of the chemical industry. Particularly instructive is the disconnection of the C–O bond in **TM 7.2b** as *retro*-cyclization with participation of the NH_2 group (Scheme 7.3).

This disconnection suggests that in the synthetic direction, the OH group of oxime **TM 7.2d** approaches the cyano group, completing the cyclization. Presentation of both functional groups as acyclic cyano-oxime suggests interconversion to 3-cyanobutan-2-one and then to 3-chlorobutan-2-one, easily available by chlorination of butan-2-one. Acetanilide is an available commodity used in the production of *para*-chlorosulfonyl derivative **TM 7.2c**.

A proposal for the synthesis is given in Scheme 7.4.

Chlorination of butanone is industrially performed in the gas phase with continuous separation of isomeric 1-chlorobutanone, followed by substitution of the chloro for the cyano group. Production of *para*-acetylaminobenzensulfonyl chloride

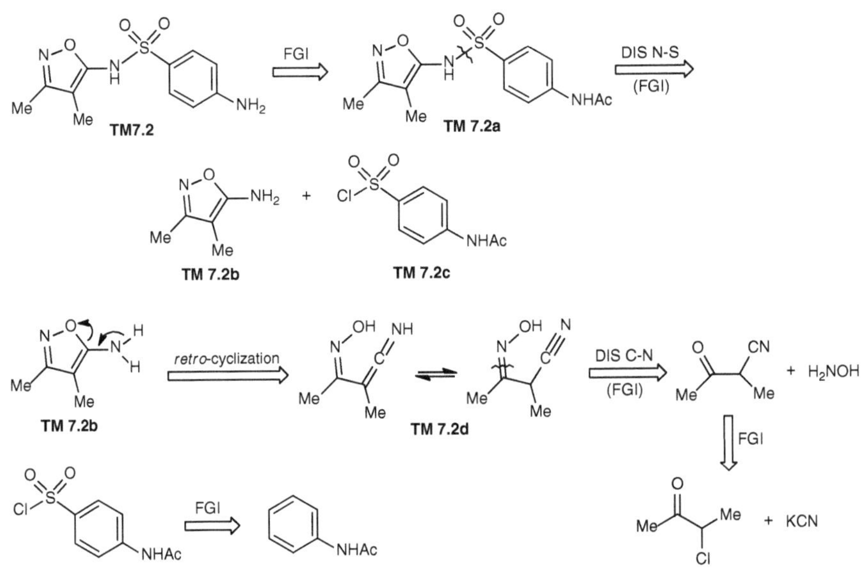

Scheme 7.3 Retrosynthetic analysis of *sulfisoxazole* **TM 7.2**

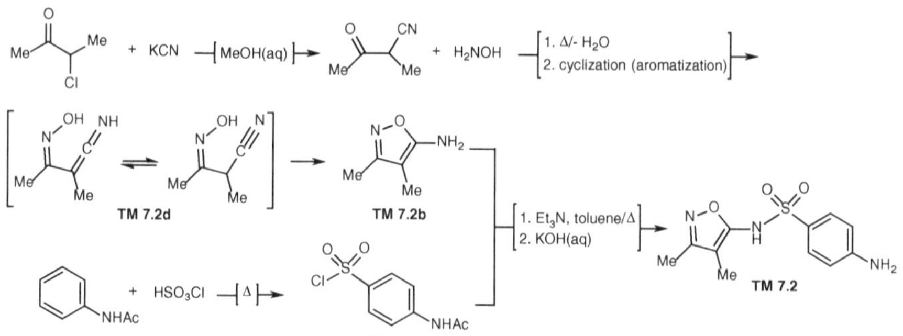

Scheme 7.4 Proposal for the synthesis of *sulfisoxazole* **TM 7.2**

(PAS) uses an excess of chlorosulfonic acid, which also serves as the solvent. Sulfonation of heterocyclic amine **TM 7.2b** is performed in an organic solvent in the presence of amine as acid scavenger. The industrial method for the sulfonation of amines prefers *Schotten-Baumann* conditions, where the reaction is completed in aqueous medium, an acetone-water or methanol-water solvent mixture, with continuous addition of one mole of hydroxide.

The next examples nicely illustrate how changes of certain substituents on the heterocyclic ring require a completely different approach to structurally related target molecules.

Example 7.3 Propose the retrosynthesis and then synthesis of the antimicrobial agent *sulfamethoxazole* **TM 7.3**.

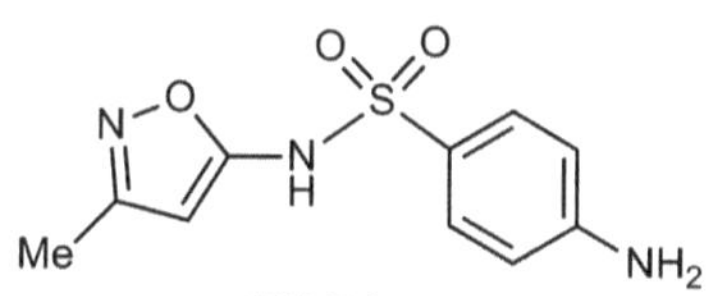

TM 7.3
Sulfamethoxazole

The first two disconnection steps correspond to those for **TM 7.2** and afford **TM 7.3b** and **TM 7.3c** (Scheme 7.5).

For the next two retrosynthetic steps, we need new knowledge related to the *Hofmann rearrangement* of amides to amines with the loss of one C atom, discussed in Sect. 8.3. This rearrangement enables the introduction of an amino group in **TM 7.3b** starting from the corresponding carboxylic amide. The high value of this retrosynthetic approach becomes obvious on the next disconnection steps, *retro*-cyclization to β-ketoxime and its interconversion to easily available diketo-carboxylic acid.

Prompting by retrosynthesis following the synthesis of **TM 7.3** is proposed (Scheme 7.6).

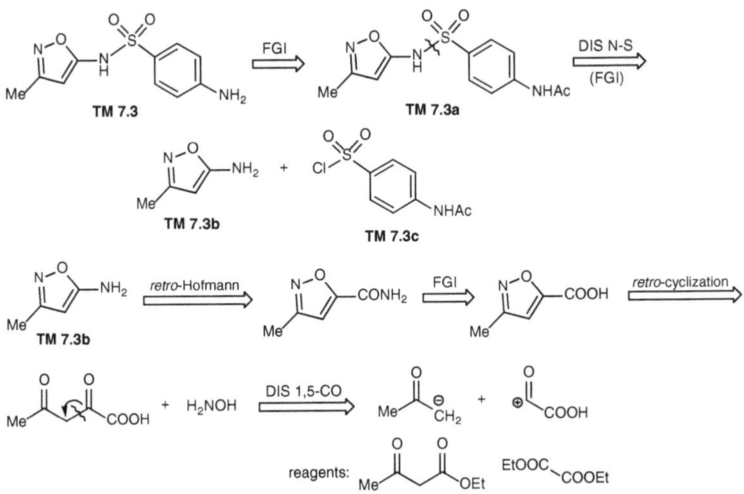

Scheme 7.5 Retrosynthetic analysis of *sulfamethoxazole* **TM 7.3**

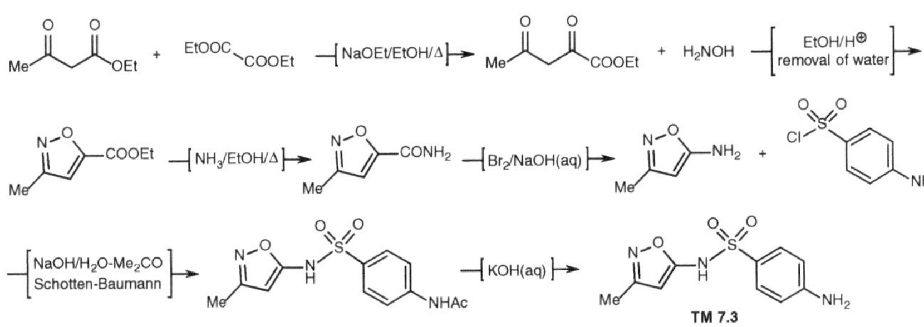

Scheme 7.6 Proposal for the synthesis of *sulfamethoxazole* **TM 7.3**

In the first steps, ester condensation to diketo acid, preparation of oxime and cyclization are completed under standard conditions. Ester ammonolysis is followed by Hofmann rearrangement of amide to heterocyclic amine. Sulfonation of amine under Schotten-Baumann conditions, in a water-organic solvent mixture, is based on an interesting "trick;" hydroxide is added at a controlled rate to scavenge the hydrochloric acid formed in sulfonation and to maintain the neutral pH of the aqueous solution of heterocyclic amine as a stronger nucleophile than water, avoiding the hydrolysis of sulfonyl chloride!

Note It is important to observe that the ester of α,γ-diketo acid chemoselectively affords only γ-oxime since the α-keto group is completely enolized (Scheme 7.7). It is also worth mentioning that inversion of the reactivity of two carbonyl groups is possible on lactonization. Within stable enol-lactone, the α-carbonyl group then reacts.

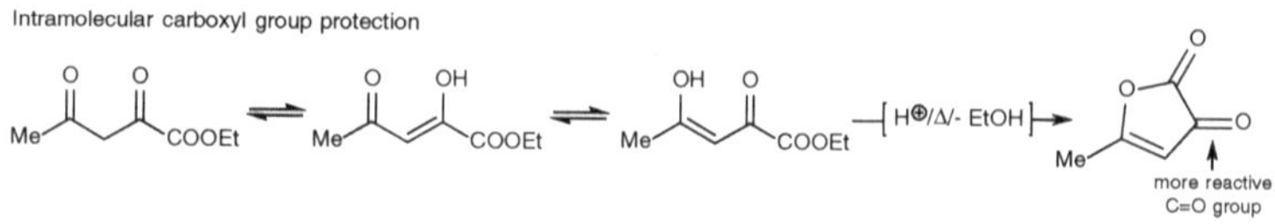

Scheme 7.7 Inversion of the reactivity of the carbonyl groups in enol-lactone

In the next examples, we consider the retrosynthetic analysis of some biologically active compounds that comprise two N atoms in the six-membered ring.

Example 7.4 Perform the retrosynthetic analysis and then suggest the synthesis of the insecticide *primicarb*, **TM 7.4**.

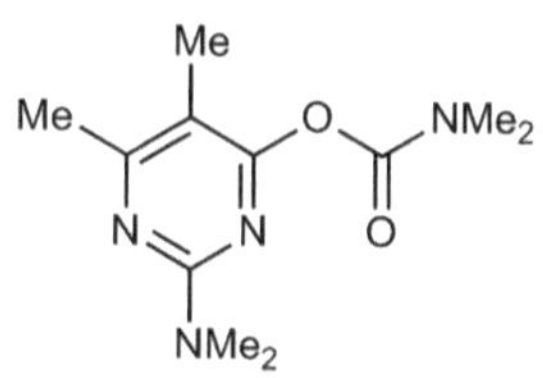

TM 7.4
Primicarb

The target molecule characterizes the presence of the carbamate group on the dimethylaminopyrimidine ring. This suggests first FGI, C–O disconnection of carbamate to **TM7.4a** and chlorodimethyl formamide (*N,N*-dimethlychloromethanamide). This highly reactive building block is available in multi-kilo quantities from phosgene and dimethylamine (Scheme 7.8).

In the target molecule of the next generation **TM 7.4a**, two C–N disconnections lead to the available building blocks, the Me-derivative of ethyl acetoacetate **TM 7.4b** and *N,N*-dimethyl guanidine **TM 7.4c**. The first one is available by C-methylation and the second by the addition of dimethylamine to cyanamide.

According to this analysis, the synthesis is presented in Scheme 7.9 where the usual reaction conditions for industrial-scale production are indicated. The process

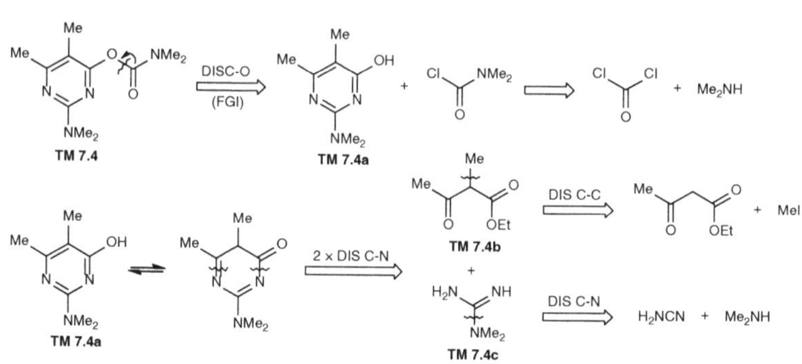

Scheme 7.8 Retrosynthetic analysis of *primicarb*, **TM 7.4**

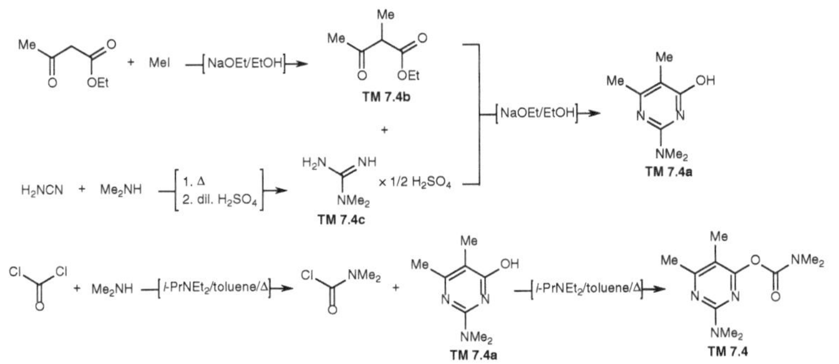

Scheme 7.9 Proposal for the synthesis of *primicarb*, **TM 7.4**

is convergent starting from the methylation of ethyl acetoacetate to **TM 7.4b** and the thermally promoted addition of dimethylamine on cyanamide in solvent-free medium and crystallization of sulfate of **TM 7.4c**.

In the penultimate step, these two building blocks cyclize to **TM 7.4a**.

Example 7.5 Complete the retrosynthesis and then propose the synthesis of **TM 7.5**, the odor principle of green paper.

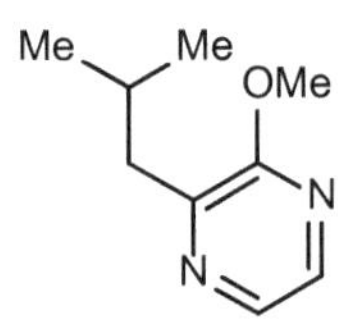

TM 7.5

First, we recognize the methyl derivative of the enolic form of pyridiazin-2-one (Scheme 7.10). Assuming O-methylation in the last step of the synthesis, we start the retrosynthetic analysis by Me–O disconnection (FGI). The enolic form of the six-membered heterocycle is in equilibrium with the less stable keto form **TM 7.5a**, amenable to disconnection of the two C–N bonds. This step affords two acyclic precursors, **TM 7.5b** and glyoxal. **TM 7.5b** is the amide of 2-aminohomovaleric acid (2-amino-4-methylpentanoic acid), a homolog of natural 2-aminovaleric acid, which is not available in nature. It is conveniently prepared by Strecker synthesis starting from isobutyraldehyde. This method for the production of α-amino acids on an industrial scale characterizes the use of ammonium carbonate and the intermediary formation of the hydantoine derivative (Sect. 5.2.1, Scheme 5.8). Under controlled basic conditions, the hydantoine derivative hydrolyzes to the amide of α-amino carboxylic acid **TM 5.7b**. Due to the limited reactivity of the amide, thermal cyclization with glyoxal requires continuous elimination of water by either azeotropic distillation with a Dean-Stark water separator or irreversible binding to molecular sieves. Selective C-methylation of enole is preferably completed by diazomethane to avoid N-alkylation with MeI or Me_2SO_4.

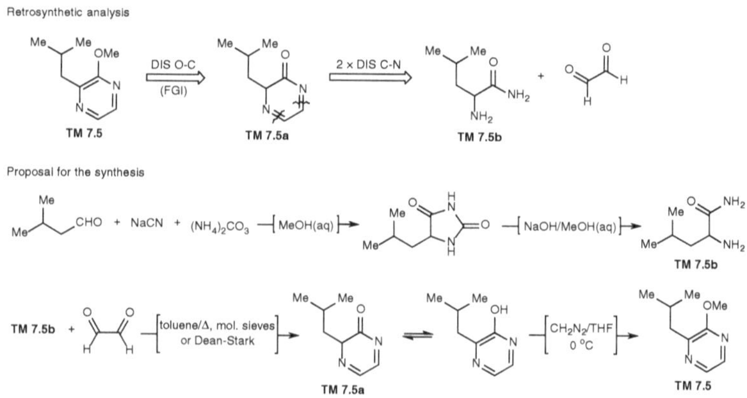

Scheme 7.10 Retrosynthetic analysis and proposal for the synthesis of **TM 7.5**

Example 7.6 Consider the retrosynthesis and then propose the synthesis of the antihypotensive agent *hydralazine* **TM 5.8**.

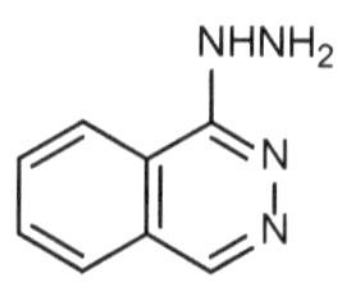

TM 7.6
Hydralazine

The basic heterocyclic ring is benzo-1,2-diazine or benzopyrazine, which obviously incorporates one mole of hydrazine. The second mole of hydrazine appears as a substituent on C-1, suggesting the first step, C–N disconnection of the hydrazine group (Scheme 7.11).

In the next step, disconnection of two C–N bonds in **TM 7.6a** leads to "extrusion" of the second mole of hydrazine and formation of a monoaldehyde of phthalic acid **TM 7.6b**. To end up with phthalanhydride as the available starting material, we need three FGIs, first of aldehyde to primary alcohol, second to the carboxylic group and a final cyclization to anhydride.

The synthetic proposal in the same scheme suggests the two-step transformation of phthalanhydride to aldehyde **TM 7.6b**, first by standard hydride reduction and then by partial oxidation. Condensation with hydrazine passes over aldimine as a stable intermediate, and then the cyclization requires heating and continuous water elimination. Activation of the heterocyclic ring by chlorination of the enolic form is required before substitution by hydrazine to **TM 7.6**.

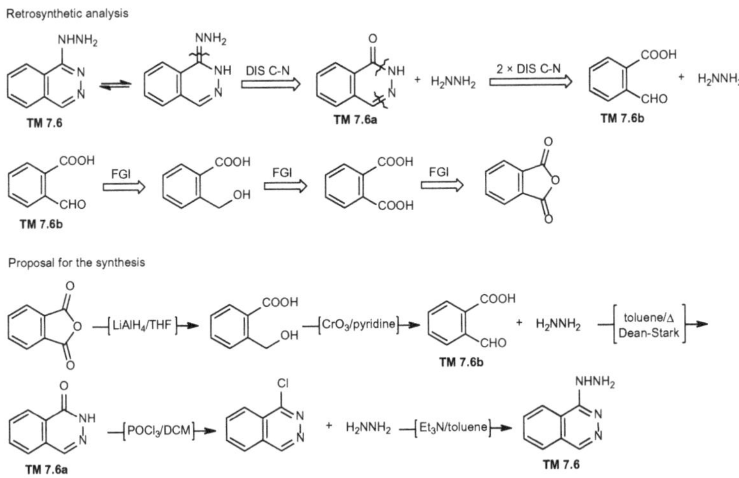

Scheme 7.11 Retrosynthetic analysis and proposal for the synthesis of *hydralazine* **TM 7.6**

7.3 Preferred Route to Non-aromatic Cyclic Molecules; the Baldwin Rules

So often synthetic or medicinal chemists experience that a certain reaction anticipated in analogy to a known example either does not work at all or affords miserable yields, or even unexpected products. The reason for such failures should be investigated concerning the unexpected mechanism of the reaction, usually due to improper reaction conditions. The failure of a certain cyclization reaction, although the disconnection of the selected bond leads to the available acyclic precursor or reagent, is also an often-encountered situation. A systematic analysis and explanation of this issue are offered by J. E. Baldwin, who proposes a series of empiric rules that we will discuss in this section.

The Baldwin rules indicate which cyclizations of acyclic precursors *are favored and which are not* [6–8]. It is important to note the substantial difference to the Woodward-Hoffmann rules [9–11], based on the theoretical concept stating which reactions are "*allowed*" and which are "*forbidden*" by the orbital symmetry. The Baldwin rules are based on experimental observations and indicate the *relative preference* of the acyclic precursor to form cyclic compounds. Therefore, the expressions "*favored*" and "*disfavored*" indicate energetically convenient and inconvenient cyclization. These rules refer predominantly to nucleophilic intramolecular reactions where the nucleophile approaches the C atom, which can be tetrahedral (*tet*, sp^3), trigonal (*trig*, sp^2) or diagonal (*dig*, *sp*). Based on experimental experience, they provide a particularly important guide for synthetic organic chemists.

According to the concept of this book, we introduce the Baldwin rules analyzing cyclizations to five- or six-membered heterocyclic rings with the O or N atom, or their failure!

Example 7.7 Does the cyclization of γ-hydroxyester **1** in the presence of a strong base result in transesterification to γ-lactone **TM 7.7**, or is five-membered cyclic ether **TM 7.8** formed by Michael addition to the terminal C-atom of enone (Scheme 7.12)?

Answering this question helps in the analysis of the steric approach of the alkoxy anion to the trigonal C atom in C=C or C=O bonds. The Baldwin rules state that such an approach is defined by the *trajectory* (a straight line or curve that a certain body describes by moving through space) between reacting atoms. The orientation of the reacting double bond determines the preferred approach of the nucleophile. It either remains inside the forming ring (*endo*) or stays outside of the ring (*exo*). These elements define the stereoelectronic control of the cyclization process, based on the Baldwin rules. The steric courses of the possible cyclization paths from Scheme 7.12 are presented in Scheme 7.13.

The first route is assigned as *exo*- and the second as *endo*-cyclization. *exo*-Cyclization to **TM 7.7** characterizes the approach of the alkoxy anion at an angle of approximately 110°, whereby the C=O bond remains outside of the emerging ring (Scheme 7.14). Note that this angle is close to that of the tetrahedral intermediate in transesterification!

On the other hand, cyclization to **TM 7.8** requires the approach of the nucleophile under a small angle, approximately 90°, and the *endo*-position of the double bond, both stereoelectronically unfavorable. Generally, the Baldwin rules state that cyclization to a five-membered ring is favored when the reacting double bond on the trigonal C-atom is *exo* and disfavored when it is *endo*. In Table 7.1, the Baldwin

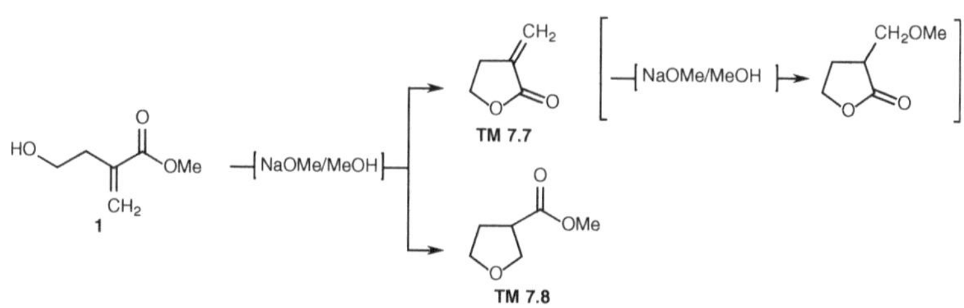

Scheme 7.12 Two possibilities for the cyclization of ester **1**

Scheme 7.13 Steric course of the *exo*- and *endo*-cyclization of α-methylene-γ-hydroxyester **1**

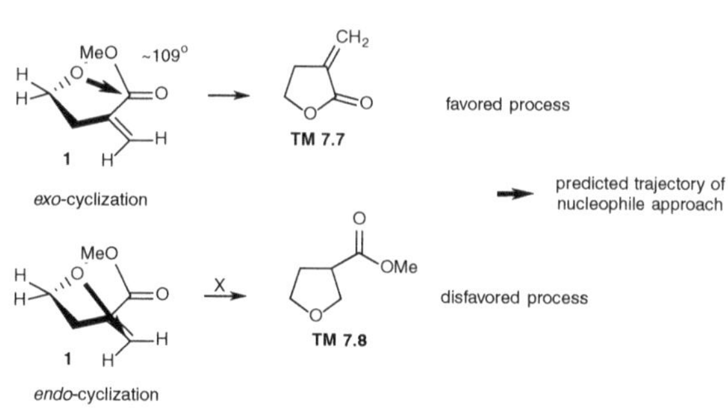

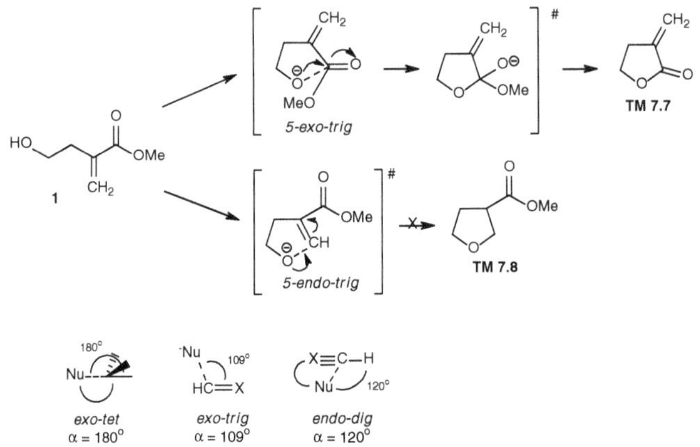

Scheme 7.14 The effect of the angle size on the favored and disfavored cyclization

Table 7.1 Baldwin rules for favored and disfavored cyclizations

Ring size	3	4	5	6	7
Type of cyclization	Exo/endo	Exo/endo	Exo/endo	Exo/endo	Exo/endo
Tet	+/−	+/−	+/−	+/−	+/+
Trig	−/−	+/−	+/−	+/+	+/+
Dig	−/+	−/+	+/+	+/+	+/+

rules for cyclization to small and medium-large rings, those with three to seven atoms in the ring, are summarized.

These rules are based on differences in the activation energy for cyclization to the rings of different sizes and are in accord with the kinetic control of these processes. It is therefore not surprising that cyclization to the highly strained four-member ring in **TM 7.9** (Scheme 7.15), although formally favored as a 4-*exo-trig* reaction, does not compete with 5-*exo-trig* cyclization to **TM 7.7** (Scheme 7.14).

Disfavored cyclization needs to surmount a high-energy, four-member transition state characterized by a high steric strain and mismatching negative charge on the internal C atom of the conjugated C=C bond in enone. The Baldwin rules reveal the stereoelectronic properties of the transition state as a determinant for the cyclization process, along with the ring size and position of the electrophile and nucleophile.

Example 7.8 Cyclization of *tert* alcohol **2** into **TM 7.10**, although on the first sight a disfavored 5-*endo-trig* process, proceeds in acidic but not in basic media.

Scheme 7.15 Disfavored cyclization to **TM 7.9**

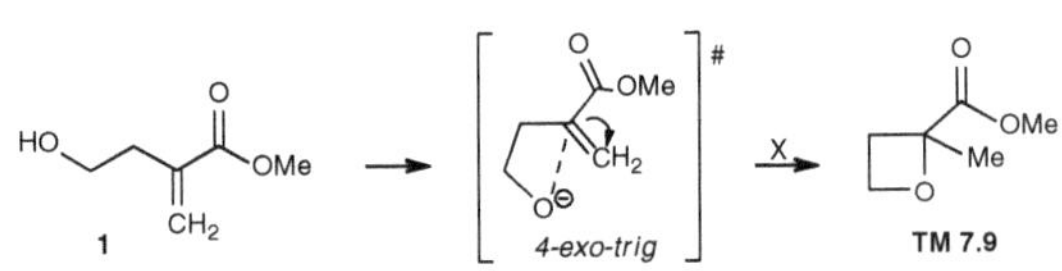

Scheme 7.16 Cyclization to **TM 7.10** is disfavored according to the Baldwin rules

The approach of the nucleophile to the C=C bond in **2** is presented in Scheme 7.16. How do you explain this outcome?

Since the Baldwin rules are not exclusive, disfavored cyclization can take place when the reaction conditions are properly changed. In basic medium the enone form of **2** reacts, and the alkoxy anion approaches the β-C atom of the C=C bond in the disfavored 5-*endo-trig* mode. In the acidic medium, protonation of the carbonyl group promotes the *formation of enol* in equilibrium with enone. This reactive species enters cyclization in 5-*exo-trig* mode (Scheme 7.17). **TM 7.10** is isolated in 84% yield on prolonged heating in dichloroethane in the presence of a catalytic amount of *p*-TsOH.

This example leads to an important conclusion: when the reaction conditions affect the cyclization reaction, they determine the stereoelectronic properties of the transition state and hence the mechanism of reaction. Generally, when different reaction conditions lead to different products, two reactions follow different mechanistic paths.

Example 7.9 Different outcomes of cyclization with changed reaction conditions are well evidenced in the recyclization reactions of γ-hydroxyepoxide **3** to either a derivate of tetrahydropyrane **TM 7.11** or tetrahydrofurane **TM 7.12** (Scheme 7.18). Suggest the possible mechanistic, i.e., kinetic reasons for these two routes.

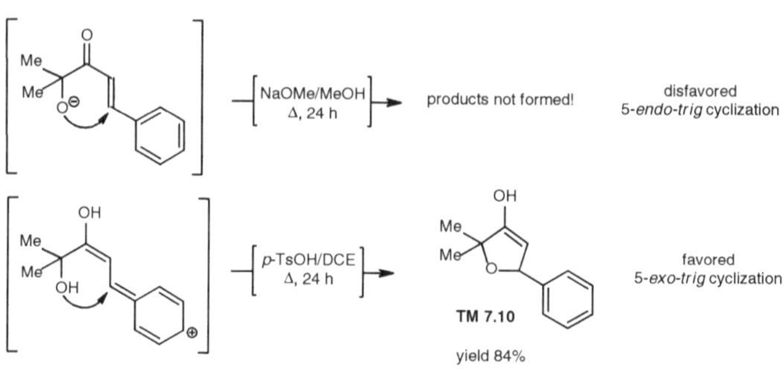

Scheme 7.17 Cyclization to **TM 7.10** depends on the reaction conditions

Scheme 7.18 Conditions for recyclizations of **3** in **TM 7.11** or **TM 7.12**

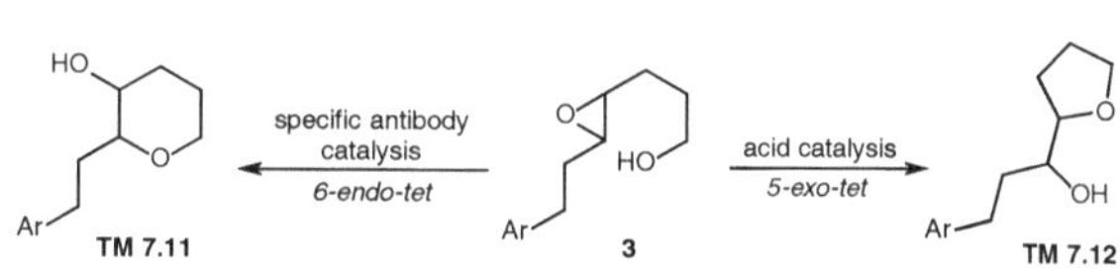

An acid catalysis leads to protonation of the basic O atom of epoxide and favors the 5-*exo-tet* approach of the OH group with formation of **TM 7.12**. On the contrary, a specific antibody developed as a biocatalyst directs the recyclization of **3** by the disfavored 6-*endo-tet* route and exclusive formation of **TM 7.11**. This outcome is explained by the preferred conformation of **3** in the active site of the antibody, which closely resembles those on the 6-*endo-tet* route. Calculations revealed the energy of the transition state for the 5-*exo-tet* route to **TM 7.12** 1.8 kcal/mol is lower than for the 6-*endo-tet* route to **TM 7.11**. Selective lowering of the TS energy on the route to **TM 7.11** is the result of preorientation of bound substrate **3** in the active site of the antibody, an important aspect of the mechanism of many enzymatic reactions. Remember that a less than 3 kcal/mol difference in the transition-state energy on the routes to enantiomeric products assures nearly 100 % e.e. (Sect. 3.4). Similarly, a small energy difference in activation energies determines the direction of chemoselective reactions.

The next three examples refer to syntheses of generic drugs developed by technological processes by chemists at the PLIVA Co. (Croatia). In all three examples, the heterocyclic unit is formed on cyclization according to the Baldwin rules. Since they were not postulated at the time these processes were developed, the authors might only speculate why they are "favored"!

Example 7.10 Suggest the bases for the retrosynthetic analysis of *glutethimide* **TM 7.13**, a compound with hypnotic and sedative activity, and then we consider the industrial method of production.

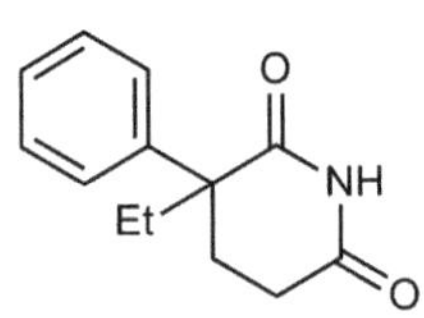

TM 7.13
Glutethimide

The six-membered imide seems to be the result of cyclization of a C_5 dicarboxylic acid derivative, most probably diamide **TM 7.13a**. This consideration suggests the first retrosynthetic step (Scheme 7.19).

Diamide **TM 13a** is available by controlled hydrolysis of dinitrile **7.13b**, and now this second-generation target molecule can be disconnected in *retro*-Michael mode to 2-phenylbutyronitrile and acrylonitrile.

All synthetic steps and reaction conditions are presented on an industrial scale in Scheme 7.20.

The first two steps are performed in one reactor since both are PTC-catalyzed alkylations of benzylic carbon in benzyl cyanide. The order of alkylation is determined by the reactivity of the alkylating agent; acrylonitrile is more reactive in Michael addition and therefore is preferred for alkylation of the less reactive *tert* carbanion. Less reactive ethyl bromide is preferred for alkylation of the more reactive

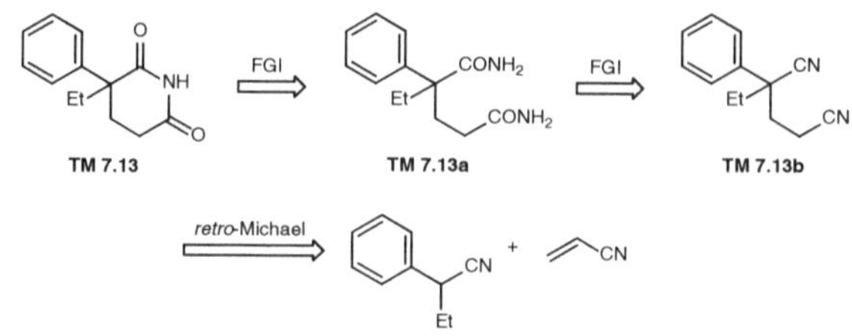

Scheme 7.19 Retrosynthetic analysis of *glutethimide* **TM 7.13**

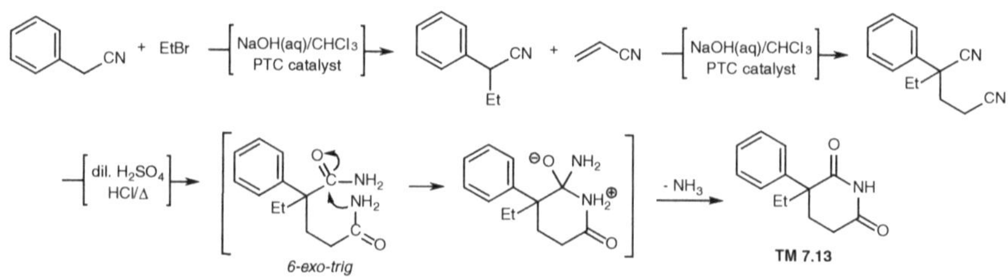

Scheme 7.20 Industrial synthesis of *glutethimide* **TM 7.13**

carbanion of benzyl bromide in the first step. This order follows the well-known principle for the sequence of reactions in which the more reactive reagent is used in the reaction with a less reactive partner, and *the energetically more demanding reaction is completed first*. Partial hydrolysis of dinitrile is controlled by the acidity of the medium and reaction temperature. On completed formation of diamide, the temperature is enhanced, and cyclization takes place with elimination of one mole of ammonia. According to the Baldwin rules, cyclization follows the favored 6-*exo-trig* approach of amide. Although the amino group in amide is weak, the nucleophile intramolecularity of the process and six-membered transition state favor the formation of a tetrahedral intermediate and elimination of ammonia.

Example 7.11 Starting from the easily available raw materials, a kind of "total" synthesis of *sulfathiazole* **TM 7.14**, a broad-spectrum antibiotic, is developed. A summary of this process is presented in Scheme 7.21.

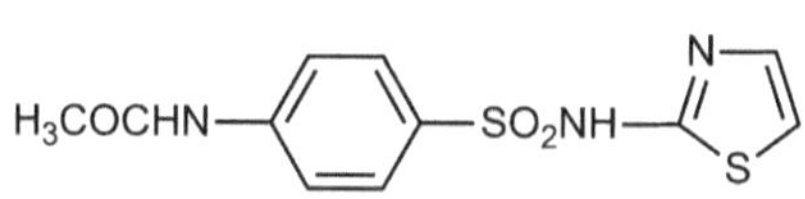

TM 7.14
Sulfathiazole

A key step in this synthesis is the cyclization of the thiazole ring by double alkylation of thiourea with 2,3-dichloropropionic acid (Scheme 7.22). Cyclization follows the preferred 5-*exo-tet* route, and the unstable intermediate undergoes fast aromatization triggered by hydrolysis and decarboxylation.

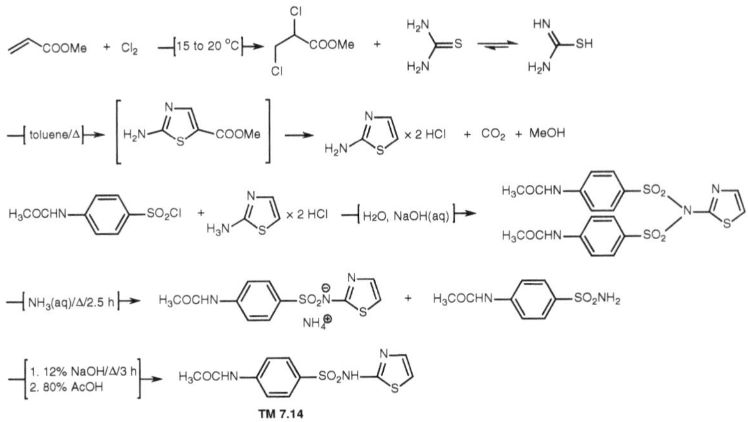

Scheme 7.21 Industrial method for production of *sulfathiazole* **TM 7.14**

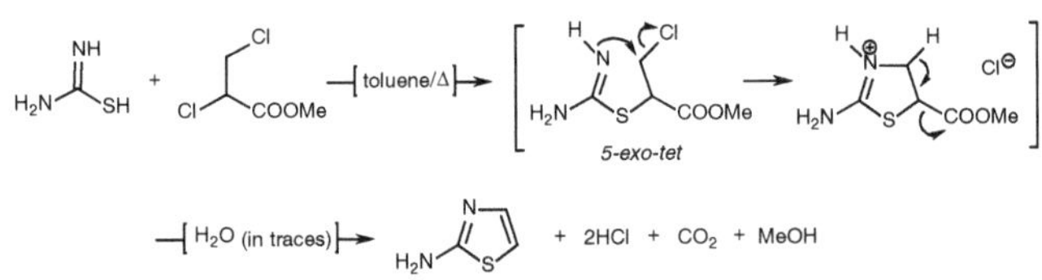

Scheme 7.22 Mechanism of formation of 2-aminothiazole

Preparation of *para*-acetylaminobenzensulfonyl chloride (PAS) was described in Example 7.2. Sulfonation of 2-aminothiazole is performed in aqueous medium under controlled addition of hydroxide according to the already-discussed Schotten-Baumann process. As the result, double-sulfonated amine is formed, which does not represent a serious technological issue since on heating with ammonia to the reaction pot, selective ammonolysis is achieved and one mole of sulfonamide obtained. This is a useful side product that on separation is recycled in the production of biologically active sulfonamides. The ammonium salt of the *sulfathiazole* precursor is submitted to hydrolysis of the *N*-acetyl group and *sulfathiazole* isolated by crystallization.

Let us now reconsider the cyclization of the isoxazole derivative on the route to *sulfisoxazole* **TM 7.2**, Scheme 7.4. Cyclization to **TM 7.2b** runs under slightly basic conditions and follows the Baldwin rules. The intermediary anion of ketoxime approaches as a nucleophile cyano group in the favored 5-*exo-dig* path (Scheme 7.23).

Scheme 7.23 Mechanism of cyclization to **TM 7.2b**

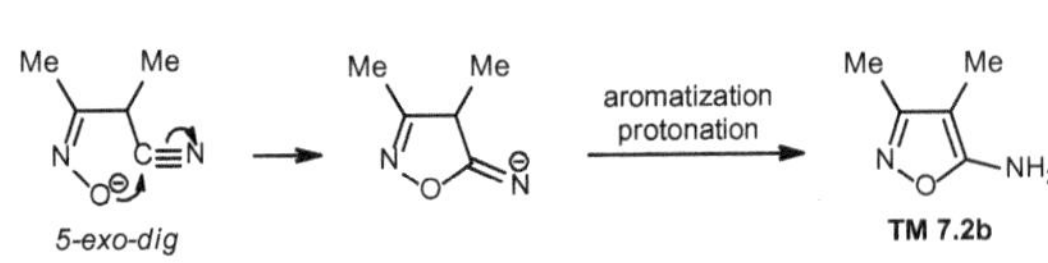

In conclusion, the Baldwin rules are a useful tool for explaining the cyclizations and planning the synthesis of small and medium rings. They reveal that the preferred cyclization not only depends on the ring size and reactivity of the terminal groups, but also on their proper orientation. These rules do not offer a "yes-no" answer for the workability of certain reactions since they are based on a kinetic argumentation; hence, disfavored reactions can be promoted under specific reaction conditions because of a change in the reaction mechanism. Knowledge of these rules can help retrosynthetic analysis and suggest the original synthesis of carbocyclic and heterocyclic structures, important building blocks in natural and biologically active compounds.

References

1. Taylor EC, Wipf P, Weissberegr A (eds) (1950–2005) Chemistry of heterocyclic compounds, vol 1–64. Wiley, New York
2. Katritzky AR (ed) (1963–2010) Advances in heterocyclic chemistry, vol 1–99. Elsevier, San Diego
3. Joule JA, Mills (2010) J Heterocyclic Chemistry, 5th ed. Wiley-Blackwell, Hoboken
4. Katritzky AR, Ramden OA, Joule JA (eds) (2010) Handbook of heterocyclic chemistry. Elsevier, San Diego
5. (1973) Org Synth Col, 5:801–804
6. Baldwin JE (1976) J Chem Soc Chem Comm 18:734–736
7. Nicolaou KC, Prasad CVC, Somers PK, Hwang CK (1989) J Am Chem Soc 111:5330–5334
8. Correy FA, Sundberg RJ (1990) Advanced organic chemistry. Part A; Structures and mechanism. Plenum Press, New York, pp 165–167
9. Woodward RB, Hoffmann R (1969) Angew Chem Int Ed 8:781–853
10. Hoffman R (2004) Angew Chem Int Ed 43:6586–6590
11. Geerlings P, Ayers PW, Toro-Labbé A, Chattaraj PK, De Proft F (2012) Acc Chem Res 45:683–695

Chapter 8
Rearrangements—Synthetic Reactions "Not Liable" to Retrosynthetic Analysis

Abstract Molecular rearrangements are not amenable to retrosynthesis because of their complex mechanisms. Still, in some cases *retro*-rearrangements are a conceivable and useful approach to selected target molecules. In this chapter, arguments for the retrosynthetic approach to some well-known rearrangements, Beckmann, Hofmann, Arndt-Eistert, Favorskii, pinacol and Bayer-Villiger, are presented. The mechanism of these rearrangements is explained. Retrosynthesis and synthesis, which include a specific rearrangement in the key step, are proposed for selected target molecules, among them *paracetamol*, *dinestrol* and spasmolytic biphenyl carboxylic acid.

8.1 Introduction

A rearrangement reaction is a broad class of organic reactions where the carbon skeleton of a molecule is rearranged to give a structural isomer of the original molecule or a substituent moves from one atom to another, without or with elimination of carbon atom(s). Rearrangements are characterized by the split and formation of C–C and C–X bonds in the same process.

Many rearrangements, such as Arndt-Eistert, Beckman or Hofmann rearrangements, have long been known to organic chemists. The industrial application of some important rearrangements is outlined [1]. The mechanism and stereochemistry of these reactions are well defined.

The observation of the target molecule imposing *retro*-rearrangement requires experience and skill. In the retrosynthetic direction it is not obvious how to disconnect certain C–X bonds with contemporaneous formation of a new C–Y bond (X, Y=C or heteroatom) whereby synthons of the specific electronic structure appear, such as carbocations or carbenes. In this chapter, some examples of the most important rearrangements for organic synthesis are presented, underlining their mechanistic characteristics and common properties to enable the reader to propose *retro*-rearrangement of selected target molecules.

V. Šunjić and V. Petrović Peroković, *Organic Chemistry from Retrosynthesis to Asymmetric Synthesis*, DOI 10.1007/978-3-319-29926-6_8

8.2 Beckmann Rearrangement

In this reaction, oximes rearrange to amides in the presence of protic or Lewis acids. Scheme 8.1 presents the mechanism for the Beckmann rearrangement of acyclic oxime **1** and cyclic oxime **2**. The actual methods for the production of oximes are discussed in Sect. 1.5, Scheme 1.20

Rearrangement of oxime **1** leads to acyclic acetanilide **3**, while rearrangement of cyclic oxime **2** results in lactame **4** (Sect. 1.3.2). Rearrangement is triggered by protonation of the OH group in oxime followed by C to N migration *anti* to the hydroxyl group. In this step, an unstable carbiminium cation is formed, which on hydration forms enol amide on the route to a very stable carbonyl form [2, 3].

Example 8.1 Propose the retrosynthetic analysis and synthesis of the analsetic *paracetamol* **TM 8.1**.

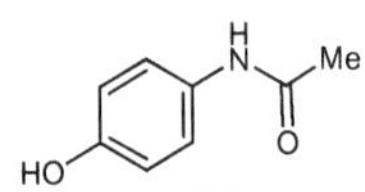

TM 8.1
Paracetamol

The classic approach to *N*-acetylated *para*-aminophenols comprises the nitration of phenol, reduction of the nitro to amino group and *N*-acetylation. The acetylation step proceeds with limited chemoselectivity, however, accompanied by the formation of *N,N*-diacetylated and *O*-acetylated products. The proposed retrosynthesis represents an elegant solution: *retro*-Beckmann to oxime as the key step, followed by FGI to ketone and *retro*-Friedel-Crafts disconnection to phenol (Scheme 8.2).

Scheme 8.1 Mechanism of the Beckmann rearrangement

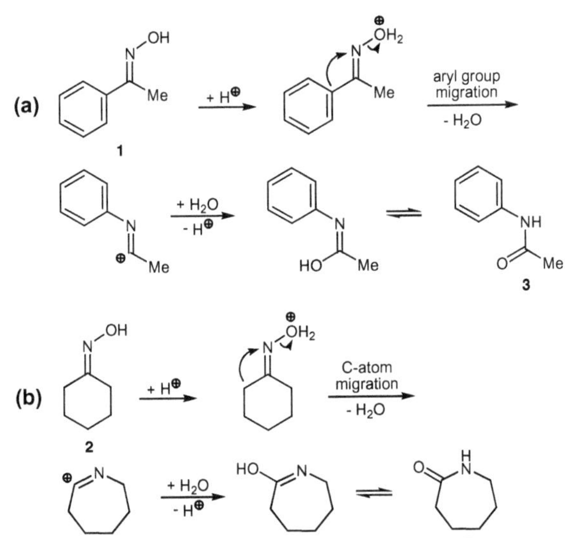

4

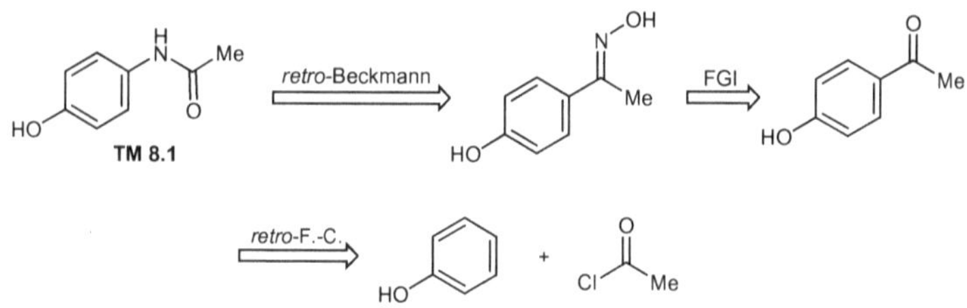

Scheme 8.2 Proposal for the retrosynthesis of *paracetamol* **TM 8.1**

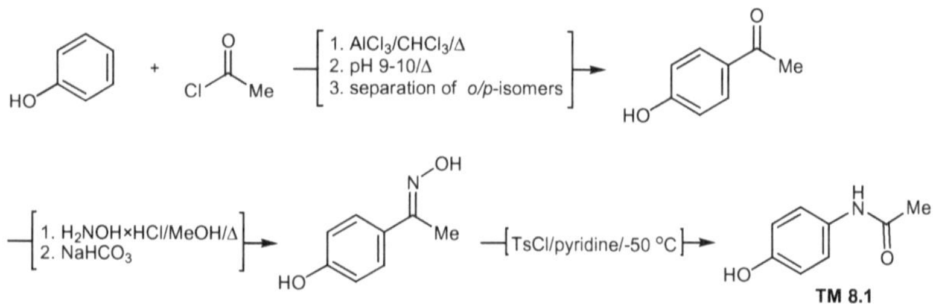

Scheme 8.3 Proposal for the synthesis of **TM 8.1**

The reaction conditions are outlined in Scheme 8.3, and it is worth mentioning that separation of *ortho/para* isomers cannot be avoided if either acetylation or nitration of phenol is performed.

The cited conditions for the preparation and rearrangement of oxime have been selected among many others reported in the literature and patents. For specific oximes, the Beckmann rearrangement is catalyzed by zeolites or nanoporous materials in ionic liquids in the presence of Co(III), Mn(III) or Fe(III) salts [4].

8.3 Hofmann Rearrangement

Using the Hofmann rearrangement, the primary amides are converted to amines with one C atom less than in the starting amide (Scheme 8.4) [5, 6].

Key steps of this rearrangement represent *N*-bromination, followed by deprotonation of *N*-bromamide and C to N migration of the alkyl group inside acyl-nitrene with the formation of isocyanate [7, 8]. In aqueous medium, isocyanate is immediately hydrated to unstable carbamic acid, which decarboxylates to primary amine. In alcohols stable carbamate esters are formed.

Synthetic application of this reaction extends from rearrangements of aliphatic amides to aromatic and heterocyclic amides. Results of the Hofmann rearrangement are primary amines or their *N*-acyl derivatives.

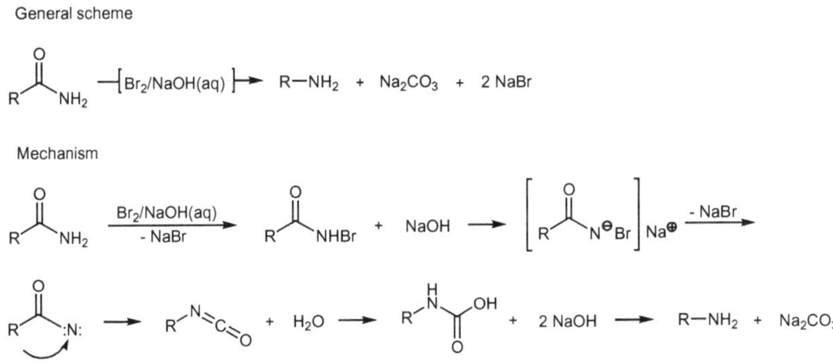

Scheme 8.4 General scheme and mechanism of the Hofmann rearrangement

Example 8.2 Complete retrosynthetic analysis and proposed synthesis of **TM 8.2**.

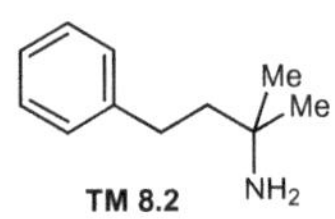

TM 8.2

In the target molecule, the amino group is bound to the *tert*-C-atom and conveniently available via Hofmann rearrangement of the corresponding amide. Another possibility for the introduction of this functionality is discussed in Sect.2.5 , Scheme 2.38.

Retrosynthesis is outlined in Scheme 8.5. The *retro*-Hofmann step results in amide **TM 8.2a**, and then FGI leads to a carboxylic group at the branching point in **TM 8.2b**. We already described the proper introduction of this functionality by the reaction of CO_2 with the Grignard reagent at the branching point. This suggests the next generation **TM 8.2c** and **TM 8.2d**. Two *retro*-Grignard disconnections lead over carbionol **TM 8.2e** to ester **TM 8.2f**.

The Kast building block is conveniently available by hydrogenation of cinnamic acid ester **TM 8.2g**, a commodity produced on the large scale. An old technology is

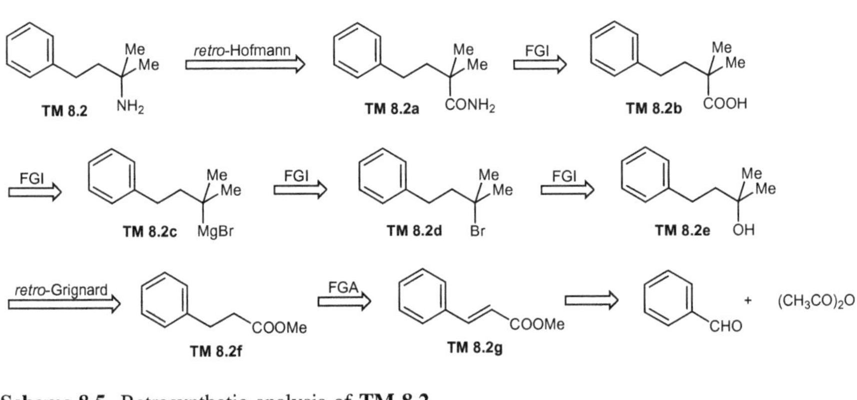

Scheme 8.5 Retrosynthetic analysis of **TM 8.2**

based on the Perkin reaction, the condensation of benzaldehyde with acetic anhydride in the presence of sodium acetate at ca. 100 °C [9, 10]. Competitive technologies are based on Knoevenagel condensation [11–13], Claisen condensation [14, 15] and Heck coupling of aryl and vinyl halides [16, 17].

Based on retrosynthetic considerations, the complete synthesis of **TM 8.2** is proposed in Scheme 8.6.

The next example discovers the importance of the Hofmann rearrangement on the route to a derivative of a heteroaromatic amine not available by other methods.

Example 8.3 Consider the retrosynthetic analysis and then propose the synthesis of **TM 8.3**.

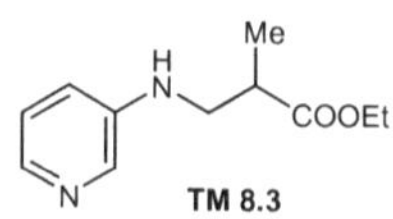

TM 8.3

The first retrosynthetic step introduces a great simplification of **TM 8.3** by disconnection of the C–N bond in a *retro*-Michael mode resulting in 3-aminopyridine **TM 8.3a** and ethyl methacrylate. The heterocyclic amine is not available by either the nitration of pyridine and reduction or bromination and ammonolysis. In both cases the pyridine nucleus proves resistant to electrophilic aromatic substitution.

Note There is a great difference in the availability of 2-amino and 3-aminopyridine. The former is produced on an industrial scale via the amination of pyridine by sodium amide, known as the *Chichibabin reaction* (Scheme 8.7) [18, 19].

Sodium amide in liquid ammonia reacts as a nucleophile at the two or six position of the pyridine ring. In a sequence of reversible steps, the first hydride ion is eliminated, and then aromatization followed by protonation takes place.

The reactivity of single positions of the pyridine ring is governed by the electron-withdrawing effect of the nitrogen atom, which is enhanced by protonation

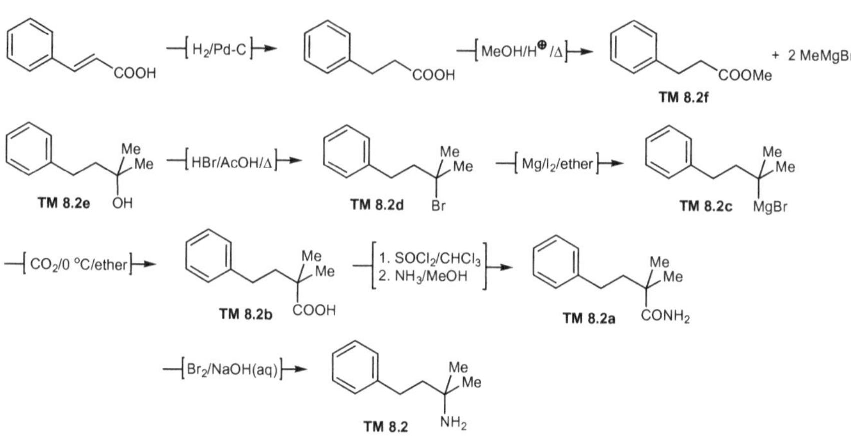

Scheme 8.6 Proposal for the complete synthesis of **TM 8.2**

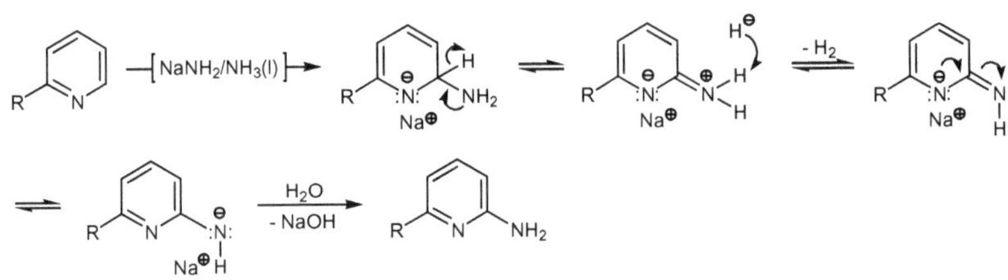

Scheme 8.7 Mechanism of the Chichibabin reaction, the amination of pyridine

or coordination. This explains the poor reactivity of pyridine with electrophiles and selective reactivity at positions 2 and 6 with nucleophiles.

The retrosynthetic analysis of **TM 8.3** presented in Scheme 8.8 involves a proposal for the retrosynthesis of 3-aminopyridine **TM 8.3a**.

The *retro*-Hofmann consideration is based on the availability of pyridine-3-carboxamide on a large scale. Actually, this compound, known as nicotinamide or vitamin B3, is produced in multi-ton quantities from 3-methylpyridine [20, 21]. For this building block, the catalytic industrial process is based on the one-pot cyclization of acrolein, ammonia and propanal in the presence of oxides of Sb(II), Ti(IV) or V(V) as catalysts. 3-Methylpyridine is submitted to *amoxidation*, the combined action of oxygen and ammonia to obtain 3-cyanopyridine (Scheme 8.9) [22].

In the last step, the partial hydrolysis in a continuous process affords nicotinamide [23]. This approach to nicotinamide supports the proposal for the complete synthesis of **TM 8.3** via Hofmann rearrangement.

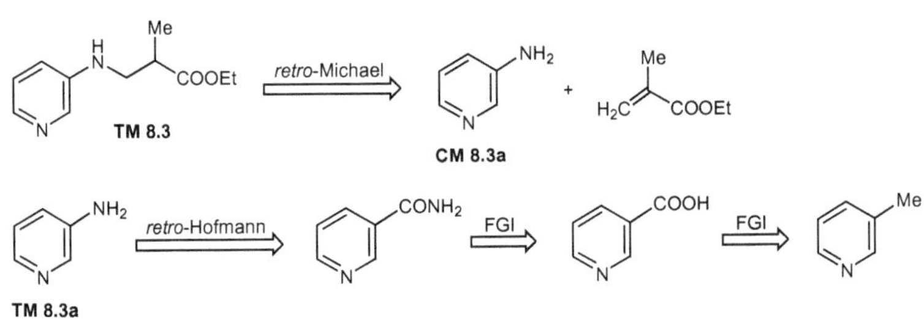

Scheme 8.8 Retrosynthetic analysis of **TM 8.3**

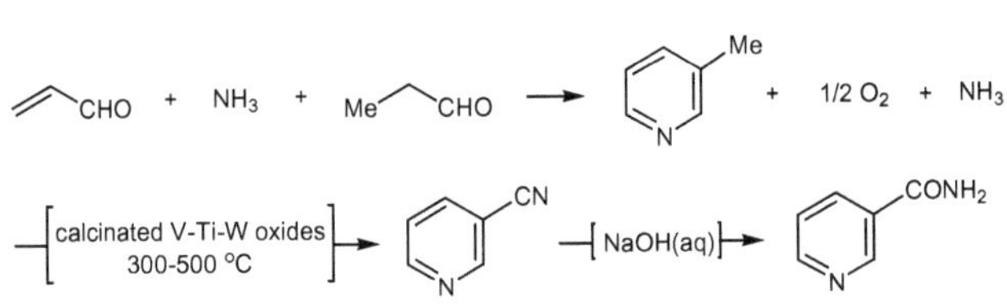

Scheme 8.9 Industrial method for the production of nicotinamide

8.4 Arndt-Eistert Rearrangement

The *Arndt-Eistert rearrangement* is a series of reactions designed to convert a carboxylic acid to a higher carboxylic acid homolog and is considered a *homologation process*. This rearrangement is particularly useful when a certain 1,*n*-difunctional compound with one carboxylic function is available, and we need a target molecule with a 1,*n* + 1 difunctional pattern. In other words, by Arndt-Eistert rearrangement, we can prolong the carbon chain of carboxylic acid for one C atom. This rearrangement can be formally presented in the retrosynthetic direction by extrusion of the α-methylene group (Scheme 8.10).

From the retrosynthetic scheme it is not clear how this reaction proceeds and where the rearrangement takes place. To understand this process, the mechanism is presented in Scheme 8.11.

Acyl azide, available on the heating of acyl chloride with diazomethane in the absence of a carbene scavenger, forms an unstable carbene that rearranges to ketene on the migration of the alkyl group promoted by Ag(I) ions or photochemically. This reactive intermediate adds a nucleophilic solvent, usually alcohol, affording ester. The final result is chain elongation of the starting carboxylic acid for one methylene group, inserted in the C–COOH bond.

The reaction of acyl chlorides with diazomethane is fast and selective in the presence of a C=C bond, but not in the presence of other carbene acceptors. Knowledge of the chemoselective reactivity of carbenes is of prime importance for planning the *retro*-Arndt-Eistert disconnection of the target molecule.

Example 8.4 Consider the retrosynthesis and then propose the synthesis of **TM 8.4**.

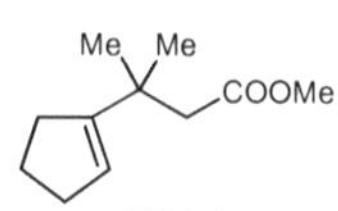

TM 8.4

At first sight the maximal simplification of **TM 8.4** can be achieved by the initial FGI introducing a carbinol OH group instead of the C=C bond in **TM 8.4a**

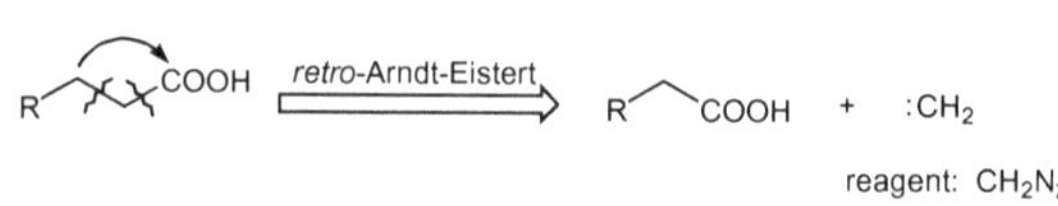

Scheme 8.10 Presentation of the *retro*-Arndt-Eistert rearrangement

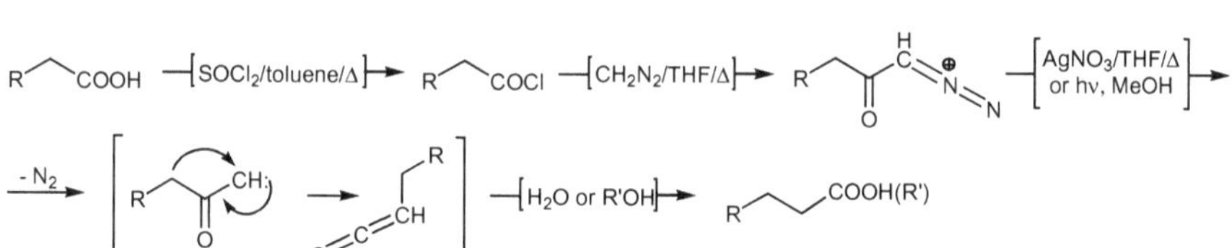

Scheme 8.11 Mechanism of the Arndt-Eistert rearrangement

followed by *retro*-Grignard disconnection *a* of the side chain (Scheme 8.12). However, **TM 8.4a** comprises a 1,4-CO pattern, and disconnection *a* is hampered by the formation of an illogical synthon, a carbanion on the tertiary β-C atom of methyl 3-methylbutanoate. This carbanion is destabilized by a double effect, the hyperconjugation of two methyl groups and "wrong" polarization due to the unfavorable distance to the carbomethoxy group. More about the mismatch of charges in the 1,4-CO pattern is presented in Chap. 5, Sect. 5.3.

By "cutting off" one methylene group in **TM 8.4a** in *retro*-Arndt-Eistert mode *b*, an unfavorable effect of the carbethoxy group in **TM 8.4a** is inverted by moving this group one bond closer to the carbanion to obtain synthon **TM 8.4b**. One FGI disconnection leads to **TM 8.4c** and *retro*-Grignard to an acceptable reagent derived from α-bromoisobutyric acid.

Complete synthesis of **TM 8.4** from cyclopentanone and isobutyric acid and selected reaction conditions are presented in Scheme 8.13.

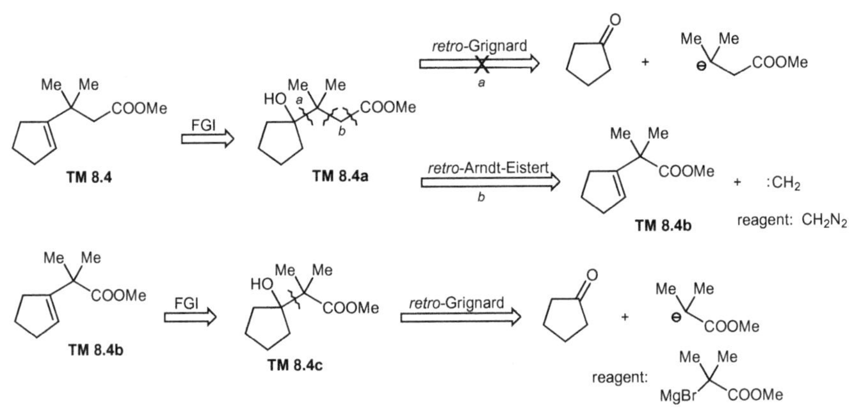

Scheme 8.12 Proposal for the retrosynthetic analysis of **TM 8.4**

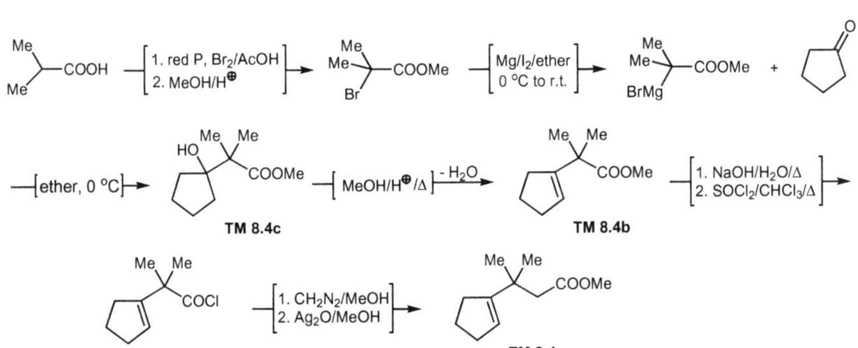

Scheme 8.13 Proposal for the synthesis of **TM 8.4**

Selective reactivity of the Grignard reagent with the ketone group and not with the sterically hindered carbethoxy group in the second molecule of the reagent is expected. Before preparation of carboxylic chloride to react with diazomethane, previous elimination of the reactive *tert*-OH group is needed to form a C=C bond less reactive with carbene.

8.5 Favorskii Rearrangement

The Favorskii rearrangement is a useful method for the preparation of carboxylic acids or their esters by rearrangement of easily available α-haloketones when the usual methods are not applicable. In the base-promoted reaction, α-haloketones cyclize to cyclopropanone derivatives and then rearrange with ring opening to carboxylic esters (Scheme 8.14).

During rearrangement, a more stable carbanion usually becomes the leaving group, affording a substantially modified carbon framework.

Example 8.5 Target molecule **TM 8.5** exhibits spasmolytic activity. Consider its retrosynthesis and then propose the synthesis.

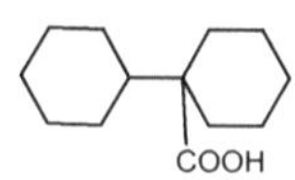

TM 8.5

Retrosynthetic imagination is needed to observe the *retro*-Favorskii disconnection, resulting in α-chlorodicyclohexyl ketone. Due to the molecular symmetry, the Cl atom can be conceived at any of two equivalent α-positions (Scheme 8.15).

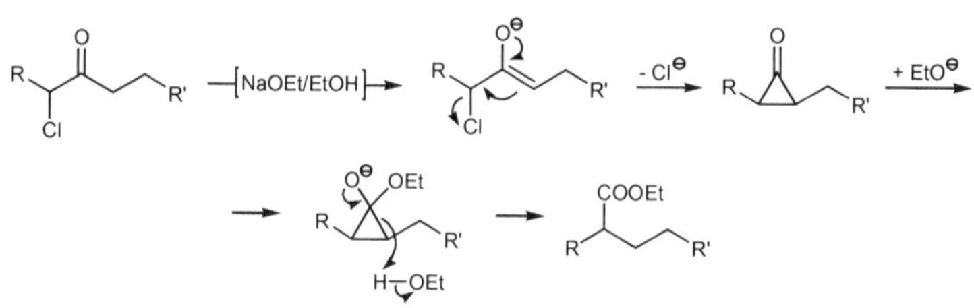

Scheme 8.14 Mechanism of the Favorskii rearrangement of α-haloketones

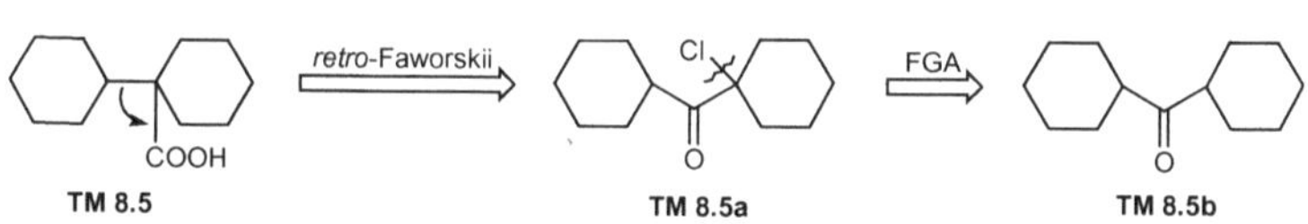

Scheme 8.15 Proposal for the retrosynthetic analysis of **TM 8.5b**

First-generation **TM 8.5a** is available by α-chlorination of dicyclohexylketone **TM 8.5b**, whose synthesis is not trivial, however. Large-scale production of **TM 8.5b** is based on the high-temperature dimerization of two moles of cyclohexanecarboxylic acid promoted by the MnO catalyst under hydrogen atmosphere (Scheme 8.16).

For this industrial method, the introduction of hydrogen into the mixture of MnO/carboxylic acid at high temperature to prevent the oxidation of Mn(II) to a higher, catalytically inactive oxidation state is characteristically controlled [24]. Chlorination, a base-promoted rearrangement and hydrolysis of the intermediary carboxylic ester are completed under standard conditions.

8.6 Pinacol Rearrangement

Pinacols are vicinal carbinols with two OH groups on *tert*-C atoms. They are available by Mg-promoted reductive dimerization of ketones involving radical-type intermediates (Scheme 8.17).

Recent technologies for the production of pinacols apply $TiCl_2$ or $TiCl_3$ as reducing agents in the presence of elementary Mn or Zn. This is a low-temperature reaction, usually performed at 0 °C in THF, and hydrolysis of the Ti complex is completed in an aqueous solution of potassium carbonate.

Pinacol rearrangement is traditionally studied by physical chemists and is not of much interest for synthetic organic chemists. Some structural features in the target molecule are elegantly available by this rearrangement, in particular *tert* alkyl ketones. Therefore, recognition of *retro*-pinacol disconnection can greatly help retrosynthetic consideration of specific target molecules. Before entering a discussion of some examples, we visit the mechanism of pinacol rearrangement (Scheme 8.18).

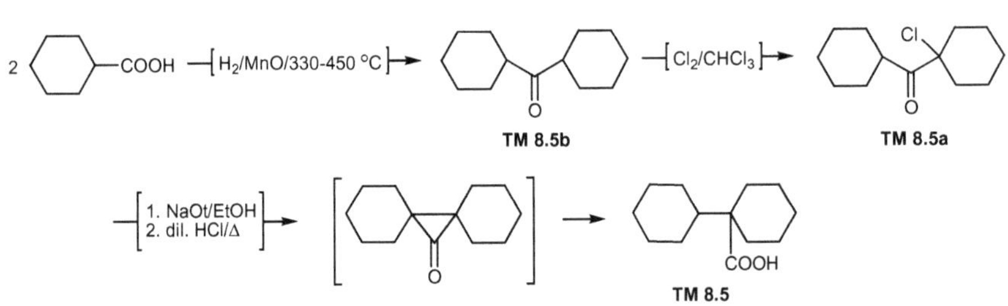

Scheme 8.16 Proposal for the synthesis of **TM 8.5**

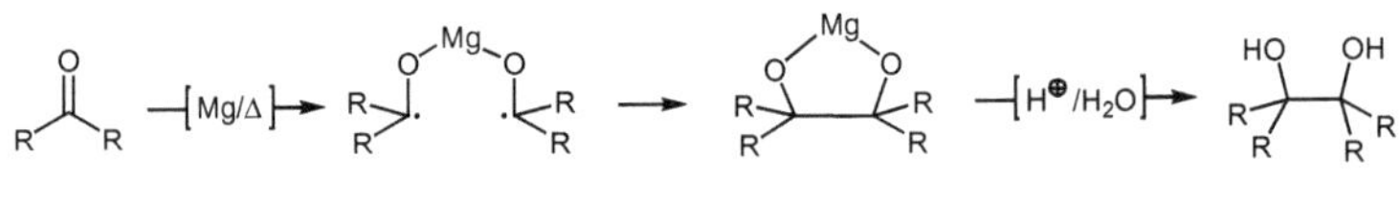

Scheme 8.17 General scheme for the dimerization of ketones to pinacols

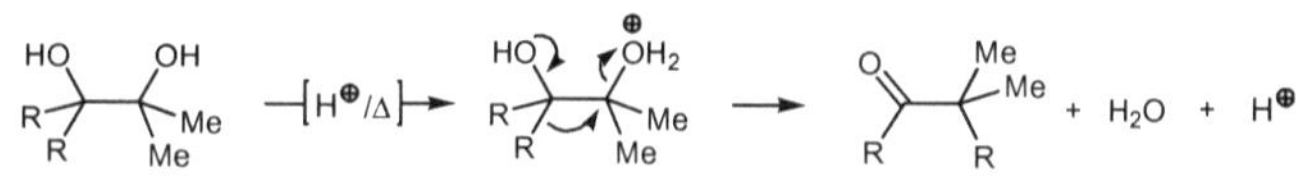

Scheme 8.18 Mechanism of pinacol rearrangement

Acid-catalyzed rearrangement leads to the migration of one alkyl group with contemporaneous elimination of water resulting in the formation of *tert* alkyl ketone.

The next example demonstrates the important role of intermediary pinacol in the route to the target molecule. The second example reveals the difficulties in planning the synthesis of specific target molecules if the *retro*-pinacol consideration is overlooked.

Example 8.6 Dienestrol **TM 8.6** exhibits estrogenic activity. Propose the retrosynthesis that in the key step conceives the formation of pinacol.

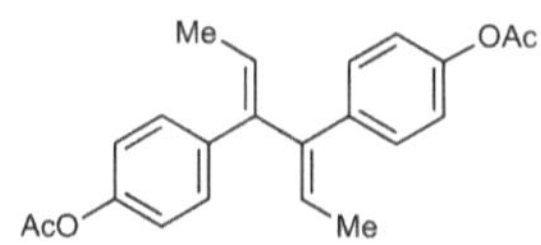

TM 8.6
Dienestrol

The pinacol unit is generated in **TM 8.6a** by FGI of the two C=C bonds to *tert* carbinol. Then *retro*-pinacol disconnection leads to two moles of *para*-acetoxypropiophenone (Scheme 8.19).

In the synthetic direction, the preparation of pinacol can be completed by heating ketone in the presence of Mg powder followed by the elimination of two moles of water by heating in acetyl chloride, presumably over the diacetylpinacol intermediate.

Example 8.7 Propose the retrosynthetic analysis and then suggest the synthesis of spiro-bicyclic ketone **TM 8.7**.

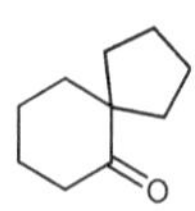

TM 8.7

The presence of the quaternary C atom in the ketone **TM 8.7** suggests formation by pinacol rearrangement. Therefore, *retro*-pinacol in the first step affords pinacol

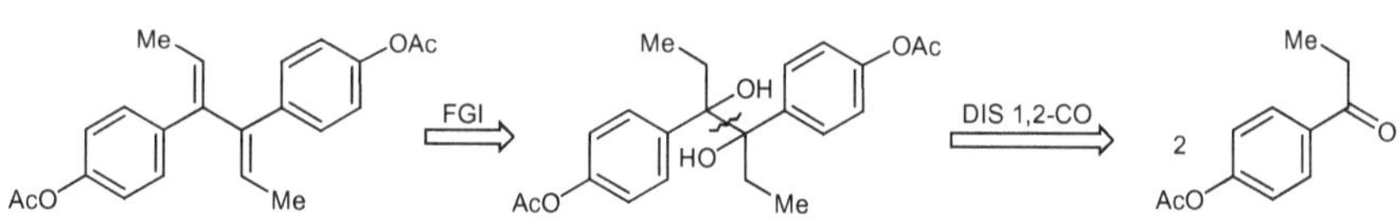

Scheme 8.19 Proposal for the retrosynthetic analysis of **TM 8.6**

TM 8.7a, which is prone to disconnection into two moles of cyclopentanone (Scheme 8.20).

It is important to note that attempts to prepare the *spiro* unit in **TM 8.7** by double alkylation of one α-C atom in cyclohexanone with 1,4-dihalobutane will result in an α,α′-bridged compound. The proposal for the synthesis of **TM 8.7** is given in Scheme 8.21.

Both former examples reveal an important limitation of this method. To complete an unambiguous pinacol rearrangement, the starting pinacol should be the product of dimerization of two moles of ketone. Two different ketones afford non-symmetric pinacol in low yield, which in turn rearranges into more products.

8.7 Baeyer-Villiger Rearrangement

This reaction was first reported in 1899 [25] and represents an oxidative rearrangement of ketones to esters of carboxylic acids, with peroxy acids or hydrogen peroxide as oxidants (Scheme 8.22) [26].

Note *meta*-Chloroperbenzoic acid (MCPBA) is a broadly used oxidant on the laboratory and industrial scale. It is produced by the oxidation of *meta*-chlorobenzoyl chloride with hydrogen peroxide in the presence of magnesium sulfate in an aqueous solution of sodium hydroxide and dioxane [27]. Since pure MCPBA

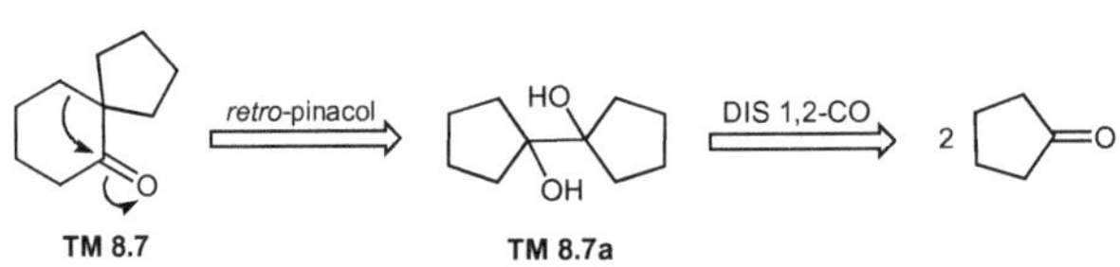

Scheme 8.20 Retrosynthetic analysis of **TM 8.7**

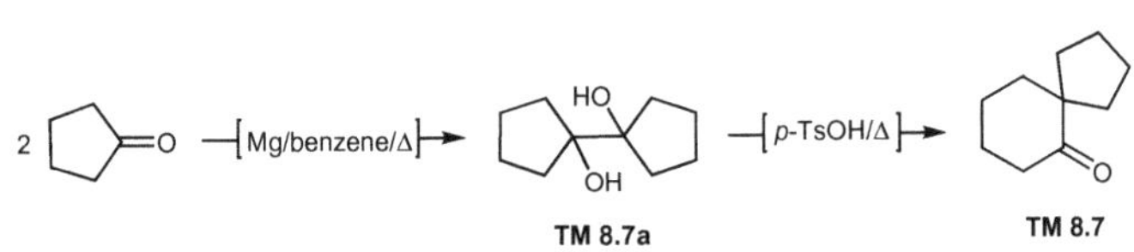

Scheme 8.21 Proposal for the synthesis of **TM 8.7**

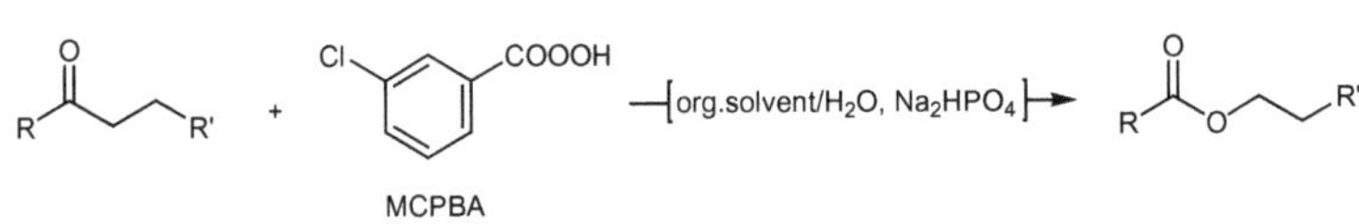

Scheme 8.22 General scheme of the Baeyer-Villiger rearrangement

can detonate on shock or spark, the commercial product contains less than 72 % of MCPBA along with 10 % *meta*-chlorobenzoic acid and wetted to10 % by water.

To consider the *retro*-Baeyer-Villiger in retrosynthesis, knowledge of the mechanism is mandatory (Scheme 8.23) [28].

In the first step, peroxy acid adds to the carbonyl group of ketone with the formation of a Criegee-type intermediate similar to that appearing in the ozonolysis of the C=C bond (Sect. 5.4, Scheme 5.55) [29]. In the next step, one carbon atom migrates to the peroxy O atom with the departure of one mole of carboxylic acid. In this process, the C atom that better stabilizes the positive charge migrates, i.e., possesses higher *migratory aptitude*. This property diminishes in the order H > *tert* alkyl > cyclohexyl > *sec* alkyl > aryl > *prim* alkyl > methyl.

Example 8.8 Consider retrosynthetic analysis and then propose the synthesis of **TM 8.8**.

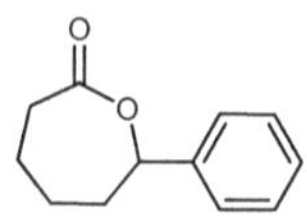

TM 8.8

Conceiving the *retro*-Baeyer-Villiger in the first step, we arrive at the convenient TM of the next generation, 2-phenylcyclohexanone **TM 8.8a** (Scheme 8.24).

This intermediate is available by an interesting detour, the reaction of cyclohexane epoxide with phenyl-Grignard reagent, followed by oxidation of *sec* alcohol. The reader should propose a synthetic scheme to **TM 8.8** and suggest workable reaction conditions.

Note There is fascinating chemistry, however, behind cyclohexene, a precursor of cyclohexene epoxide! Although a valuable chemical intermediate, it is difficult to

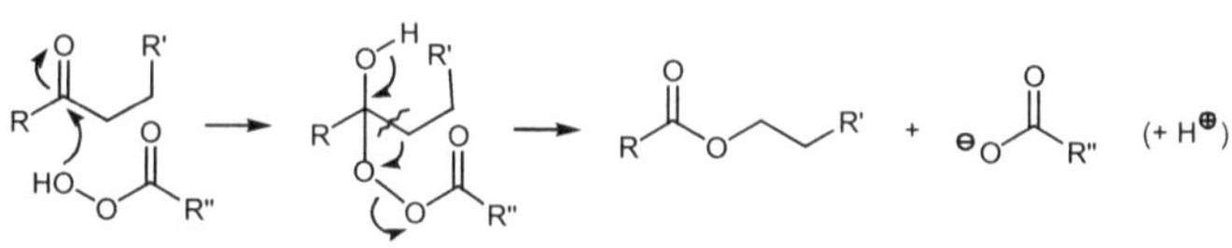

Scheme 8.23 Scheme of the Baeyer-Villiger rearrangement

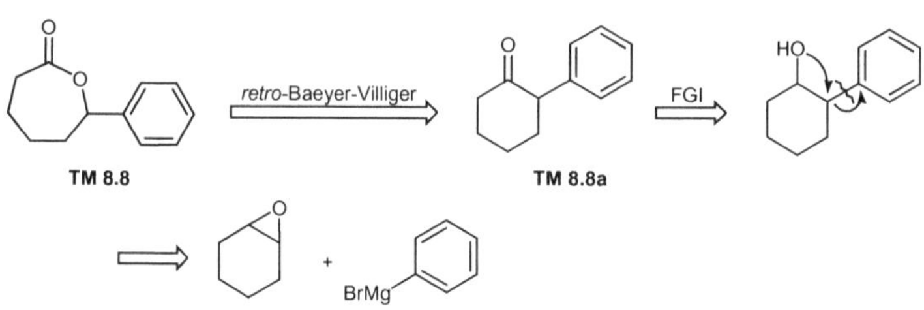

Scheme 8.24 Retrosynthetic analysis of **TM 8.8**

manufacture. Essentially, any catalyst active enough to hydrogenate benzene to cyclohexene hydrogenates cyclohexene to cyclohexane. One way to increase the yield of cyclohexene is to shape the selective carbon molecular sieve (CMS)-based catalyst to shift the product distribution to cyclohexene by exploiting the difference in size and shape of benzene, cyclohexene and cyclohexane. The selectivity of the catalyst to cyclohexene increases with decreasing pore diameter, water concentrations and benzene/hydrogen ratio. Selective catalysts are prepared by reducing $RuCl_2$ dissolved in water and alcohol and deposited in CMS pores at 250–300 °C. This procedure results in Ru crystals of proper size in the pores of CMS [30]. Other methods are claimed for the preparation of Ru-based catalysts with crystallite size of 200 A or less and doped with Zn salts as promoters [31, 32].

Quite often, *retro*-Baeyer-Villiger consideration enables synthetic chemists to propose surprising retrosynthesis of target molecules where this rearrangement is not obvious. Anticipation of Baeyer-Villiger rearrangement of ketone to lactone "hidden" in the remote retrosynthetic step is elaborate and requires considerable retrosynthetic imagination. The next example serves to support this deliberation.

Example 8.9 Propose the retrosynthetic analysis of cyclic α-amino acid **TM 8.9** and then the synthetic route that includes Baeyer-Villiger rearrangement in the key step.

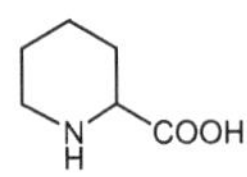

TM 8.9

An obvious and apparently attractive synthetic route to this target molecule starts from 2-methylpyridine (α-picoline) followed by oxidation to carboxylic acid and catalytic hydrogenation of the heterocyclic ring. Hydrogenation is accompanied, however, by the formation of side products apparently by cleavage of the C–N bond, alkylation of the *sec* amine formed and condensation of the side reactions.

We therefore start the retrosynthesis with the apparently illogical (ugly!) opening of the piperidine ring by two interconversions of C–N bonds leading to acyclic 1,5-dibromo derivative **TM 8.9a** and ammonia (Scheme 8.25).

The target molecule of the next generation seems an unacceptably complex derivative of *n*-hexanoic acid. However, the six position of one bromine atom suggests α-bromo lactone **TM 8.9b**. This intermediate is available on selective

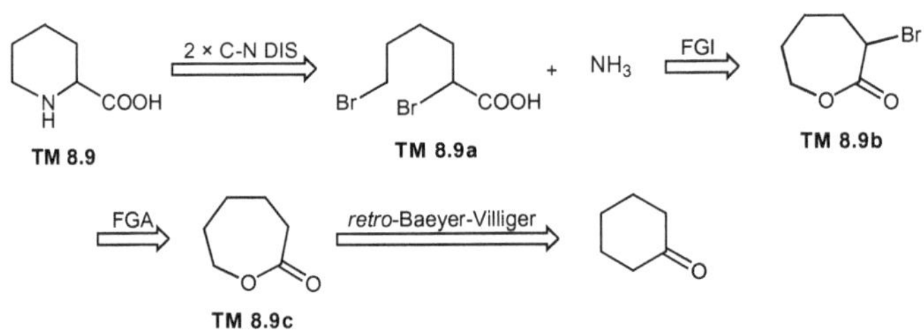

Scheme 8.25 Proposal for the retrosynthetic analysis of **TM 8.9**

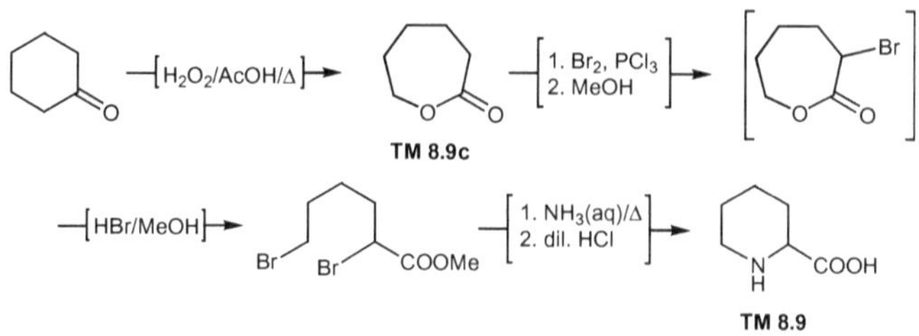

Scheme 8.26 Proposal for the synthesis of **TM 8.9**

α-bromination of caprolactone **TM8.9c**, and then the *retro*-Baeyer-Villiger step leads to cyclohexanone as an available commodity.

The surprisingly short synthesis of **TM 8.9** is presented in Scheme 8.26.

Some synthetic steps differ slightly from the retrosynthetic scheme. Namely, using Br_2 activated by PCl_3 in methanol for α-bromination of lactone, one mole of HBr is liberated and serves as an acid catalyst for contemporaneous transesterification with ring opening to the intermediary methyl ester of 2,6-dibromohexanoic acid. In the final stage, two separate synthetic steps are needed, first cyclization by ammonia and then hydrolysis of ester.

With the somewhat surprising retrosynthesis and proposal for the synthesis of pyperidine-2-carboxylic acid, we conclude this chapter with some well-known rearrangements that are demonstrated as a retrosynthetically feasible concept. In conclusion, retrosynthetic consideration of rearrangements requires the creative skill of synthetic chemists but rewards them with the discovery of short, unexpected synthetic routes to selected target molecules.

References

1. Moulay S (2002) Chem Educ Res Pract Eur 3:33–64
2. Gawley RE (1988) The Beckmann reactions: rearrangement, elimination-additions, fragmentations and rearrangement-cyclizations. Org React 35:14–24
3. Yamabe Y, Tsuchida N, Yamazaki S (2005) J Org Chem 70:10638–10644
4. Gou S, Du Z, Zhang S, Li D, Li Z, Deng Y (2006) Green Chem 8:296–300
5. Hofmann AW (1881) Ber Dtsch Chem Ges 14:2725–2736
6. Shioiri T (1991) Comp Org Syn 6:800–806
7. Clayden J (2007) Organic chemistry. Oxford University Press, pp 1073–1078
8. Zagulayave AA, Banek CT, Yusubove MS, Zhdankin VV (2010) Org Lett 12:4644–4647
9. Rosen T (1991) Comp Org Synt 2:395–408
10. Bansel RK (1988) Organic reaction mechanism, 3rd edn. McGrow Hill, pp 199–201
11. Knoevenagel E (1898) Ber Dtsch Chem Ges 31:2596–2619
12. Wilk BK (1997) Tetrahedron 53:7097–8002
13. Smith MB (2007) J. March's advanced organic chemistry, 6th edn. J. Wiley, pp 1358–1363
14. Claisen L (1890) Ber Dtsch Chem Ges 23:977–986
15. Smith MB (2007) J. March's advanced organic chemistry, 6th edn. J. Wiley, pp 1364–136
16. Heck RF (1982) Org React 27:345–390

17. Beletskaya IP, Cheprakov AV (2000) Chem Rev 100:309–3066
18. Chichibabin AE, Zeide OA (1914) Russ Zhr Obsch Khim 46:1212–1215
19. Pozharskii AF, Simonov AM (1978) Doron’kin VN. Russ Chem Rev 47:1042–1060
20. Shimizu S, Watanabe N, Kataoka T, Shoji T, Abe N, Morishita S, Ichimura H (2002) Pyridine and pyridine derivatives. Ullmann’s encyclopedia of industrial chemistry
21. Scriven EFV, Murugan R (2005) Pyridine and pyridine derivatives. Kirk-Othmer encyclopedia of chemical technology, XLI
22. Beschke H (1984) US Pat 4,447,612 (Degussa-Huls Ag)
23. Grendze M, Vorhies SL (1999) US Pat 6(218):543
24. SNIA, Viscosa (1967) GB Pat 1.063,267
25. Baeyer A, Villiger V (1899) Chem Ber 32:3625–3633
26. Renz M, Meunier B (1999) Eur J Org Chem 4:737–750
27. McDonald RN, Streppel RN, Dorsey E (1988) Org Synth Col. 6:276–278
28. Alvarez-Idaboy JR, Reyes L, Mora-Diez N (2007) Org Biomol Chem 5:3682–3689
29. Criegee R (1948) Liebigs Ann Chem 560:127–132
30. Srinivas G, MacQueen DB, Dubovik M, Mariwala RK (1998), Shape selectiv hydrogenataion of benzene to cyclohexene. In: Herkes FE (ed) Catalysis of organic reactions. Marcel Decker, Inc, pp 577–582
31. US 4734536 A (1988) for Asahi Kasei Kogyo Kabushiki Kaisha
32. US 5589600 (1992) for BASF AG

Chapter 9
Retrosynthetic Considerations and Syntheses of Complex, Biologically Active Molecules

Abstract Retrosyntheses and related syntheses of four selected products and biologically active compounds are presented. Retrosynthesis of (±)-*menthol* and synthesis of (−)-*menthol* are presented in the first section. Synthesis of (±)-*chloramphenicol* and asymmetric synthesis of the antibiotic (−)-*chloramphenicol* are compared in the second section. The third section deals with (−)-*sertraline*, a compound with antidepressive and anxiolytic activity. The example of (−)-*sertraline* illustrates diverse approaches to biologically active compounds in the optically pure form, which is regularly investigated and used in a certain phase of the development of new drug entities (NDEs) in the pharmaceutical industry. In the last section, the retrosynthesis, stereoselective synthesis and asymmetric synthesis of diastereomeric α-, β- and γ-*lycoranes*, members of an important class of alkaloids and a goldmine for new drugs, are discussed.

9.1 Introduction

Natural and synthetic products used in medicine and agriculture or as materials in modern technologies are frequently complex organic or organometallic molecules. Syntheses of natural compounds, known as *total syntheses*, start from the available building blocks and construct target molecules over many synthetic steps [1–3]. The multistep synthetic route to biologically active compounds not available from natural sources is also called total synthesis.

The retrosynthetic approach to complex molecules of natural compounds is not easy. This approach is more convenient for analysis of key intermediates of less complex structures. Total syntheses of natural compounds were supported by contemporaneously developed computer programs [4–6]. It seems, however, that the trial-and-error approach to the synthesis of natural products and other complex molecules still prevails in many academic and industrial synthetic laboratories.

We have already presented the retrosynthetic approach to the biologically active compounds (±)-*fluoxetine* **TM 4.15**, (−)-*fenpropimorph* **TM 4.16** and (−)-*frontalin* **TM 5.15**. In this chapter, four additional topics related to retrosynthesis and the

V. Šunjić and V. Petrović Peroković, *Organic Chemistry from Retrosynthesis to Asymmetric Synthesis*, DOI 10.1007/978-3-319-29926-6_9

stereoselective and asymmetric synthesis of natural products and drug molecules are discussed.

9.2 Synthesis of Racemic and (−)-*Menthol*

(−)-*Menthol* is a natural compound present in mentha and rose oil. It acts as a mild inhibitor of inflammation processes and a promoter of drug penetration through the skin [7]. The menthol skeleton has three chiral centers and therefore $2^3 = 8$ stereoisomers, i.e., four diastereomeric racemates. In nature, both (+)- and (−)-*menthol* are present (Fig. 9.1).

It is worth of noting that rose oil contains a mixture of enantiomers consisting of up to 80 % (−)-*menthol*. Since natural sources of (−)-*menthol* are limited and the requirements of the fragrance and pharmaceutical industries are enormous, industrial methods for the production of (−)-*menthol* have been developed over the years.

Example 9.1 Complete the retrosynthesis and then propose the synthesis of (±)-*menthol* **TM 9.1**, without considering the stereostructure of this target molecule.

First, we observe a trisubstituted cyclohexane ring, whose construction by the Diels-Alder reaction in the absence of an electron-accepting group is rather inconvenient.

Note The reader is prompted to consider a retrosynthesis that includes the *retro*-Diels-Alder disconnection of the precursor of **TM 9.1** to isoprene and nitro-alkene as a dienophile. The retrosynthesis presented in Sect. 2.5, Example 2.20, provides a hint.

Hydrogenation of an aromatic precursor emerges as an attractive retrosynthetic approach (Scheme 9.1).

For the next generation aromatic target molecule **TM 9.1a**, preferably disconnected, is at an isopropyl unit, affording easily available *meta*-methylphenol (*meta*-cresol) and propylene as convenient Friedel-Crafts alkylating agents.

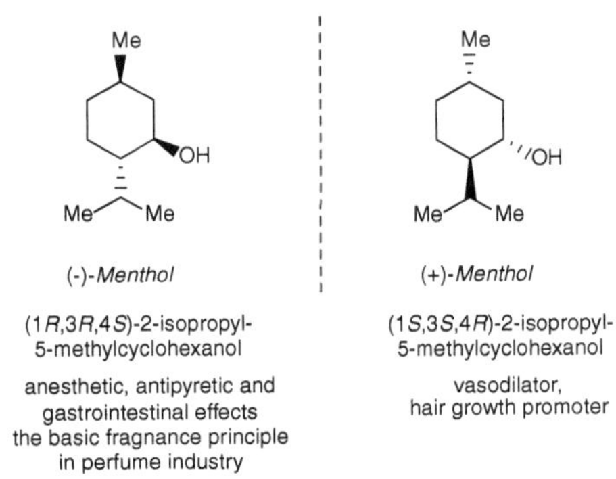

Fig. 9.1 Structural formulae and some properties of enantiomers of *menthol*

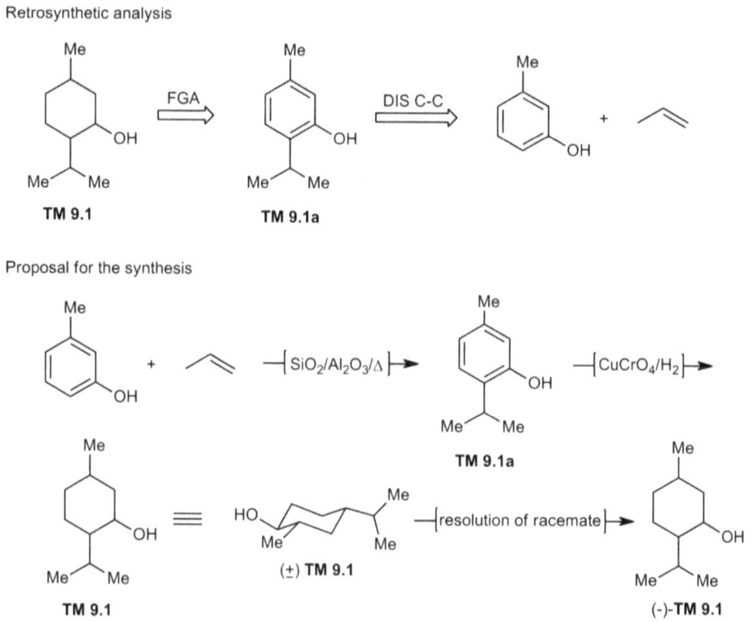

Scheme 9.1 Retrosynthetic consideration and proposal for the synthesis of (±)-*menthol* **TM 9.1**

Stereoselective synthesis of racemic menthol possessing *all-equatorial* substituents on the cyclohexane ring is surprisingly simple (Scheme 9.1). The alkylation step is completely *ortho*-selective since the phenolic OH group is *ortho/para* directing and the *para* position is sterically hindered by the methyl group. Alkylation by propylene is catalyzed on the industrial scale by weak Lewis acids and hydrogenation by the number of heterogeneous catalysts, among them Cu (II) chromate. On the industrial scale, the approach to enantiomers used to be completed via separation of diastereomeric esters with optically pure carboxylic acids and subsequent hydrolysis to (−)-*menthol* [8]. This method was abandoned in 2010 when the demand for (−)-*menthol* increased to 4500 tons/year.

Details of an ingenious industrial process for the production of (−)-*menthol* by asymmetric synthesis are presented in the next example.

Example 9.2 The industrial method for large-scale production of both enantiomers of menthol is based on an asymmetric catalytic reaction in the key step promoted by the Rh(II) complex of the well-known chiral ligand BINAP, available in both enantiomeric forms. This process was developed by Takasago Research Institute, Tokyo, in collaboration with Nobel laureate R. Noyori (Scheme 9.2) [9].

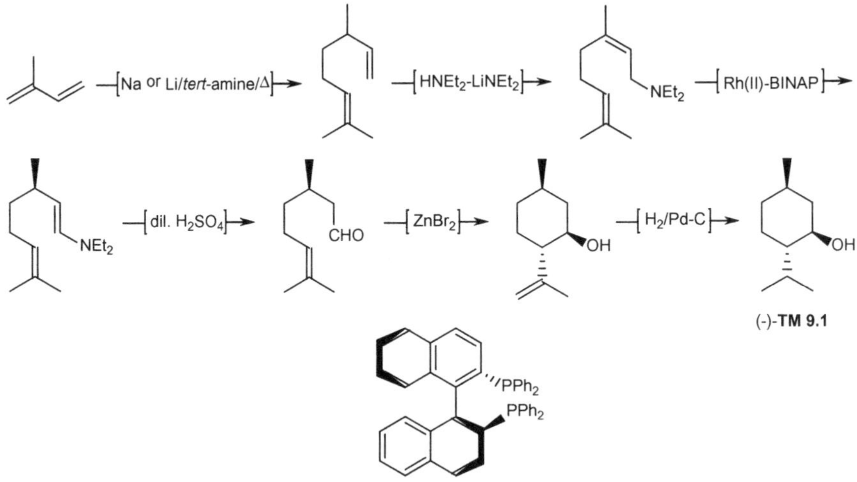

Scheme 9.2 Industrial method for production of (−)-*methol* **TM 9.1**

The process starts with isoprene, nowadays available at 800,000 tons/year, as a by-product of the thermal cracking of naphtha in the production of ethylene. In the first step, dimerization to myrcene is promoted by a strong base, which in the next step reacts with the strongly nucleophilic diethylamide anion to the prochiral allyl amine derivative. In the key step, this intermediate undergoes *asymmetric allylic migration* catalyzed by the Ru(II) complex of (*S*)-BINAP [10]. The asymmetric reaction results in a chiral enamine of (*R*) configuration with 98 % e.e. Aldehyde, obtained from the mild hydrolysis of enamine, is submitted to $ZnBr_2$-catalyzed cyclization to trisubstituted cyclohexane with two new chiral centers with (3*R*,4*S*) configurations. This remarkably stereoselective process is outlined in Scheme 9.3.

Asymmetric induction by the chiral center on C(1) is effected in the acyclic precursor coordinated to Zn via the carbonyl group. The C–H acid allylic Me group forms a hydrogen bond in the transition complex with $ZnBr_2$ and triggers cyclization. It is important to note that cyclization represents a 6-*exo-trig* process, allowed according to the Baldwin rules (Sect. 7.3). In the last step, the hydrogenation of the terminal C=C bond affords (−)-*menthol* with 100 % e.e.!

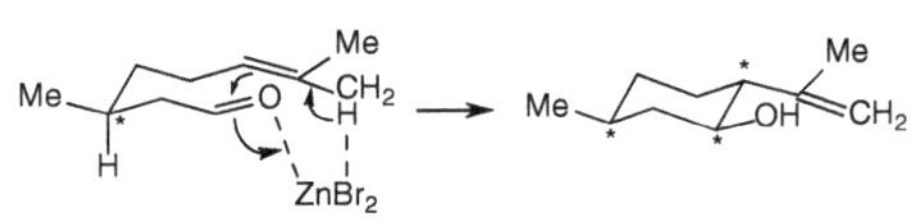

Scheme 9.3 Mechanism of $ZnBr_2$ promoted cyclization

9.3 Synthesis of Racemic and (1*R*,2*R*)-*Chloramphenicol*

Example 9.3 Structure and stereochemistry of the antibiotic (−)-*chloramphenicol* is presented in (1*R*,2*R*)-**TM 9.2**. Propose the retrosynthesis of racemic **TM 9.2** neglecting absolute configurations at stereogenic centers and then suggest the synthesis from benzaldehyde and 2-nitroethanol and propose the simple synthesis of 2-nitroethanol. Indicate a reaction step that can be completed diastereo- and enantioselectively!

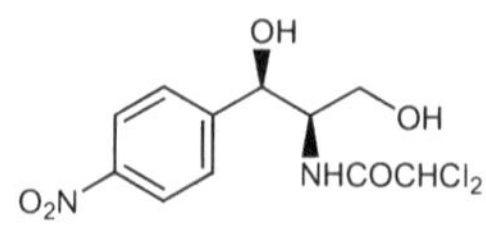

(1*R*,2*R*)-**TM 9.2**
(1*R*,2*R*)-*Chloramphenicol*

Retrosynthetic analysis starts with the observation that the dichloracetylamino group appears on the acylation of the amino group, available by reduction of the nitro group, which originates from nitroethanol. The nitro group in the aromatic ring, instead, should be introduced only on completed reduction of the nitro group in the aliphatic chain and hence retrosynthetically eliminated first (FGE). The next two interconversions of **TM 9.21** result in **TM 9.2b** (Scheme 9.4).

First, *retro*-aldol disconnection greatly simplifies the structure of **TM 9.2b** to benzaldehyde and 2-nitroethanol, amenable to a second *retro*-aldol disconnection to nitromethane and formaldehyde.

The proposal of the synthesis needs careful selection of the reaction step order (Scheme 9.5).

On completed aldol reactions in the first two steps, the nitro group is reduced before the introduction of the second one in the aromatic ring. To this aim, protection of hydroxyl groups is needed and can be completed by acetylation. Nitration is followed by deprotection of all acetyl groups since selective acetylation of amino groups can be performed by dichloroacetyl chloride in the presence of *tert*-amine.

As already mentioned, this retrosynthesis and proposed synthetic route do not take into account absolute configurations on the two chiral centers of the biologically active (−)-enantiomer. Both chiral centers are formed in aldol reactions;

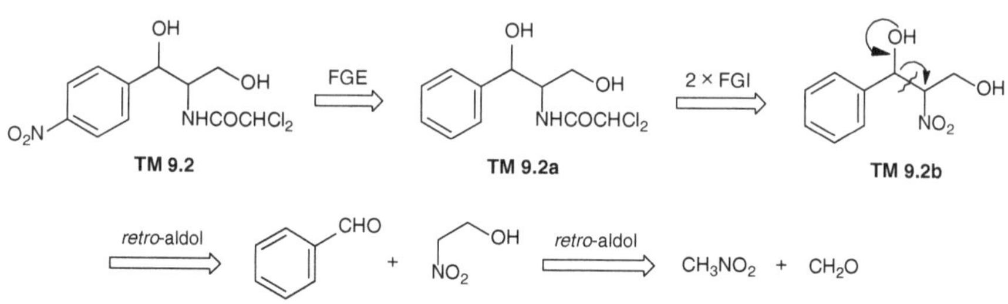

Scheme 9.4 Retrosynthetic analysis of racemic *chloramphenicol* **TM 9.2**

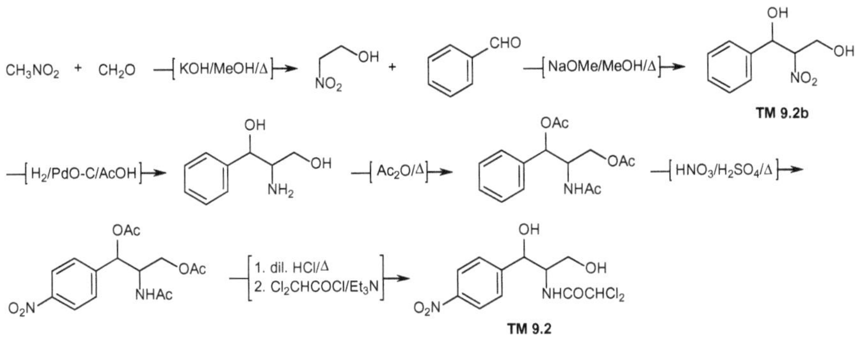

Scheme 9.5 Proposal for the synthesis of racemic *chloramphenicol* **TM 9.2**

therefore, the chiral variant of this reaction (Sect. 4.3.1.3, Scheme 4.13) suggests the possibility to prepare one enantiomer. Suggest the reaction scheme and catalyst!

In the next example, a fascinating solution for the asymmetric synthesis is presented.

Example 9.4 Complete asymmetric synthesis of (−)-*chloramphenicol*, (1*R*,2*R*)-**TM 9.2** is outlined in Scheme 9.6. An exemplary application of chiral organocatalysts is present in symmetric catalytic cyclization to aziridine **4** in the key step of the synthesis [11, 12]. Scheme 9.7 presents the detailed mechanism of chemo- and stereoselective rearrangement of chiral aziridine derivative **4** in the final intermediate **5**.

In the first key step of this reaction, prochiral aldimine **3** reacts with diazoacetic ester promoted by 1 mol% of (*S*)-VAPOL and affords chiral aziridine derivative **4** with 99:1 *cis/trans* diastereoselectivity and 96.4 % e.e. of *cis* product.

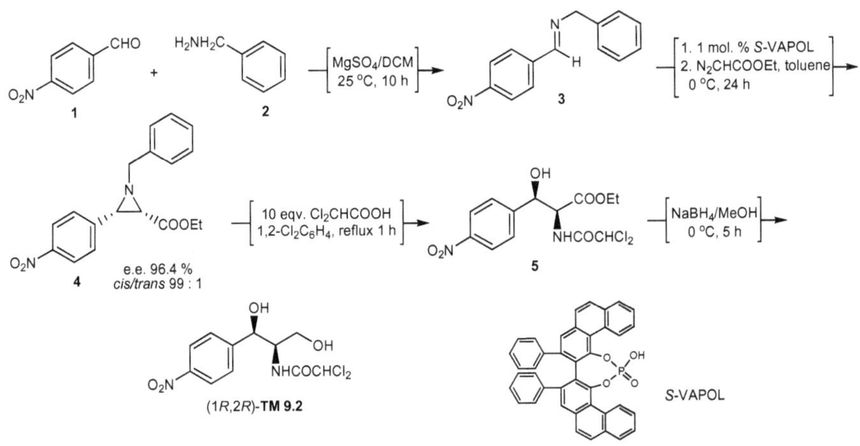

Scheme 9.6 Asymmetric synthesis of (−)-*chloramphenicol*, (1*R*,2*R*)-**TM 9.2**

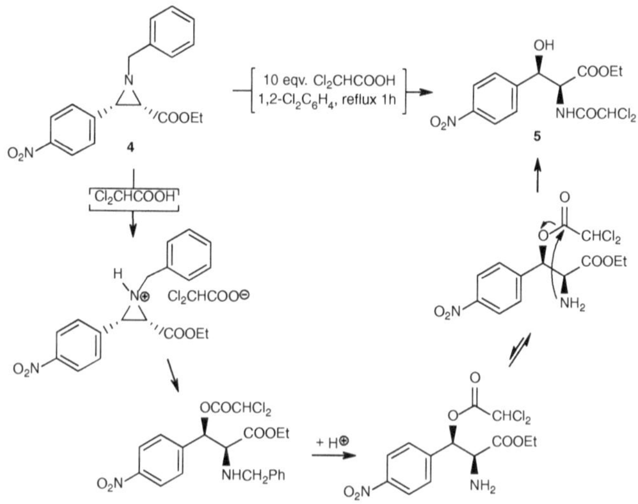

Scheme 9.7 Mechanism of the ring opening and rearrangement of **4**

Note (*S*)-VAPOL is an organic derivative of phosphoric acid that acts as an exceptional chiral organocatalyst promoting Diels-Alder and aldol reactions as well as aziridination of imines. The mechanism of its catalytic activity has been studied by physico-chemical and computational methods, but is still not completely understood [13].

In the second key step, a great molar excess of dichloroacetic acid promotes the opening of aziridine derivative **4** into the immediate precursor of (−)-*chloramphenicol,* compound **5**, at elevated temperature. The proposed mechanism of this fascinating reaction cascade is outlined in Scheme 9.7.

On protonation of the N atom in aziridines, the dichloroacetate anion acts as a strong nucleophile and opens the aziridine ring with inversion of the configuration approaching the benzylic C atom from the less hindered side. The second mole of this strong acid promotes hydrolysis of the benzylic group on the N atom. The free amino group now acts as an internal nucleophile triggering 1,3-O,N-migration of the dichloroacetyl group and formation of ester **5**. In the last step, chemoselective reduction of the ester group in the presence of the amide and nitro group affords (−)-*chloramphenicol* **TM 9.2**.

Observe that the former presented retrosynthesis and non-stereoselective synthesis of racemic *chloramphenicol* have no relation to this asymmetric synthesis of (−)-*chloramphenicol.* It is an often-encountered situation in planning the synthesis of complex chiral molecules, although it does not detract from the value of retrosynthetic consideration. Unavailability of the *chiral variant* of certain reactions prompts synthetic organic chemists to explore other routes besides the one suggested by retrosynthesis and to discover new, original solutions. Analysis of the *sertraline* molecule in the next section offers impressive support of this statement.

9.4 (+)-*Sertraline*, Interplay of Non-stereoselective and Asymmetric Syntheses

9.4.1 Non-stereoselective Synthesis of Sertraline

(+)-*Sertraline*, (1*S*,4*S*)-**TM 9.3**, (1*S*,4*S*-1-methylamino-4-(3′,4′-dichlorophenil)-tetrahydronaphthalene hydrochloride) is a widely used antidepressant and anxiolytic drug with the trade names *Zoloft* or *Lustral*. The carbon framework characterizes the 1,4-disubstituted tetrahydronaphthalene and the C–C bond to C4 of the dichlorophenyl unit. Two chiral centers are in the 1,4-position on the annulated cyclohexane unit.

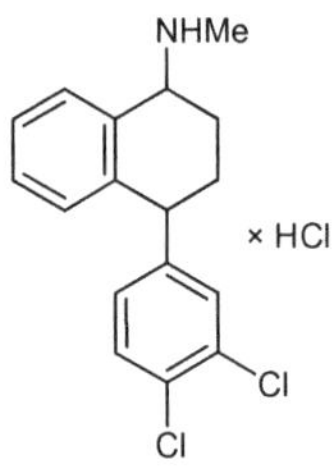

TM 9.3
Sertraline

We start the retrosynthetic consideration of **TM 9.3** by FGI of the methylamino to phenol group and continue with aromatization to the naphthalene ring to arrive at an α-naphthol derivative and complete C–C disconnection between two aromatic units (Scheme 9.8).

In the synthetic direction, coupling of the dichlorobenzene unit to α-naphthol emerges immediately as a key synthetic issue and is discussed in the next example.

Example 9.5 Consider synthetic routes to **TM 9.3** neglecting any stereochemical aspect of the synthetic steps.

To couple two aryl units in the first synthetic step, *new knowledge* is needed. α-Naphthol is known to be in equilibrium with the non-aromatic, unstable and reactive keto form. This form coordinates two moles of $AlCl_3$, affording a highly electrophilic complex (Scheme 9.9).

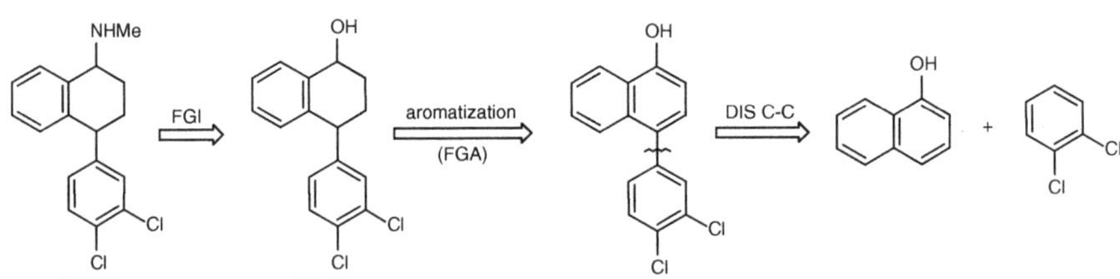

Scheme 9.8 Retrosynthetic consideration of racemic *sertraline*, **TM 9.3**

Regioselective electrophilic attack of the *vinylogous acyl cation* in the $AlCl_3$ complex on 1,2-dichlorobenzene affords **TM 9.3a** controlled by steric perturbation at both *ortho* positions to chlorine atoms.

Note The *vinylogy rule* states that in α,β-unsaturated carbonyl compounds, and generally in alkenes with EWG at the α-C atom, the H atom in the γ position (**II**) exhibits similar acidity as in the α position (**I**) to the carbonyl group (Fig. 9.2).

This rule can be extended to other conjugated systems and applied to the reactivity of other groups in the vinylogous position to the EWG group.

Based on original synthesis of ketone **TM 9.3a**, we can now easily propose the last steps on the route to racemic *sertraline* (Scheme 9.10).

The industrial method is developed based on the preparation of ketimine **TM 9.3b** followed by reduction. It should be observed that ketone **TM 9.3a** and imine **TM 9.3b** are racemic compounds. By non-stereoselective reduction of imine to methylamine, diastereomeric 1,4-*cis* and 1,4-*trans* racemates are formed.

9.4.2 Stereoselective Routes to (+)-Sertraline

Enantioselective synthesis of (+)-*sertraline*, (1*S*,4*S*)-**TM 9.3**, can be completed starting from either achiral building blocks, completing the key reaction in an enantioselective fashion, or starting from chiral raw materials, available from the chiral pool of nature, and completing the key reaction in a diastereoselective mode.

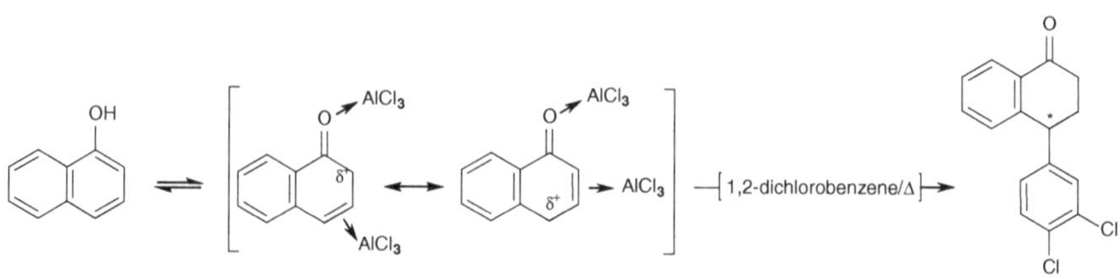

TM 9.3a

Scheme 9.9 Proposed mechanism of formation of intermediary ketone **TM 9.3a**

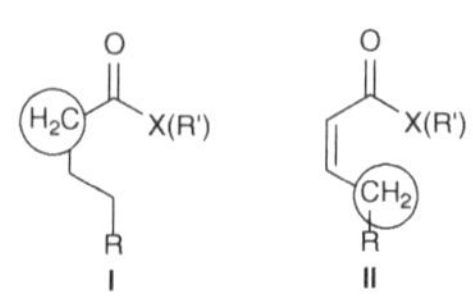

Fig. 9.2 Vinylogous methylenic group **I** and **II**

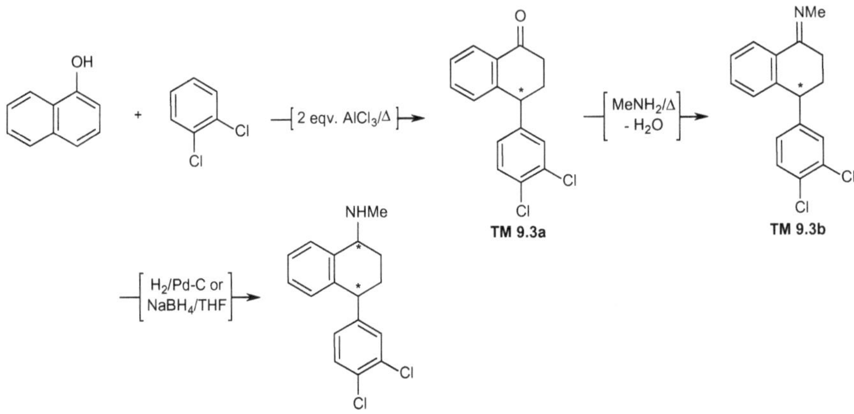

Scheme 9.10 Synthetic route to the mixture of stereoisomers of *sertraline* **TM 9.3**

Example 9.6 According to Scheme 9.10, **TM 9.3a** and **TM 9.3b** are obtained as racemic compounds. How can they be used on the route to optically pure (1*S*,4*S*)-*sertraline*, **CM 9.3**?

Both racemic compounds can be separated to enantiomers by *chiral chromatography* based on chiral stationary phases (CSPs). Chromatographic columns for laboratory- and industrial-scale separation of enantiomers are available [14, 15]. However, on the industrial scale, chiral chromatography has two serious drawbacks: the low-capacity of the process (kg racemate/kg CSP, h) and the "wrong" (*R*)-enantiomer as waste. The first issue is solved by the *simulated moving bed* (SMB) technology, successfully applied in large-scale separation of racemic ketone **TM 9.3a**. More about this technology, continuous chromatographic separation on multi-column equipment, can be found in the suggested reading [16]. The economy of the process is solved by coupling continuous SMB separation and racemization to one process by continuous treatment of (4*R*)-**TM 9.3a** with a weak base and recycling of racemate into chromatographic separation [17].

Diastereoselective reduction of optically pure imine (4*S*)-**TM 9.3b** to (+)-*sertraline* (1*S*,4*S*)-**TM 9.3** is completed with achiral heterogeneous catalysts for hydrogenation, RaNi/H_2 or Pd/C/H_2.

Example 9.7 One enantioselective synthesis of (−)-*sertraline* **TM 9.3** starts from the easily available oxabenzonorbornadiene **2**. This *meso*-compound is *desymmetrized* by the catalytic action of chiral complex Ni(II)-(*S*)-BINAP and diisobutylaluminium hydride as the reducing agent (Scheme 9.11) [18, 19].

Note Meso-compounds are prochiral structures possessing an internal symmetry plane and hence two enantiotopic groups. Two chiral centers separated by a symmetry plane are of the opposite configurations. Modification of one of the

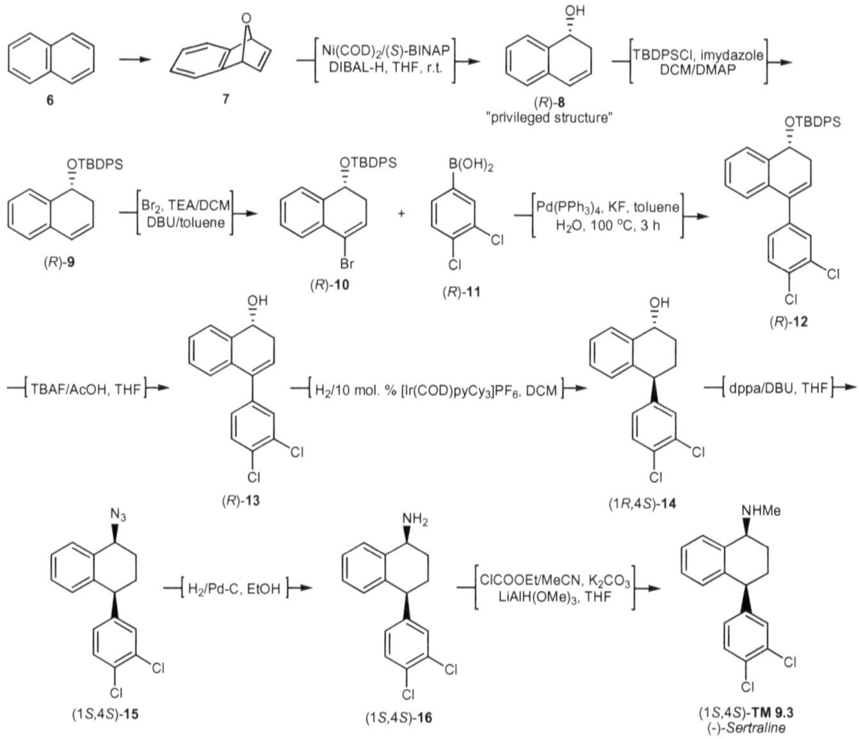

Scheme 9.11 Enantioselective synthesis of (−)-*sertraline* (1*S*,4*S*)-**TM 9.3**

enantiotopic groups leads to one of two possible enantiomers. This enantiostereoselective transformation is known as *desymmetrization*.

Chiral dihydronaphthole, *sec*-alcohol (*R*)-**8**, is a ubiquitous structure present in numerous natural compounds. It belongs to the family of "*privileged structures,*" the starting point in the synthesis of libraries of compounds for testing as potential drugs (Sect. 9.5). Coupling of chiral/achiral reagent (*S*)-BINAP/DIBAL-H enables hydrometallation-elimination from one of the *enantiotopic* C atoms in the C–O–C bridge of prochiral **7**. Broad application of this reagent is reported in a review article [20]. Chiral dihydronaphthole (*R*)-**8** is obtained in 98 % e.e. and 84 % chemical yield and protected as the *tert*-butyldiphenylsilyl derivative (TBDPS) before the preparation of vinyl bromide (*R*)-**10**. Selective bromination of (*R*)-**9** is an interesting reaction involving addition-elimination steps as outlined in Scheme 9.12.

Addition of bromine is promoted by *tert*-amine, but elimination of hydrogen bromide requires a strong organic base, DBU. Regioselective elimination of HBr deserves comment. The kinetically preferred product is formed on the elimination of bromine from the benzylic C atom and hydrogen from the vicinal position. Then, the 1,2-shift of bromine leads to (*R*)-**10** with the Br atom in the benzylic position,

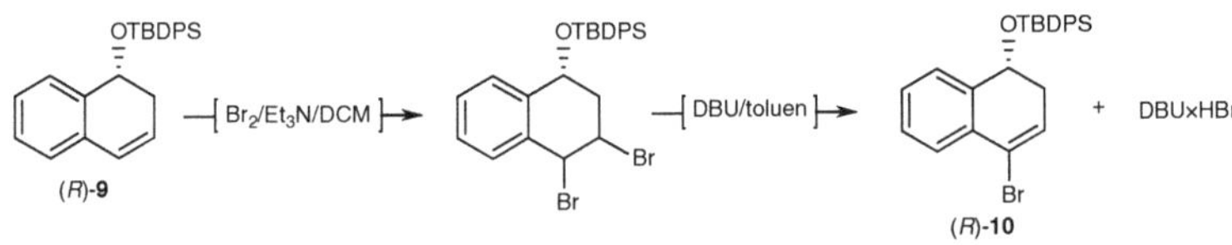

Scheme 9.12 Regioselective bromination of (*R*)-**9** to vinylbromide (*R*)-**10**

triggered by the high stability of the benzylic cation as an intermediate. Regioselective formation of (*R*)-**10** is therefore the result of thermodynamic control!

Coupling of (*R*)-**10** and (*R*)-**11** to (*R*)-**12** is completed by the well-known Suzuki-Miyaura reaction where Pd(0) complex catalyzes the formation of the C–C bond (Sect. 6.3, Example 6.4). In the next step, the protecting group is eliminated and the C=C bond reduced by achiral Ir(I) complex to *trans*-(1*R*,4*S*)-**14**. It is important to note the "wrong" *R* configuration at the C(1) atom in this and the previous intermediate. Inversion of the configuration in (1*S*,4*S*)-**15** is achieved by the Mitunobu reaction with diphenylphosphorylazide (dppa) as the source of nucleophilic azide ions in the presence of DBU. This reaction is the method of choice for the transformation of alcohols in many other functionalities, azides, esters, alkyl-aryl ethers, imides, sulfonamides, etc., and its mechanism is explained in considerable detail [21, 22].

Example 9.8 This is an example of the creative use of a chiral building block from nature in the synthesis of (+)-*sertraline* (1*S*,4*S*)-**TM 9.3**. Asymmetric synthesis starts from D-glyceraldehyde available by oxidative splitting of mannose or ascorbic acid [23]. Protected as acetal **17**, this aldehyde is coupled to *E*-alkene **19** with phosphonate **18**, a reagent in the Horner-Wadsworth-Emmons reaction (Scheme 9.13).

This reaction is a variant of the Wittig reaction (Sects. 2.2.1–2.2.3). The phosphonate group stabilizes the carbanion and can be easily cleaved on the completed reaction as a water-soluble side product [24, 25]. DMPU is an aprotic polar solvent that has replaced the toxic hexamethylphosphortriamide (HMPTA) formerly used in the laboratory and industry. The mechanism of this reaction is analogous to the Wittig reaction (Sect. 2.2.2), with enhanced thermodynamic control of the formation of the *E* isomer.

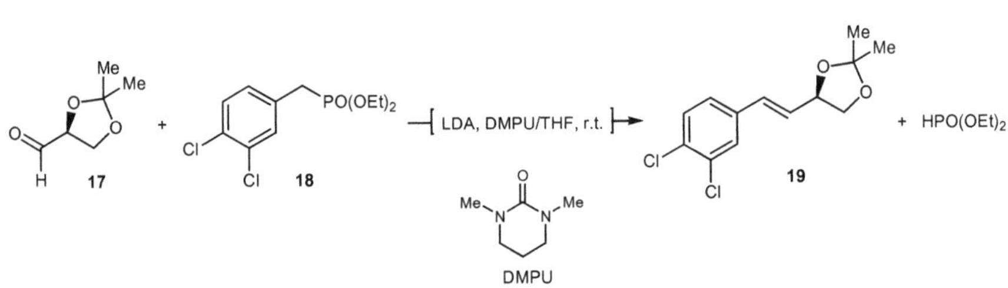

Scheme 9.13 Horner-Wadsworth-Emmons preparation of chiral *E*-alkene **19**

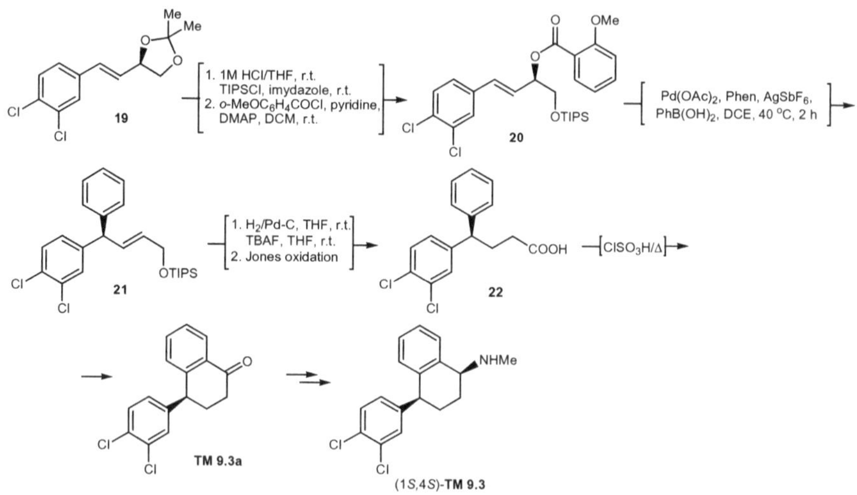

Scheme 9.14 Synthesis of (−)-*sertraline,* (1*S*,4*S*)-**TM 9.3**

The final steps on this route to (1*S*,4*S*)-*sertraline* are presented in Scheme 9.14. In the above scheme, the detailed reaction conditions are presented for any synthetic steps. Vicinal diol **19** is de-protected, and then the *prim*-OH group is protected as the triisopropylsilyl (TIPS) group and *sec*-OH group activated as benzoate ester **20**. The key event represents allyl-aryl coupling between the intermediary allylic ester and phenylboronic acid to **21**, known as Tsuji-Trost coupling [26, 27]. The reaction is promoted by catalytic complex Pd(II)Phen (1,10-phenantroline) and runs with 1,3-*syn*-transfer of chirality determining the *S* configuration on the future C(4) atom of (−)-*sertraline*. In the final steps, unsaturated alcohol **21** is hydrogenated and oxidized to carboxylic acid **22** and then cyclized to ketone **TM 9.3a**, whose transformation to racemic *sertraline* **TM 9.3** is presented in Scheme 9.10.

9.5 Lycoranes, a Goldmine of Pharmaceutical Candidates

9.5.1 Polycyclic Framework of Lycoranes

Total syntheses of three tetraacyclic diastereomeric alkaloids, α-*lycorane* **TM 9.4**, β-*lycorane* **TM 9.5** and γ-*lycorane* **TM 9.6**, have attracted the attention of chemists and pharmacologists for many years. Absolute configurations indicated in the formulae are present in natural products.

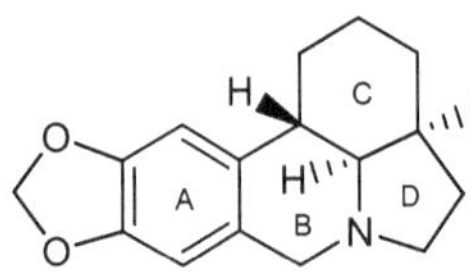

TM 9.4
α-Lycorane
BC *trans*
CD *cis*

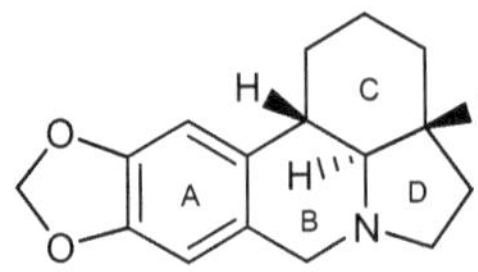

TM 9.5
β-Lycorane
BC *trans*
CD *trans*

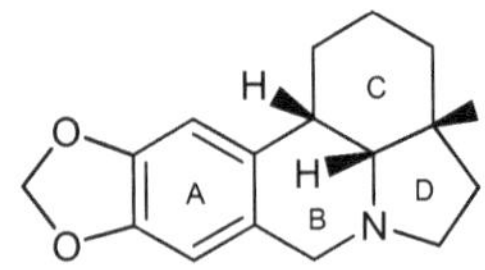

TM 9.6
γ-Lycorane
BC *cis*
CD *cis*

Lycoranes belong to the family of *Amaryllidaceae* alkaloids, structures defined as a goldmine of potential new drugs. Their derivatives and congeners exhibit antiviral activity, cholinesterase inhibition and promising results in the treatment of Alzheimer's disease.

We start the retrosynthetic analysis of structures **1–3** related to lycoranes and then discuss some examples of non-asymmetric and asymmetric total synthesis of these target molecules.

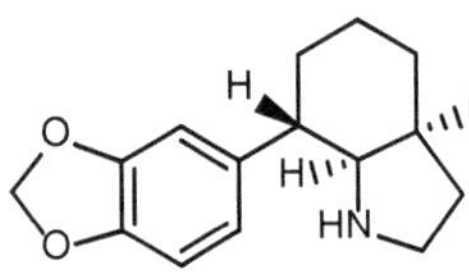

trans,cis-**1**

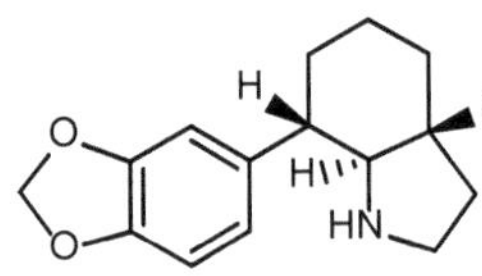

trans,trans-**2**

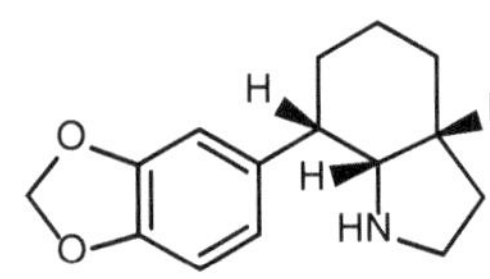

cis,cis-**3**

Precursors to lycoranes **1–3** are initially regarded as the main synthetic target since their cyclization into α-, β- or γ-lycorane **TM 9.4–TM 9.6** seemed feasible by some standard methods. However, they offer more opportunities for parallel synthesis of the libraries of analogs with targeted biological profiles.

Compounds **1–3** comprise a perhydroindole ring connected to the (3,4-methylenedioxo)aryl unit. This unit is annulated to a tetracyclic structure in lycoranes **TM 9.4–TM 9.6**. Both sets of compounds comprise three chiral centers; in **1** and α-lycorane, their relative configuration is *trans,cis*, in **2** and β-lycorane it is *trans,trans* and in **3** and γ-lycorane *cis,cis*.

9.5.2 Stereoselective Synthesis of Racemic α- and β-Lycorane

Example 9.9 Consider the retrosynthesis and propose the synthesis of racemic **1–3**, neglecting absolute configurations at the three chiral centers but taking into account their relative *cis* or *trans* configuration.

Assuming that the perhydroindole unit in **1–3** is not available by hydrogenation of aromatic indole derivatives, we consider its construction from the proper building blocks. The proper starting point seems to be the reconnection of the aromatic moiety. This unit may originate from either *O,O*-methylene protected resorcinol (*ortho*-methylenedioxybenzene) **4** or the easily available natural aldehyde piperonal **5**.

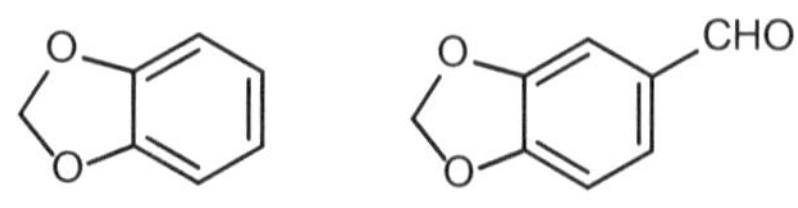

4 **5**

The plausible retrosynthetic approaches to **1–3** envisage one or two C–C bond disconnections to arrive at **4** or **5** and C_8 or C_7 building blocks (Scheme 9.15).

One-bond disconnection *a* affords **4** (X=Hal or Li) as the proper reagent for introducing the aromatic unit, and two-bond disconnection *b* results in *piperonal* **5** as the available aromatic building block.

The remaining open question is how to construct building blocks C_8N and C_7N in the above scheme. One proper retrosynthetic approach to non-aromatic bicyclic units considers cyclization of the five-member pyrolidine ring to the cyclohexane unit in **1–3**. The five-member heterocyclic ring is available via a lactame precursor; therefore, the first two retrosynthetic steps are interconversion of the amino group in **1–3** to lactame in **6** and then C–N disconnection with opening of the five-membered ring to a γ-amino acid unit on the cyclohexane ring in **7** (Scheme 9.16).

Note For the sake of simplicity, in Schemes 9.16 and 9.17 all three stereoisomers are presented by a single number, amides as **6**, amino acids as **7** and structurally isomeric (by the position of the C=C bond) nitro acids as **8** and **9**. When these compounds result in stereoselective reactions as racemates in the next schemes, the relative configurations are presented.

To continue the retrosynthetic analysis of **7**, it helps to recall that polysubstituted cyclohexane derivatives are available via intermediary cyclohexenes, products of Diels-Alder cyclization. Before *retro*-D.-A. disconnection, an important interconversion of the amino to nitro group and addition of the C=C bond to cyclohexane in

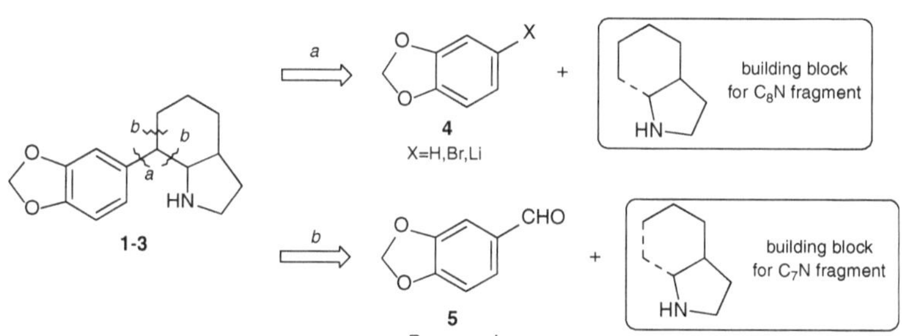

Scheme 9.15 Two possible disconnections of the central C–C bonds in **1–3**

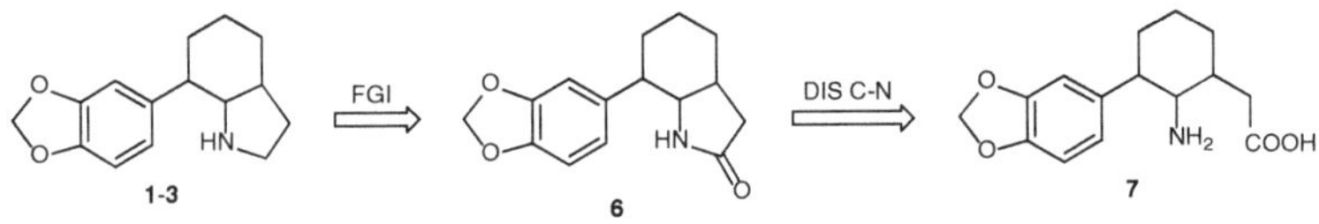

Scheme 9.16 Retrosynthetic analysis of the bicyclic unit in **1–3**

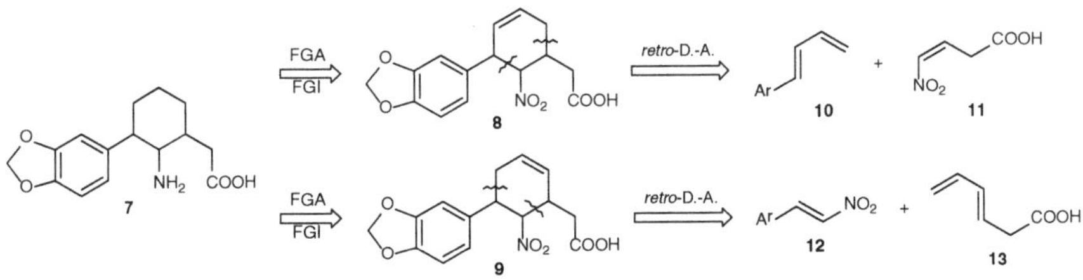

Scheme 9.17 Continuation of retrosynthesis of **7** to acyclic building blocks

one of the two preferred positions should proceed, as indicated in the retrosynthesis to **8** and **9** (Scheme 9.17).

Now we observe that both cyclohexene derivatives **8** and **9** afford surprisingly simple and easily available starting building blocks **10** and **11** and **12** and **13** on *retro*-D.-A. disconnection. The preferred couple diene/dienophile is **12/13** since their disconnection leads to the most convenient starting materials (Scheme 9.18).

The first synthesis of **1** was completed according to the presented retrosynthesis [28]. It deviates somewhat from the retrosynthetic analysis in some details, as presented in Scheme 9.19. Additional steps to racemic α-*lycorane* **TM 9.4** are included in the scheme.

2,3-Dimethylenedioxy-β-nitrostyrene **12** as a dienophile and diene methyl hexa-3,5-dienoate, the ester of **13**, are formed in a stable *E* configuration. Diels-Alder cyclization is completed in refluxing toluene and results in the *trans,cis* juncture of the BC/CD rings. Hydrogenation of the nitro group and C=C double bond results in spontaneous cyclization of the intermediary γ-amino ester to the pyrrolidone derivative. Reduction of lactame to pyrrolidine derivative **1** followed by Pictet-Spengler cyclization to ring B completes the synthesis of **TM 9.4**. Note that the Diels-Alder product and all other intermediates and the final product α-*lycorane* **TM 9.4** are racemic since no chiral intervention is implemented in the reacting system.

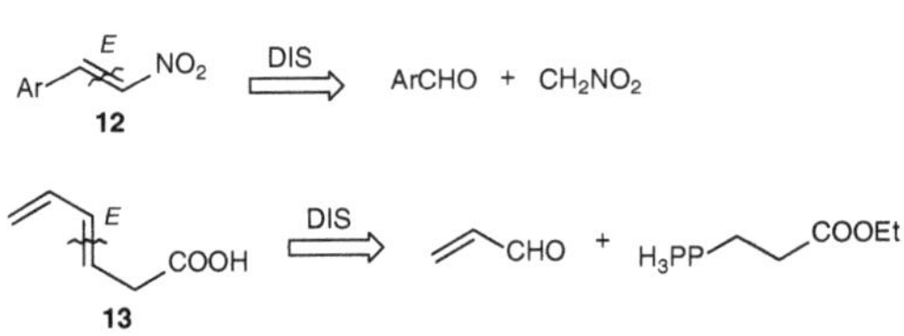

Scheme 9.18 Disconnections of **12** and **13**, final steps in the retrosynthesis of **1–3**

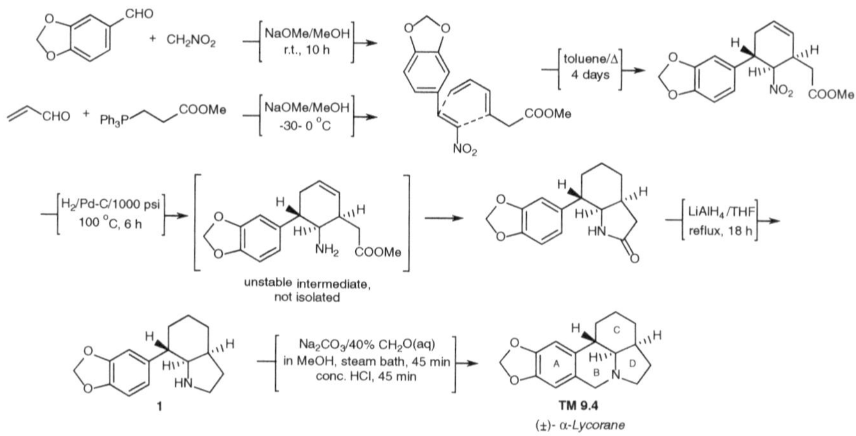

Scheme 9.19 Total synthesis of (±)-α-*lycorane* **TM 9.4**

Note In the *Pictet-Spengler* reaction, β-arylethylamines undergo ring closure on condensation with an aldehyde or ketone. Usually an acid catalyst is employed and the reaction mixture heated, but some reactive compounds give good yields even at physiologic conditions. The Pictet-Spengler reaction can be considered a special case of the Mannich reaction [29, 30].

An alternative diastereoselective synthesis of (±)-α-*lycorane* was reported in 1977 (Scheme 9.20) [31].

The Grignard reaction of cyclohexanone **15** with 2,3-methylenedioxyphenyl magnesium bromide **14** in THF followed by dehydration and hydroboration-oxidation produced cyclohexanol **16**, which afforded cyclohexanone derivative **17** by Jones oxidation.

Note Jones reagent is a solution of chromium trioxide or potassium dichromate in dilute sulfuric acid and acetone. Acetone markedly affects the properties of chromic acid. Oxidation is very rapid, usually exothermic, and the yields are typically high. This reagent rarely oxidizes unsaturated bonds.

Alkylation via enamine **18** and successive alkaline hydrolysis furnished 2-oxo-cyclohexylacetic acid derivative **19**. Refluxing keto acid with benzylamine in *ortho*-xylene and then in 87 % formic acid yielded unsaturated lactame **20**. Catalytic hydrogenation of the C=C bond afforded *trans,cis* **21** as the sole stereoisomeric product. In the last steps, reduction of lactame by a complex hydride, hydrogenolytic debenzylation and the Pictet-Spengler reaction gave (±)-α-*lycorane* **TM 9.4** in a moderate overall yield.

Another inventive total synthesis of (±)-α- and (±)-β-*lycorane* is completed by sequential chemoselective conjugate addition-stereoselective nitro-Michael cyclization of a ω-nitro-α,β,ϕ,ω-unsaturated ester [32]. This approach is based on the following retrosynthetic analysis (Scheme 9.21).

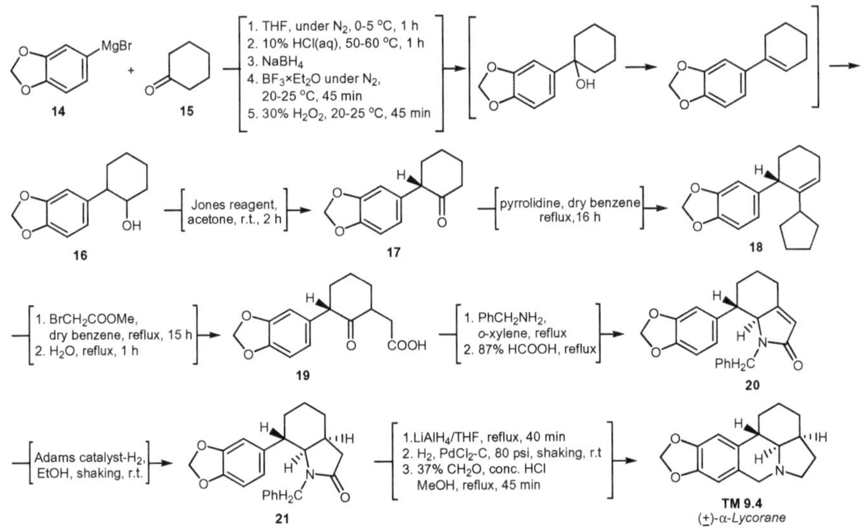

Scheme 9.20 Total synthesis of (±)-α-*lycorane* **TM 9.4**

The first FGIs lead to amino- and nitro-carboxylic acid derivatives, similar building blocks to those in retrosynthetic Schemes 9.16 and 9.17. *retro*-Michael disconnection opens the cyclohexane ring, and the second *retro*-Michael-type disconnection results in phenyl lithium and α,ω-disubstituted C_8 diene.

Intramolecular *retro*-Michael disconnection of the γ-nitro ester and opening of the cyclohexane ring provide an innovative retrosynthetic concept. Next, *retro*-Michael disconnection splits the aromatic unit from the unsaturated nitro ester.

The total synthesis starts with a four-step synthesis of 8-nitroocta-2,7-diene carboxylic acid ester **26** (Scheme 9.22).

Wittig reaction of hemiacetal **22** leads to ω-hydroxy ester **23** prevalently as *E* isomer, which is oxidized by the Pfitzner-Moffatt method to aldehyde **24**. Aldol

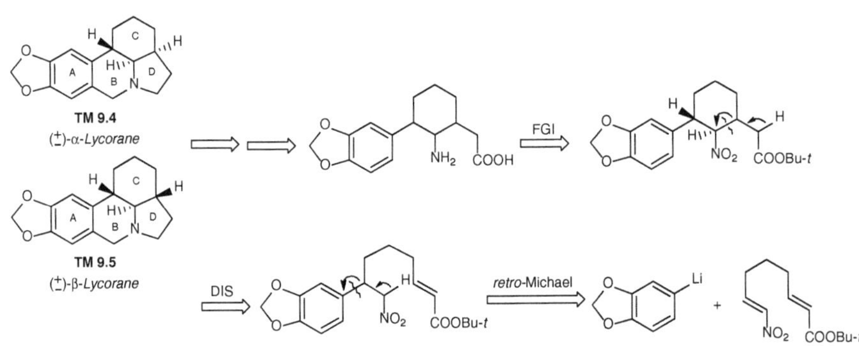

Scheme 9.21 Retrosynthetic analysis of racemic α- and β-*lycoranes*

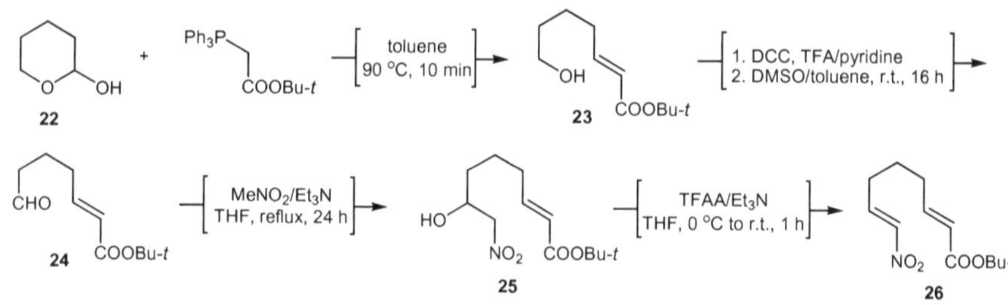

Scheme 9.22 Synthesis of unsaturated nitro-ester **26**

reaction of **24** with nitromethane to **25** is catalyzed by triethylamine and followed by dehydration with trifluoroacetic anhydride to afford **26**.

Note Pfitzner-Moffatt reaction represents oxidation of primary and secondary alcohols by dimethyl sulfoxide (DMSO) activated with carbodiimides, usually dicyclohexylcarbodiimide (DCC). Intermediary alkoxysulfonium ylides rearrange to aldehydes or ketones. This reaction yields urea as a by-product that is difficult to remove [33, 34].

The first step in Scheme 9.23 is convergent because of coupling of the two large building blocks, **26** and **27.** Aryllithium **27** is generated by treating the corresponding bromide with butyllithium in THF at −78 °C and reacts with complete chemoselectivity, forming nitro-ester **28** as the sole product.

Intramolecular nitro-Michael cyclization to **29** is completed in THF at room temperature with 2 mol of cesium fluoride and myristyltrimethylammonium bromide as promoters. Prevailing *trans,trans*-isomer **29** is separated by crystallization and the nitro derivative reduced to amine **30** in 98 % yield. Treatment with MeONa in MeOH afforded lactame **31**, which is reduced by $LiAlH_4$ to known pyrrolidine derivative **2**. Direct conversion of **2** to β-lycorane by Pictet-Spengler type cyclization using paraformaldehyde and mineral acid failed. Synthesis was accomplished in high yield by a detour, methoxycarbonylation to **32**, cyclization to **33** and final $LiAlH_4$-reduction of six-membered lactame to (±)-β-*lycorane* **TM 9.5**.

It is important to note that all presented syntheses are diastereoselective and afford one racemic diastereomer of four possible ones. In the next section, the asymmetric synthesis of (+)-γ-*lycorane* is presented, one enantiomer of eight present in four racemic mixtures! In this reaction, the chiral information resides in the chiral ligand of the organometallic catalytic system.

9.5.3 Asymmetric Synthesis of (+)-γ-Lycorane

A short total synthesis of enantiopure of (+)-γ-*lycorane*, based on the asymmetric alkylation of cyclohexene 1,4-diol derivative, was first reported in 1995 [35] and

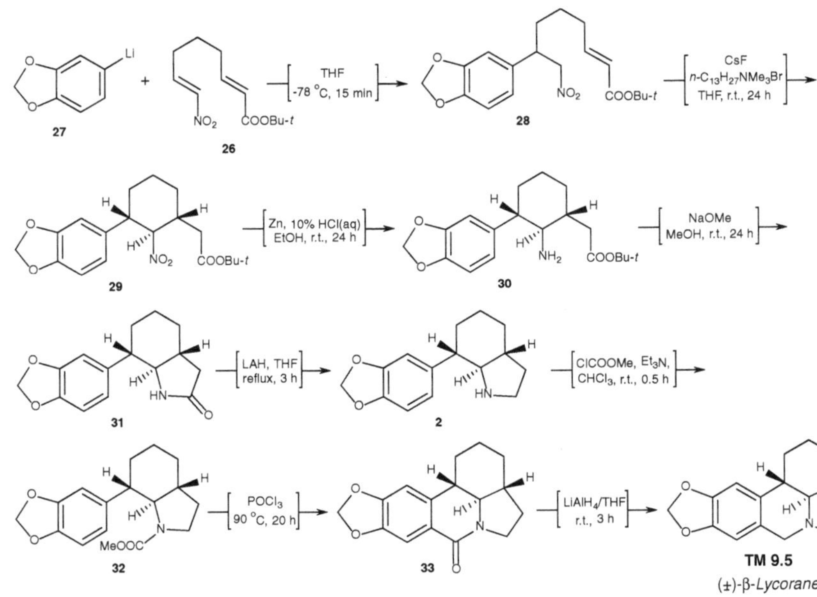

Scheme 9.23 Stereoselective synthesis of (±)-β-*lycorane*

later substantially improved [36]. Both syntheses start from piperonal as in Scheme 9.19, and the first key intermediate **37** is prepared as shown in Scheme 9.24.

Both approaches complete the *asymmetric allylic alkylation* of *meso*-1,4-di-*O*-benzylcyclohexene **38** (R=benzyl) by malonate derivative **37** catalyzed by palladium complexes of chiral phosphines. Desymmetrization of *meso*-compound **38** resembles the desymmetrization of oxabenzonorbornadiene **2** (Example 9.7 in this chapter).

In the first chapter, key chiral intermediate **39** was obtained at 40 % e.e. by catalytic allylic alkylation of **38** promoted by the palladium complex of (*S*)-BINAPO, ligand

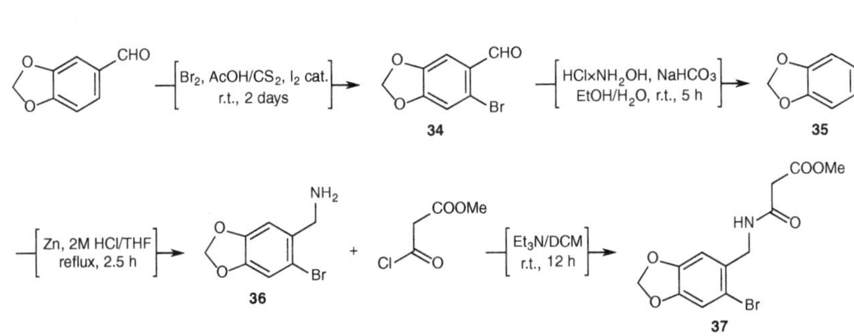

Scheme 9.24 Synthesis of achiral intermediate **37**

L1 [35]. Highly improved enantioselectivity and yield of (+)-γ-*lycorane* were achieved using the palladium complex of the unique chiral biphenol-based monodentate phosphoramidate ligand **L2** [36]. Detailed conditions for the final steps are presented in Scheme 9.25.

The authors improved the enantioselectivity and chemical yield of allylic alkylation to **39** by the screening library of the original monodentate phosphoramidate ligands [36]. The best result was obtained with ligand **L2**, 76 % chemical yield and 99.7 % e.e. of **39**, reaching 100 % conversion in 8 h. This intermediate was converted into tetracyclic oxo-lycorane **41** through one-pot tandem allylic amination-intramolecular Heck reaction.

Note The *Heck reaction* is the coupling of aryl halide with an alkene in the presence of a base and palladium catalyst to form a substituted alkene [37, 38]. The Heck reaction is of great importance as it allows substitution reactions on planar sp^2-hybridized carbon atoms catalyzed by an organopalladium catalyst, complexes of $PdCl_2$, $Pd(OAc)_2$ or $Pd(Ph_3P)_4$ with phosphines in the presence of triethylamine, potassium carbonate or sodium acetate.

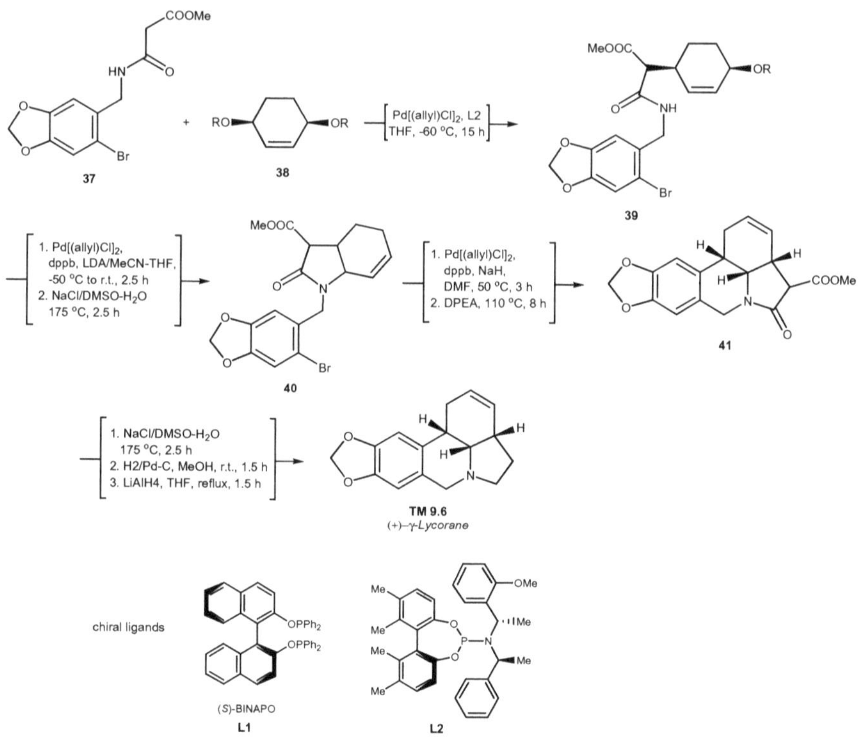

Scheme 9.25 Asymmetric synthesis of (+)-γ-*lycorane* **TM 9.6**

In the last steps, **41** was subjected to sequential demethoxycarbonylation, hydrogenation of the C=C bond and $LiAlH_4$ reduction of amide to (+)-γ-lycorane in 41 % overall and 99 % e.e from *meso*-compound **38**.

Concluding the final chapter, we suggest two recent asymmetric syntheses of *lycoranes* for independent study. The first comprises a catalytic approach to octahydroindolones and enantioselective synthesis of perhydroindole alkaloids. It is exemplified by the synthesis of (+)-α-*lycorane* and (+)-γ-*lycorine* from a common intermediate [39]. The second approach is based on asymmetric conjugate addition of organolithiums to nona-2,7-dienedioate controlled by a chiral diether ligand and subsequent intramolecular conjugate addition of the enolate intermediate affording *all-trans* trisubstituted cyclohexanes with high e.e. and yields [40].

References

1. ApSimon J, Goldsmith D (eds) (1974–1999) Total synthesis of natural products, vol 1–11. Wiley Interscience, Hoboken
2. Nicolaou KC, Vourloumis D, Winssinger N, Baran PS (2000) Angew Chem Int Ed 39:44–122
3. Nicolaou KC, Snyder SA (2003) Classics in total synthesis. More targets, strategies, methods. John Wiley-VCH, Weinheim
4. Keck GE, Giles RL, Cee VJ, Wager CA, Yu T, Kraft MB (2008) J Org Chem 73:9675–9691
5. Yamashita S, Kitajama IK, Himuro M, Hirama M (2011) J Org Chem 76:2408–2425
6. Zhou X, Xiao T, Iwama Y, Qin Y (2012) Angew Chem Int Ed 51:4909–4912
7. Kang CW, Yap PF, Lim C, Chen YZ, Ho PC, Chan YW, Wong GP, Chan SY (2007) J. Controll Release 120:211–219
8. Chaplin JA, Gardiner NL, Mitra RK, Parkinson CJ, Portwig M, Butana AM, Dickson MDE, Brady D, Marais SF, Reddy S (2002) PCT Int Appl WO 2002004384 A2, (for AECI Ltd.)
9. Tani K, Yamagata T, Akutagawa S, Kumobayashi S, Taketomi T, Takaya H, Miyashita A, Noyori R, Otsuka S (1984) J Am Chem Soc 106:5208–5217
10. Akutagawa S (1998) In: Collins AN, Sheldrake GN, Crosby J (eds) A practical synthesis of (−)-menthol with Rh-BINAP catalyst, in chirality in industry. Wiley, Hoboken, pp 313–324
11. Loncaric C, Wulff WD (2001) Org Lett 3:3675–3678
12. Hu G, Huang L, Huang RH, Wulff WD (2009) J Am Chem Soc 131:15615–15617
13. Desai A, Wulff WD (2010) J Am Chem Soc 132:13100–13103
14. Cox GB (2005) Preparative enantioselective chromatography. Blackwell, Oxford
15. Schmidt-Traub H (2005) Preparative chromatography of fine chemicals and pharmaceutical agents. Wiley-VCH, Weinheim
16. Abel S, Juza M (2007) Less common applications of enantioslective HPLC using the SMB technology in pharmaceutical industry. In: Subramanina G (ed) Chiral separation technologies, 3rd edn. Wiley-VCH, Weinheim, pp 203–273
17. Toumi A, Engeli S, Diehl M, Bock HG, Schloder J (2007) Chem Eng Process 46:1067–1084
18. Samec JSM, Lautens M, Rovis T (1997) J Org Chem 62:5246–5247
19. Lautens M, Rovis T (1999) Tetrahedron 55:8967–8976
20. Yang D, Hu P, Long Y, Zeng H, Wang H, Zou X (2009) Beilst J Org Chem 5–9
21. Kotha S, Lahiri K, Kashinath D (2002) Tetrahedron 48:9633–9695
22. Barder TE, Walker SD, Martinelli JR, Buchwald SL (2005) J Am Chem Soc 127:4685–4696
23. Ohmiya H, Makida Y, Li D, Tanabe M, Sawamura M (2010) J Am Chem Soc 132:879–889
24. Wadsworth W (1977) Org React 25:73–78
25. Blasdel LK, Myers AG (2005) Org Lett 7:4281–4283

26. Trost BM, Van Vranken DL (1996) Chem Rev 96:395–422
27. Trost BM, Crawley ML (2003) Chem Rev 103:2921–2944
28. Hill RK, John A, Joule JA, Loeffler LJ (1962) J Am Chem Soc 84:4951–4956
29. Cox ED, Cook JM (1995) Chem Rev 95:1797–1842
30. Seayad J, Seayad AM, List B (2006) J Am Chem Soc 128:1086–1087
31. Umezawa B, Hoshino O, Sawaki S, Sato S, Numao N (1977) J Org Chem 42:4272–4275
32. Yasuhara T, Nishimura K, Yamashita M, Fukuyama N, Yamada K, Tomioka K (2003) Org Lett 5:1123–1126
33. Tidwell TT (1990) Org React 39:297–572
34. Lee TV (1991) Comp Org Syn 7:291–303
35. Yoshizaki H, Satoh H, Sato Y, Nukui S, Shibasaki M, Mori M (1995) J Org Chem 60:2016–2021
36. Chapsal BD, Ojima I (2006) Org Lett 8:1395–1398
37. de Meijere A, Meyer Jr. FE (1994) Angew Chem Int Ed 33:2379–2411
38. Beletskaya IP, Cheprakov AV (2000) Chem Rev 100:3009–3066
39. Sun Z, Zhou M, Li X, Meng X, Peng F, Zhang H, Shao Z (2014) Chem Eur J 20:6112–6119
40. Nishimura K, Fukuyama N, Yasuhara T, Yamashita M, Yamamoto Y, Yamada KI, Tomioka K (2015) Tetrahedron 71:7222–7226

MIX
Papier aus verantwortungsvollen Quellen
Paper from responsible sources
FSC® C105338
FSC
www.fsc.org

If you have any concerns about our products,
you can contact us on
ProductSafety@springernature.com

In case Publisher is established outside the EU,
the EU authorized representative is:
Springer Nature Customer Service Center GmbH
Europaplatz 3, 69115 Heidelberg, Germany

Printed by Libri Plureos GmbH
in Hamburg, Germany